MATERIALS SCIENCE AND TECHNOLOGIES

MAX PHASES

MICROSTRUCTURE, PROPERTIES AND APPLICATIONS

MATERIALS SCIENCE AND TECHNOLOGIES

Additional books in this series can be found on Nova's website
under the Series tab.

Additional E-books in this series can be found on Nova's website
under the E-book tab.

MATERIALS SCIENCE AND TECHNOLOGIES

MAX PHASES

MICROSTRUCTURE, PROPERTIES AND APPLICATIONS

IT-MENG (JIM) LOW
AND
YANCHUN ZHOU
EDITORS

Nova Science Publishers, Inc.
New York

Library of Congress Cataloging-in-Publication Data

MAX phases : microstructure, properties, and applications / editors, It-Meng (Jim) Low, Yanchun Zhou.

p. cm.

Includes bibliographical references and index.

ISBN 978-1-61324-182-0 (hardcover : alk. paper) 1. Ceramic metals. 2. Transition metal carbides. 3. Transition metal nitrides. 4. Phase rule and equilibrium. I. Low, It-Meng. II. Zhou, Yanchun.

TA479.6.M39 2011

669--dc22

2011010131

Published by Nova Science Publishers, Inc. † New York

CONTENTS

PREFACE

This book deals with a new class of ceramic materials, the *MAX* Phases, which are unique new ternary carbides and nitrides. These materials exhibit a unique combination of characters of both ceramics and metals. Like ceramics, they have low density, low thermal expansion coefficient, high modulus and high strength, and good high-temperature oxidation resistance. Like metals, they are good electrical and thermal conductors, readily machinable, tolerant to damage, and resistant to thermal shock. The unique combination of these interesting properties enables these ceramics to be a promising candidate material for use in diverse fields, especially in high temperature applications.

MAX phases are nano-layered ceramics with the general formula $M_{n+1}AX_n$ (n = 1–3), where M is an early transition metal, A is a group A element, and X is either carbon and/or nitrogen. These carbides or nitrides represent a new class of solids that combine some of the best attributes of metals and ceramics. Similar to metals, they are electrically and thermally conductive, easy to machine, ductile at high temperatures, and exceptionally damage and thermal shock resistant. Like ceramics they are elastically rigid, lightweight and oxidation resistant. These phases have ultra-low friction properties and are also resistant to creep, fatigue, and corrosion. With proper alignment of the grains, it may also exhibit plasticity at room temperature due to the mobility of dislocations and multiplicity of shear-induced deformation modes. In addition, these materials have been found to exhibit "reversible plasticity" such that they can be compressed repeatedly and fully recover, whilst absorbing considerable mechanical energy. The ability of these materials to absorb vibrations is vastly superior to that of other structural ceramics, and thus they offer engineers many of the thermal, chemical and electrical advantages of ceramics, with few of their drawbacks, such as brittleness. A material that can absorb mechanical vibrations, and yet remain stiff and lightweight, should find additional applications as precision machine tools, quiet, vibration-free machinery and transportation equipment, industrial robots, and low-density armour. Other potential uses include automobile and aircraft engine components, rocket engine nozzles, aircraft brakes, heating-elements, and racing car brake pads and discs.

Hitherto, there is an enormous but fragmented amount of research papers on MAX phases published in various journals in recent years. A dedicated book on this topic is required to bring all the scattered research findings into a single volume which will provide an invaluable resource for both students and researchers in this field.

Fourteen peer-reviewed Chapters are presented in this book. Each chapter has been written by a leading researcher of international recognition on *MAX* phases. This book is concerned with the synthesis, characterisation, microstructure, properties, modelling and potential applications of MAX phases. The synopsis of each Chapter is as follows:

Chapter 1: The sintering mechanism of spark plasma sintering (SPS), also known as pulse discharge sintering (PDS), is introduced. It is shown that SPS is an effective method to fabricate *MAX* phases at a lower temperature and in shorter time when compared to conventional hot pressing. The processing parameters to sinter/synthesize *MAX* phases using SPS such as initial composition of powder mixture, sintering temperature, heating rate, holding time, and applied pressure are summarized. It is shown that the most important factors to obtain pure and fully dense *MAX* phases are initial composition, annealing temperature, and soaking time. For Ti_2AlN, Ti_2AlC, and Cr_2AlC, the optimized composition, sintering temperature, and holding time are Ti/Al/TiN/1200°C/10 min, 2Ti/1.2Al/C/1100°C/8 min, and 2Cr/1.1Al/C/1400°C/5 min respectively. For Ti_3SiC_2 and Ti_3AlC_2, the optimized sintering conditions are 3Ti/Si/0.2Al/2C/1250°C/10 min and Ti/Al/2TiC/0.1-0.2Si/1200-1250°C/8 min respectively. In the case of Nb_4AlC_3, the optimized sintering conditions are 4Nb/1.5Al/2.7C/1650°C/2 min.

Chapter 2: Thermite reactions are inorganic oxydo-reduction reactions, based on the difference of affinity for oxygen with different metals. These reactions are performed with either aluminum or magnesium, whose affinity for oxygen is amongst the highest. An overview of the many possibilities of thermite reactions from the literature of the past 10 years is presented. Details of some experiments are described to compare and contrast the direct synthesis of MAX phases, from the elements and the aluminothermic reduction of titanium dioxide.

Chapter 3: Owing to the unique nano-laminate crystal structure, the *MAX* phases with A = Al and Si offer superior mechanical properties, which make them a potential reinforcement for brittle ceramic matrix materials. This chapter focuses on the processing of Ti_3AlC_2 and Ti_3SiC_2-based composites by reactive melt infiltration technique. Capillary driven infiltration of a reactive melt allows near-net shape manufacturing of *MAX* phase reinforced composites with high flexibility in component geometry at low production costs. Reaction infiltration formation of *MAX* phase in ceramic-matrix composites such as fibre-reinforced ceramic-matrix composites and interpenetrating phase composites may extend the application fields of advanced ceramic composite materials.

Chapter 4: $Ti_3Sn_{(1-x)}Al_xC_2$ solid solutions are successfully synthesized from different reactant mixtures by using hot-isostatic pressing and their structures have been carefully characterised using Rietveld refinements. The octahedrons and trigonal prims distortion parameters and the Ti-C-Ti and Ti-A distances are calculated as a function of the Al content. Mechanical properties of solid solutions are studied from nano-indentation experiments. It is shown that solid solution hardening is not operative in this system. Hardness and elastic modulus are found to vary non-monotonically from Ti_3SnC_2 to Ti_3AlC_2, such a result is discussed in terms of Ti-A and Ti-C-Ti bond stiffness.

Chapter 5: Spark plasma sintering (SPS) is a versatile technique to rapidly sinter a number of materials which include metals, ceramics, polymers, and composites in a matter of minutes. In this technique, as-received powders are placed in a graphite die, pressed uni-axially, and then heated by passing a high pulsed current through the powders and/or the die. Sintering and densification of the material occurs within a few minutes. It is suggested that electro-discharging among particles can activate the particle surface and assist sintering. SPS synthesized Ti_3SiC_2 samples can be obtained by one of three reactions; (1) $3Ti + SiC + C \rightarrow$

Ti_3SiC_2; (2) 3Ti + Si + 2C → Ti_3SiC_2 and (3) Ti + Si + 2TiC → Ti_3SiC_2. To improve the purity of samples, aluminium can be used as a sintering aid. The results showed that addition of aluminium can accelerate the synthesis reaction of Ti_3SiC_2 significantly and fully dense, essentially single-phase polycrystalline Ti_3SiC_2 could be successfully obtained by sintering 2TiC/1Ti/1Si/0.2Al powders at 1250~1300°C. The process parameters in the sintering process revealed that addition of aluminium decreased the temperature for the synthesis reaction of Ti_3SiC_2. The sintered Ti_3SiC_2 was 97.8% dense. It was shown that spark plasma sintering is an effective method to synthesis Ti_3SiC_2.

Chapter 6: The applications, potential advantages and challenges in synthesizing *MAX* phase coatings by thermal spray techniques are reviewed. *MAX* phases are known for their unique combination of metallic and ceramic properties. *MAX* phases are expected to be promising coating materials for applications such as jet engines, aircrafts, and petrochemical installations with their ability to withstand harsh service conditions including high temperature and repeated thermal shock. However, the development of thermal sprayed *MAX* coatings on large engineering components has been very slow when compared with bulk *MAX* phases. Although thermal spray techniques are scaled-up processes for coating deposition, the commercial non-availability of pure *MAX* phase powder feedstock appears to be the main reason for lack of such coatings. This chapter summarizes examples of thermal sprayed *MAX* phase coatings in the literature and also provides future research directions in this field.

Chapter 7: The so-called *MAX* phases have shown unusual set of properties, ranging from the best of metallic to ceramic properties. In this chapter, we summarize the main results of structural and electronic properties of these fascinating materials. A description of the primitive unit cell of M_2AX, M_3AX_2 and M_4AX_3, and the pressure effect on the lattice parameters are described. The structural stability against increasing pressure and temperature is presented. Experimental and theoretical studies show that *MAX* phases are good electrical conductor where the conductivity is assured by the *d* electrons of the transition metal atoms, the stiffness of the *MAX* is attributed to the strong ionic, covalent bonds, and the remarkable machinability is related to its richness of metallic bonds and the nano-laminate nature of these materials.

Chapter 8: The two-dimensional nature (2D) of the M_2AX lamellar structure compounds is shown to be favorable for the existence of superconductivity in this new class of the compounds. In spite of there being numerous theoretical and experimental studies devoted to M_2AX phases, the physical properties of only 6 compounds were reported as having superconducting behavior. Experimental and theoretical results suggest enormous prospects to explore the possibility of other high T_c superconductors with lamellar structure. It is also shown that nitride compound (Ti_2InN) possesses more than twice superconducting critical temperature of the carbide compound (Ti_2InC). Our results suggest that other nitride compounds may also exhibit high superconducting critical temperature which may represent a new class of high temperature superconducting.

Chapter 9: An overview of the behaviour of the Ti_3SiC_2 under ionic irradiation is presented, from both structural and microstructural points of view. It appears that Ti_3SiC_2 is relatively resistant towards irradiations; even for a large number of displacements per atom, this material remains crystallized. Nevertheless, a loss of its nano-lamellar structure and an anisotropic variation of its lattice parameters, have been observed. Moreover, the swelling induced by the irradiation of the ternary compound appears as anisotropic, but seems relatively low compared to other materials. Concerning the effect of electronic interactions, they seem to enhance the oxidation of Ti_3SiC_2. Eventually, the presence of an oxide layer would cause the formation of some reliefs that seem not to have ever been observed in other materials.

Chapter 10: Layered ternary transition-metal carbides (so-called nano-laminates, or *MAX* phases) attract much attention of physicists and material scientists owing to their unique physical and chemical properties and their technological applications as high-temperature and ultrahigh-temperature structural materials. The *ab initio* approaches are very helpful in optimization and prediction of properties of these materials. Here, recent achievements in the theoretical studies of electronic band structure, chemical bonding and some properties of Ti_3SiC_2 as a basic phase of a broad family of nano-laminates are reviewed.; Besides, the peculiarities of electronic properties of non-stoichiometric and doped Ti_3SiC_2 - based species are described, and Ti_3SiC_2 surface states and hypothetical Ti_3SiC_2–based nano-tubes are discussed.

Chapter 11: The synthesis of Ti_2SC and its high pressure behaviour is addressed. A brief description of high pressure technique, requirements for conducting high pressure experiments, getting good data and obtaining the bulk modulus is also presented. The high pressure results are shown to provide insights for tailoring the physical properties of Ti_2SC and other similar compounds.

Chapter 12: The latest comprehensive studies on the mechanical behaviours of Cr_2AlC, such as fracture strength and deformation behaviour under compressive and tensile stress conditions, in a broad temperature range from room temperature to 1000°C and under various strain rates is reviewed. The fundamental mechanical properties, such as hardness, damage tolerance, thermal shock resistance and solid solution hardening are also presented. It was found that the room-temperature mechanical properties of Cr_2AlC are usually comparable to other Al-containing *MAX* phases, such as Ti_2AlC and Ti_3AlC_2, and strongly dependent on the microstructure developed. The good machinability of Cr_2AlC is attributed to low hardness, layered microstructure and good thermal conductivity. However, compared with Ti_3AlC_2 and Nb_2AlC, the thermal shock resistance property of Cr_2AlC is inferior due to the relatively smaller thermal conductivity and higher coefficient of thermal expansion. The compressive strength of Cr_2AlC decreases continuously from room temperature to 900 °C when tested at a constant strain rate. The ductile-to-brittle transition temperature is measured to be in the range of 700~800°C. When tested at different strain rates, Cr_2AlC fails catastrophically at room temperature and the deformation mode changes with strain rate at 800°C. In addition, the compressive strength increases slightly with increasing strain rate at room temperature and it is less dependent on strain rate when tested at 800°C. The variation of flexural strength of Cr_2AlC with testing temperature exhibits similar decreasing tendency to that of compressive strength and the plastic deformation occurs at temperatures higher than 800°C. The plastic deformation mechanism of Cr_2AlC at elevated temperatures is discussed on basis of microstructural observations.

Chapter 13: The tribological behaviours and the relevant mechanisms of highly pure bulk Ti_3SiC_2, Ti_3AlC_2 and the influence of TiC impurities are described. The highly pure Ti_3SiC_2 exhibits a decreasing friction coefficient (0.53-0.09) and an increasing wear rate ($0.6\text{-}2.5\times10^{-6}$ mm^3/Nm) with the sliding speed increasing from 5 to 60 m/s. The normal pressure, in the range of 0.1~0.8 MPa, also has a complex but relatively weak influence on them. The changes can be attributed to the presence and the coverage of a frictional oxide film consisting of an amorphous mixture of Ti, Si and Fe oxides on the Ti_3SiC_2 friction surface. TiC impurities cause the friction coefficient and the wear rate to increase significantly when compared with the highly pure Ti_3SiC_2 which may be attributed to an interlocking action and the pulling-out of TiC grains. When Ti_3AlC_2 slides against low carbon steel disk in the sliding speed range of 20~60 m/s, the friction coefficient is as low as 0.1~0.4 and the wear rate of Ti_3AlC_2 is only (2.3~2.5) $\times 10^{-6}$ mm^3/Nm. Such unusual friction and wear properties were confirmed to be dependant predominantly upon the presence of a frictional oxide film consisting of amorphous Ti, Al, and Fe oxides on the friction surfaces. The oxide film is in a fused state during the sliding friction at a fused temperature of 238-324°C, so it takes a significant self-lubricating effect.

Chapter 14: The successful deployment of Cr_2AlC requires the understanding of the high-temperature oxidation characteristics of Cr_2AlC such as the oxidation kinetics, the mechanism, and the oxide scales formed. Hence, the high-temperature isothermal and cyclic oxidation behavior of Cr_2AlC is described. It is well known that Cr_2AlC oxidizes according to the following equation during the isothermal oxidation between 900 and 1300 °C in air, *i.e.*, $Cr_2AlC + O_2 \rightarrow$ (Al_2O_3 oxide layer) + (Cr_7C_3 sub-layer) + (CO or CO_2 gas). The oxide scale consists primarily of the Al_2O_3 barrier layer that forms by the inward diffusion of oxygen. The consumption of Al to form the Al_2O_3 leads to the enrichment of Cr immediately below the Al_2O_3 layer, resulting in the formation of the Cr_7C_3 sub-layer. At the same time, carbon escapes from Cr_2AlC as CO or CO_2 gas into the air. During the cyclic oxidation between 900 and 1100 °C in air, Cr_2AlC similarly oxidizes according to the equation; $Cr_2AlC + O_2 \rightarrow$ (Al_2O_3 oxide layer) + (Cr_7C_3 sub-layer) + (CO or CO_2 gas). However, during the cyclic oxidation at 1200 and 1300 °C, Cr_2AlC oxidizes according to the equation, $Cr_2AlC + O_2 \rightarrow$ ($Al_2O_3/Cr_2O_3/Al_2O_3$ triple oxide layers) + (Cr_7C_3 sub-layer) + (CO or CO_2 gas), because the thermo-cycling facilitates the formation of the intermediate Cr_2O_3-rich layer between the outer Al_2O_3-rich layer and the inner Al_2O_3-rich layer.

In: MAX Phases: Microstructure, Properties and Applications ISBN 978-1-61324-182-0
Editors: It-Meng (Jim) Low and Yanchun Zhou

Chapter 1

SPARK PLASMA SINTERING (SPS), OR PULSE DISCHARGE SINTERING (PDS) OF MAX PHASES

***C. F. Hu,*[1,2] *Y. Sakka,*[1,2,3] *S. Grasso,*[2,3] *H. Tanaka,*[1] *and T. Nishimura*[2]**

[1]World Premier International Research Center (WPI) Initiative on Materials Nanoarchitectonics (MANA), National Institute for Materials Science (NIMS), 1-2-1 Sengen, Tsukuba, Ibaraki, Japan

[2]Fine Particle Processing Group, Nano Ceramics Center, NIMS, 1-2-1 Sengen, Tsukuba, Ibaraki, Japan

[3]Graduate School of Pure and Applied Sciences, University of Tsukuba, 1-1-1 Tenodai, Tsukuba, Ibaraki, Japan

ABSTRACT

In this chapter, the sintering mechanism of spark plasma sintering (SPS), also known as pulse discharge sintering (PDS), is introduced. It is shown that SPS is an effective method to fabricate MAX phases at a lower temperature and in shorter time when compared to conventional hot pressing. In the present work, the processing parameters to sinter/synthesize MAX phases using SPS such as initial composition of powder mixture, sintering temperature, heating rate, holding time, and applied pressure have been systematically summarized. The most important factors to obtain pure and fully dense MAX phases are initial composition, annealing temperature, and soaking time. For 211 MAX phases, such as Ti_2AlN, Ti_2AlC, and Cr_2AlC, the optimized composition, sintering temperature, and holding time are Ti/Al/TiN/1200°C/10 min, 2Ti/1.2Al/C/1100°C/8 min, and 2Cr/1.1Al/C/1400°C/5 min respectively. For 312 MAX phases, like Ti_3SiC_2 and Ti_3AlC_2, the optimized sintering conditions are 3Ti/Si/0.2Al/2C/1250°C/10 min and Ti/Al/2TiC/0.1-0.2Si/1200-1250°C/8 min respectively. In the case of 413 MAX phase Nb_4AlC_3, those are 4Nb/1.5Al/2.7C/1650°C/2 min.

Keywords: Ti_2AlN; Ti_2AlC; Cr_2AlC; Ti_3SiC_2; Ti_3AlC_2; Nb_4AlC_3; Spark plasma sintering

1. Introduction

Transition metal ternary compounds, MAX phases, in which M is transition metal, A is A group element, and X is C or N, are the laminar ceramics due to their special layered crystal structure with hexagonal symmetry [1,2]. To date, more than fifty M_2AX, six M_3AX_2, and eight M_4AX_3 phases have been determined [3]. It is found that they have the remarkable physical and mechanical properties, such as high Young's modulus, excellent thermal shock resistance and damage tolerance, and high thermal conductivity and electrical conductivity [4-12], promising for the structural and/or functional applications. These MAX phases have been successfully prepared using chemical vapor deposition (CVD), hot isostatic pressing (HIP), hot pressing (HP), pressureless sintering (PL), and magnetron sputtering (MS), as well as spark plasma sintering (SPS), etc. [13-18]. In the present review, we systemically introduce the investigation state of MAX phases fabricated by spark plasma sintering.

2. Fundamentals of SPS or PDS

Spark plasma sintering (SPS), also called pulse discharge sintering (PDS), is a pressure and current assisted sintering method, which permits to lower the sintering temperature and to shorten the sintering time [19-21]. SPS operational temperatures are typically 200-500°C lower than conventional sintering methods. Lower temperatures and shorter holding time permit to fully densify nanometric powders and prevent the incipient grain coarsening. SPSed materials often exhibit improved physical and mechanical properties compared to those obtained by conventional methods. Noteworthy features resulting from SPS include superplastic behaviour of ultrafine ceramics, increased permittivity of ferroelectric materials, improved magnetic properties of magnetic materials, enhanced product-scale bonding, augmented thermoelectric properties, superior mechanical properties, and reduced impurities segregation at grain boundaries.

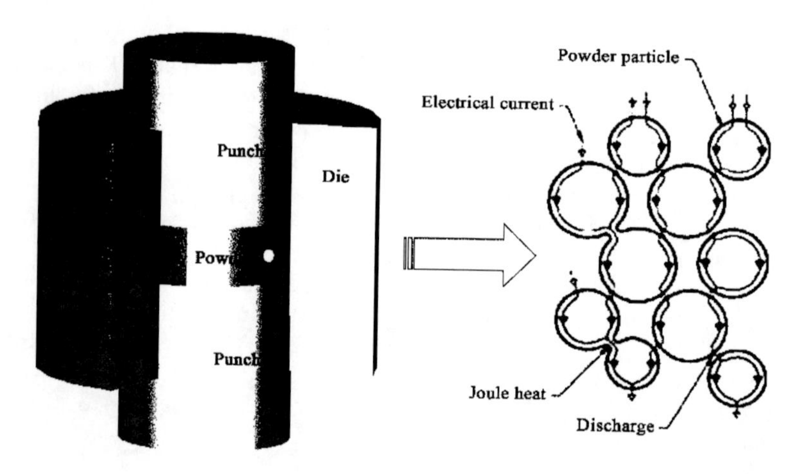

Figure 1. Sintering mechanisms among conductive powder particles during spark plasma sintering (SPS).

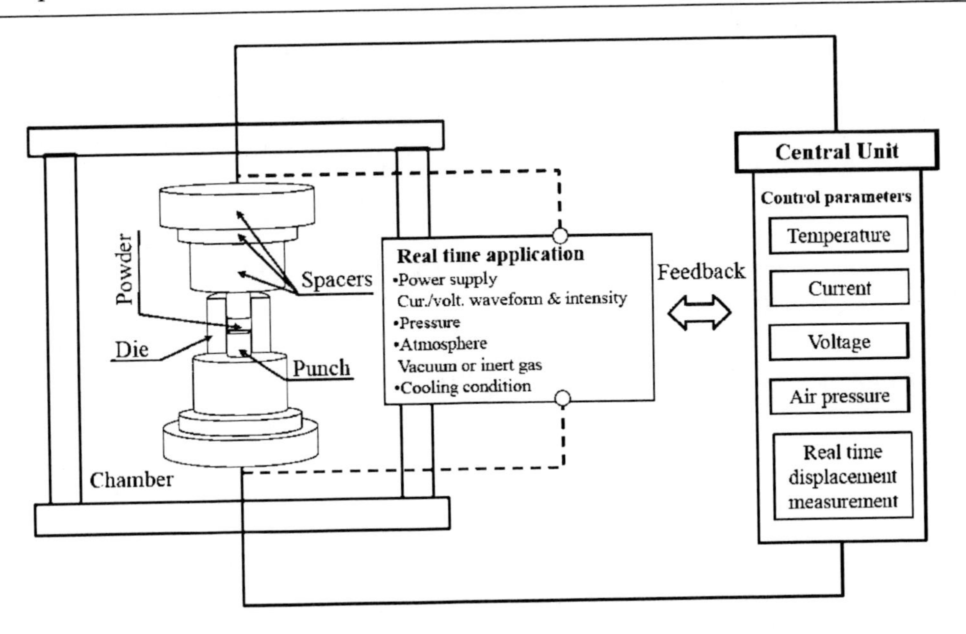

Figure 2. The sketch map of SPS facility.

SPS exploits the same punch/die system concept as the more familiar hot pressing (HP) process. A powder or green compact is placed in the die and subsequently pressed between two counter-sliding punches. Mechanical loading is normally uniaxial. It is well documented that in the SPS process the powder was mainly densified owing to a combination of thermal, pressure and current effects [22]. In the case of conductive samples, the densification is enhanced by the electrical spark discharge phenomena, including a high energy, low voltage spark generated at interparticles contact point with local temperature of several thousands degree [22]. Figure 1 shows the sintering mechanisms on the conductive powder particles. The surface of the particles are cleaned by the formation of plasma and sparking phenomena at particles contact points which cause the vaporization and/or melting due to the localized high temperature up to 10000°C. Consequently, the neck formation is greatly enhanced by current effects. At last, the densification is accelerated by plastic deformation induced by electroplasticity effect given by the combination of applied pressure and electric current. The entire process from powder to fully dense bulk sample is fast in time. In the case of nonconductive specimens, the formation of plasma is difficult to be hypothesized. The non conductive powders are heated from the die and punches [23].

The commercial SPS sintering system consists of vertical single-axis pressurization, built-in water-cooled special energizing mechanism, water-cooled vacuum chamber, atmosphere control, vacuum exhaust unit, special sintering DC pulse generator, and SPS controller. Figure 2 shows the sketch map of SPS apparatus. Apparatuses include special purpose hardware and software for multi-parameter optimization by adjusting pulse duration, duty cycle (on-time/off-time), peak current, applied voltage, repetition frequency, waveform, heating and cooling rates, primary power supply, and vacuum/atmosphere control. All the above mentioned operating parameters which affect the final properties of samples can be controlled and adjusted in real time by central control unit. Under the electric pulse energizing condition, the temperature can quickly rise at heating rate as high as 1000oC/min. In general, SPS process is energetically efficient since it can use just 20 to 50% of the energy of conventional HP [24]. Both localized heating and the reduced sintering time significantly contribute to energy saving.

3. *MAX* 211 NITRIDE AND CARBIDES

In previous investigations about 211 phases, Ti_2AlN, Ti_2AlC, and Cr_2AlC have been synthesized by SPS or PDS.

(A) Ti_2AlN

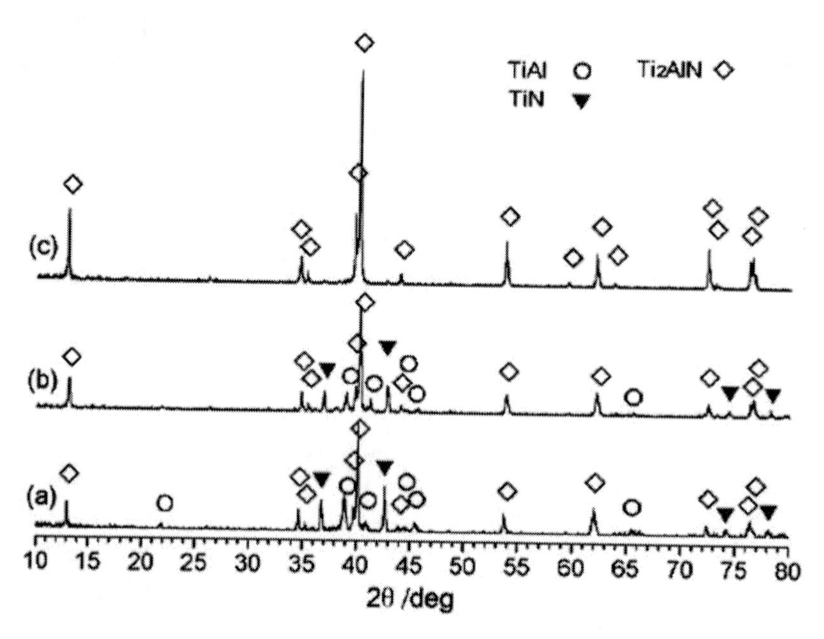

Figure 3. X-ray diffraction (XRD) patterns of samples sintered at: (a) 1000, (b) 1100, and (c) 1200oC [25].

Single phase bulk Ti_2AlN was prepared by SPS at 1200°C using Ti/Al/TiN powders [25]. The initial Ti (99%, 2.48 μm), Al (99.8%, 1.50 μm), and TiN (99.3%, 2.03 μm) powders with a molar ratio of 1 : 1 : 1 were mixed in ethanol for 24 h and then dried at 60°C for 2 h. The mixture powder was filled into a graphite die with a diameter of 16 mm and sintered by SPS (Dr. Sinter-1020, Izumi Technology Co. Ltd., Japan) in a vacuum of 0.4 Pa. The heating rate was 80°C/min, the annealing time was 10 min, and the applied pressure was 30 MPa. During the sintering, the sample densified rapidly during the temperature of 900-1100°C. Figure 3 shows X-ray diffraction (XRD) patterns of samples sintered at different temperatures. At 1000°C, the main phases were Ti_2AlN, TiAl, and TiN. When increasing the temperature as 1100°C, the content of TiN and TiAl reduced and the peak intensity of Ti_2AlN increased. It was concluded that the formation of Ti_2AlN was attributed to the reaction: $TiAl+TiN \rightarrow Ti_2AlN$. When sintered at 1200°C, single phase Ti_2AlN was obtained. In the dense sample, the grains were plate-like and have the sizes of 8-12 and 20-30 μm in width and length respectively.

(B) Ti_2AlC

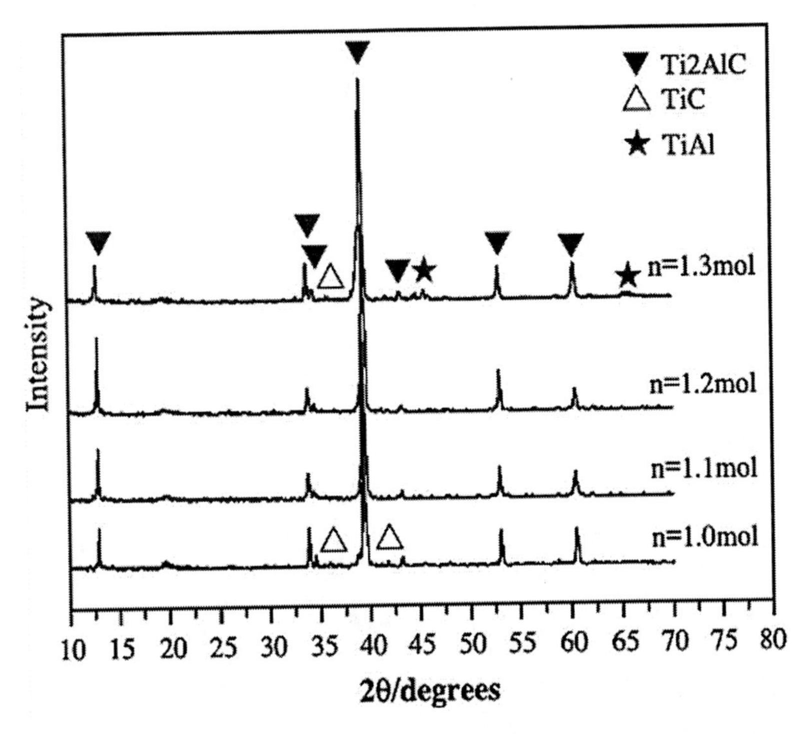

Figure 4. XRD patterns of samples sintered at 1100oC from different initial compositions in molar ratio [27].

Dense polycrystalline Ti_2AlC was successfully synthesized by SPS from the elemental Ti, Al, and C powders with a molar ratio of 2 : 1.2 : 1 [26,27]. Commercially available Ti (99%, 10.6 μm), Al (99.8%, 12.8 μm), and graphite (99%, 13.2 μm) powders were used. The powders with a designed composition of Ti : Al : C = 2 : n : 1 (n = 1, 1.1, 1.2, and 1.3) were firstly mixed in ethanol for 24 h and then dried. The mixture was put into a graphite die with an inner diameter of 20 mm and sintered in vacuum by SPS (Dr. Sinter-1050, Izumi Technology Co. Ltd., Japan). The temperature was measured by means of an optical pyrometer focused on a small hole in the die. The heating speed was 80°C/min, the vacuum degree was 0.5 Pa, and the pressure was 30 MPa. Generally, the ON/OFF ratio of pulsed current was set as 12/2 and the maximum current reached about 2000 A during sintering. The soaking time was 8 min. Figure 4 shows XRD patterns of samples sintered at 1100°C. The results confirmed that the excessive aluminum contributed to the synthesis of Ti_2AlC (n = 1.1 and 1.2). However, more aluminum was not welcomed because that the peaks of TiAl and TiC appeared again due to the reaction dynamic (n = 1.3). Additionally, the effect of sintering temperature on purity of Ti_2AlC was investigated, as shown in Figure 5. At 900-1000°C, the main phases were Ti_2AlC, TiC, and TiAl. At 1100°C, only single phase Ti_2AlC was examined. However, when the temperature was above 1200-1300°C, TiC appeared again and Ti_3AlC_2 was formed. It was considered that Ti_2AlC reacted with TiC to form Ti_3AlC_2. The density of single phase dense Ti_2AlC sample was 99.8% of theoretical value. The grain sizes were 20 μm in length and 5 μm in width. Vickers hardness tested at 1 N was 4 GPa.

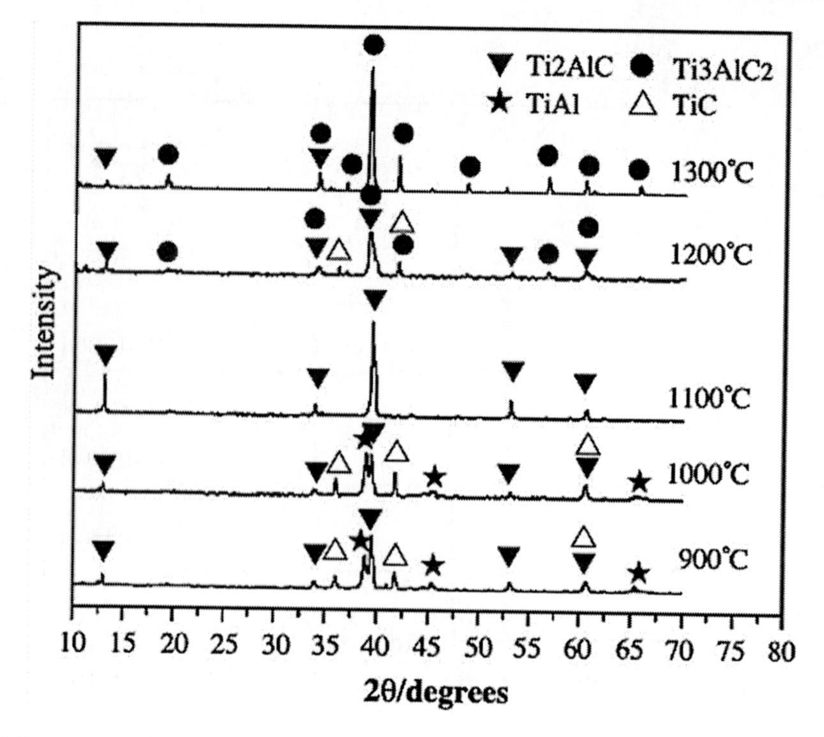

Figure 5. XRD patterns of specimens sintered at different temperatures using a molar ratio of Ti : Al : C = 2 : 1.2 : 1 [27].

(C) Cr_2AlC

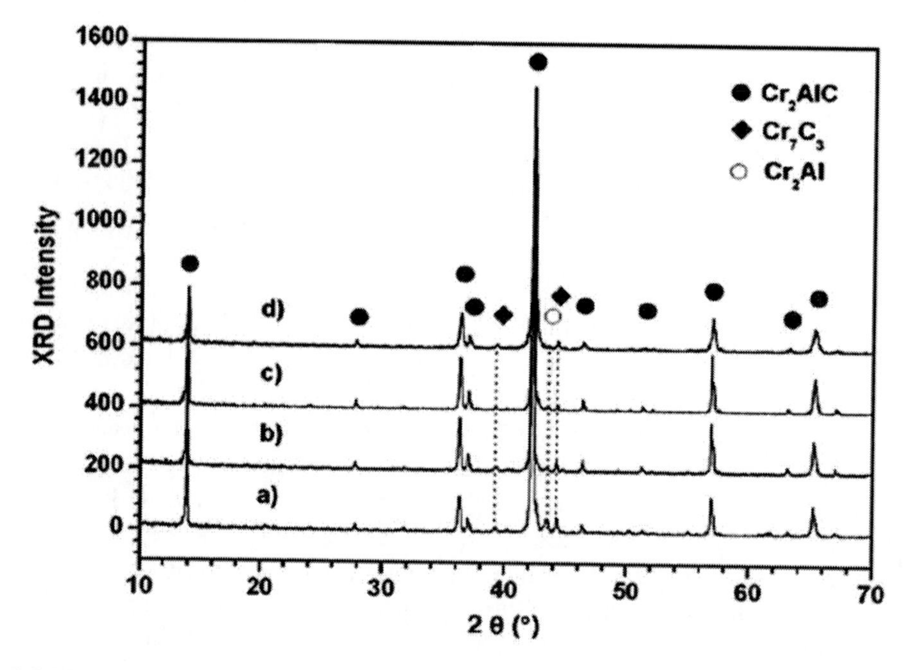

Figure 6. XRD patterns of samples sintered using coarse powders at (a) 1250, (b) 1300, (c) 1350, and (d) 1400°C [28].

Bulk Cr_2AlC ceramic was fabricated by SPS from coarse powders and fine powders in a temperature range of 1100-1400°C [28]. For the coarse powders, chromium (99.95%, 32 μm), aluminum (99.95%, 130 μm), and graphite (99%, 5 μm) powders were used. And for the fine powders, chromium (99.95%, 3 μm), aluminum (99%, 3 μm), and graphite (99%, 5 μm) powders were employed. The powders were weighed according to the composition of Cr : Al : C = 2 : 1.1 : 1 and milled in ethanol for 24 h using Si_3N_4 balls as milling media. Dried mixture was pre-pressed as pallets and sintered in SPS (FCT Systeme GmbH, Germany) at a designed temperature for 5 min under 50 MPa with a heating rate of 200°C/min. Figure 6 shows XRD spectra of samples prepared using coarse powders. For the SPSed samples, Cr_2AlC appeared as a major phase, together with a small amount of Cr_7C_3 and Cr_2Al. At 1250°C, the amount of Cr_7C_3 and Cr_2Al were comparable. When the temperature increased from 1300 to 1400°C, Cr_2Al content decreased quickly while the amount of Cr_7C_3 went down slowly. On the other hand, for the samples sintered using fine powders, there also existed the major phase of Cr_2AlC with minor amount of Cr_7C_3 and trace amount of Cr_2Al, and the amount of later two phases were declined as an increment of temperature from 1100 to 1400°C, as shown in Figure 7. By comparing, it was found that the Cr_2AlC samples synthesized at 1400°C had the high purity of 99 and 97 wt% for fine powders and coarse powders respectively. The densities were close to the theoretical value (5.23 g/cm^3). The grain size of sample sintered from fine powders was 5 μm, much less than that of sample prepared using coarse powders (20 μm).

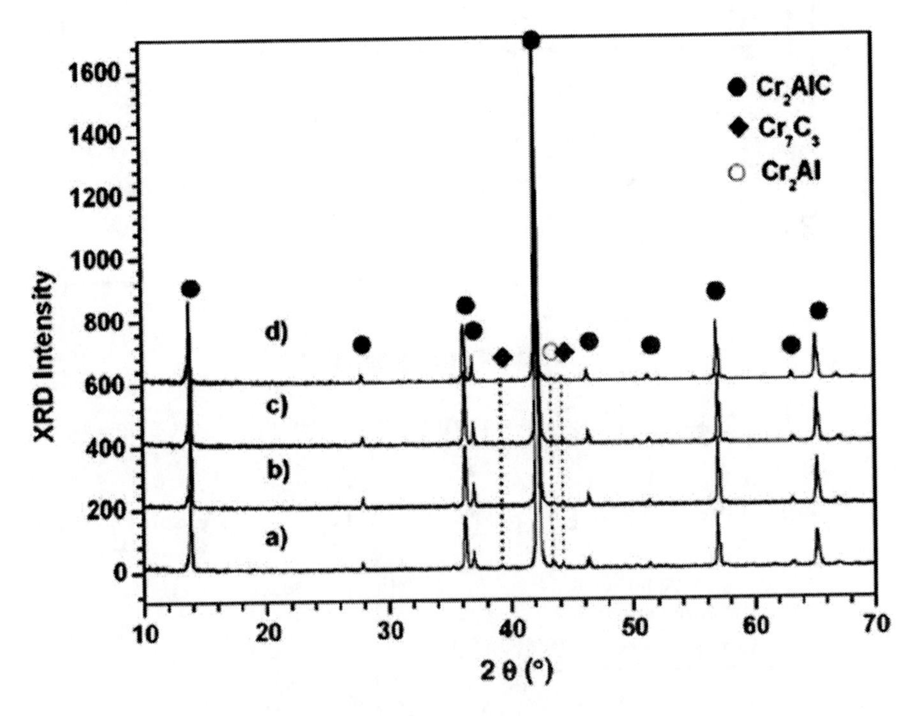

Figure 7. XRD patterns of samples sintered using fine powders at (a) 1100, (b) 1200, (c) 1300, and (d) 1400°C [28].

In addition, the reaction process of Cr_2AlC was investigated using powder mixture of Cr, Al_4C_3, and graphite by PDS [29]. The starting materials were Cr (98%, 10 μm), Al (99%, 4 μm), and graphite (99.7%, 5 μm) powders. The powders were weighed according to a molar ratio of Cr : Al : C = 2 : 1.1 : 1 and mixed by a Turbula shaker mixer for 24 h. The mixture was put into a graphite mold and sintered in vacuum by PDS (PAS-V, Sodick Co. Ltd., Japan) in a temperature range of 850 to 1350°C for 15 min with a heating rate of 50oC/min and a pressure of 50 MPa. Figure 8 shows XRD patterns of samples PDSed at different temperatures. It was found that the amount of Cr_2AlC increased significantly in the temperature range of 950-1150°C at the consumption of Cr and Al_4C_3 as well as an intermediate phase Cr_2Al. Nearly single phase Cr_2AlC was fabricated above 1250°C with a small amount of Cr_7C_3. It was considered that Cr_2AlC phase was formed near Al_4C_3 particle by the diffusion of Cr and reaction between Cr and Al_4C_3.

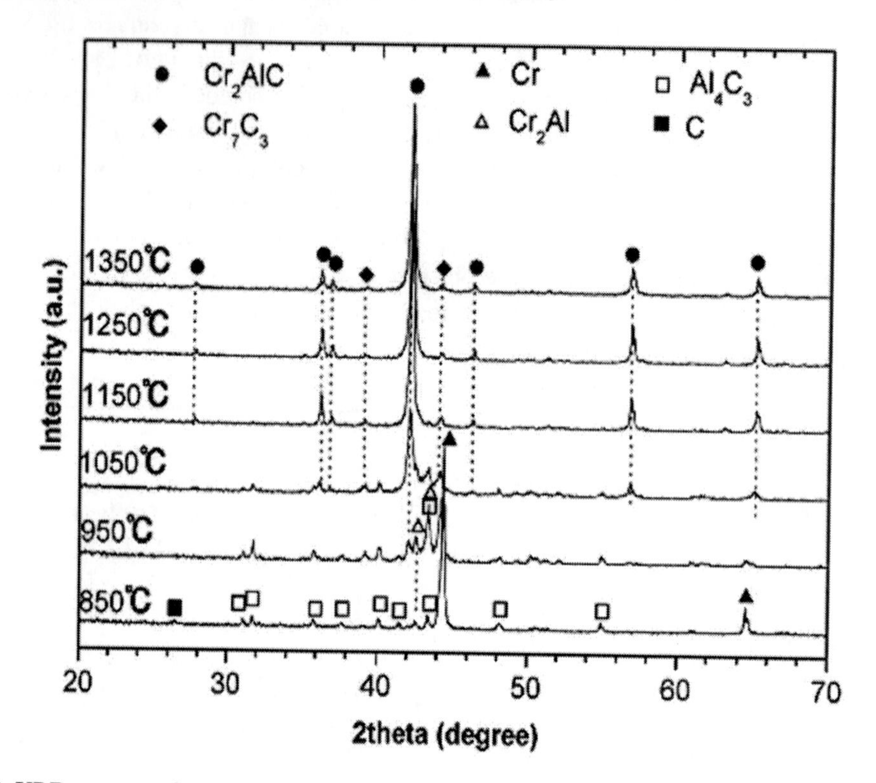

Figure 8. XRD patterns of samples PDSed from Cr-Al_4C_3-C powders at different temperatures [29].

Table 1. Compositions and sintering parameters of $(Cr_{1-x}V_x)_2AlC$ solid solutions [30]

Sample	Composition (molar ratio)				Sintering parameters
	Cr	V	Al	C	
Cr_2AlC	2	0	1.1	1	1250°C/30 min/50 MPa
$(Cr_{0.9}V_{0.1})_2AlC$	1.8	0.2	1.1	1	1250°C/30 min/50 MPa+1250°C/60 min/30 MPa
$(Cr_{0.75}V_{0.25})_2AlC$	1.5	0.5	1.1	1	1275°C/30 min/50 MPa+1275°C/60 min/30 MPa
$(Cr_{0.5}V_{0.5})_2AlC$	1	1	1.1	1	1300°C/30 min/50 MPa+1300°C/60 min/30 MPa

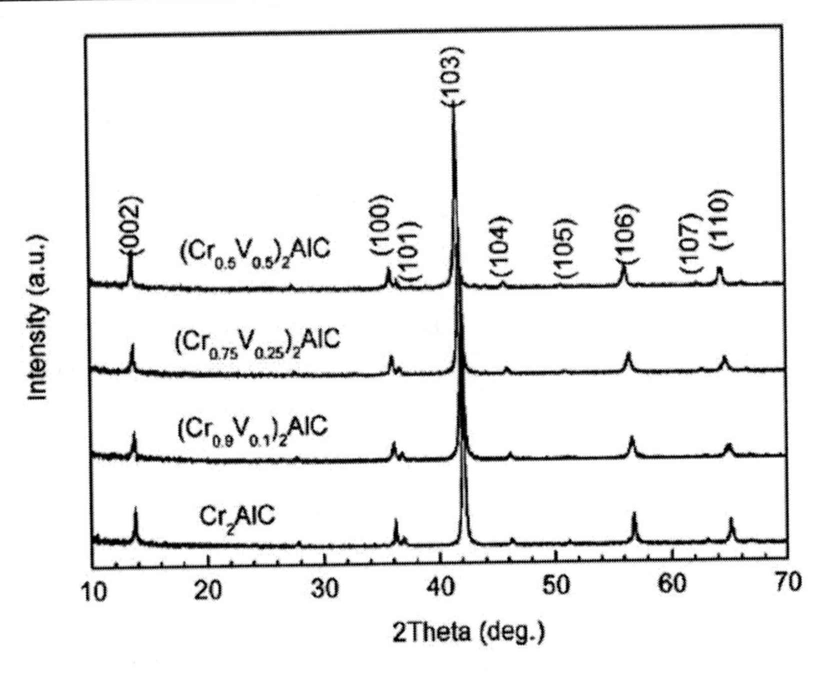

Figure 9. XRD patterns of $(Cr_{1-x}V_x)_2AlC$ solid solutions [30].

Also, $(Cr_{1-x}V_x)_2AlC$ solid solutions were synthesized by PDS [30]. The initial powders were Cr (98%, 10 μm), V (99%, 75 μm), Al (99.9%, 10 μm), and graphite (99.7%, 5 μm). The powders were weighed according to the designed compositions listed in Table 1 and mixed by a Turbula shaker mixer for 24 h. The mixture powder was put into graphite mold and sintered in vacuum by PDS (PAS-V, Sodick Co. Ltd., Japan). The optimized sintering parameters were listed in Table 1. Figure 9 shows XRD patterns of solid solutions. Nearly single phase samples were obtained by the optimized sintering processes. The diffraction peaks shift to lower angles with increasing V content, indicating the increase in lattice parameters. The larger atom size of V was responsible for the increase of the crystal parameters. The fabricated samples were nearly dense with almost the same relative densities of 98%. With the increase of V content, the grains changed from equiaxed shape to elongated morphology. The grain size of Cr_2AlC was 7.0 μm, and those of solid solutions were about 8.9-9.1 μm.

4. *MAX* 312 Carbides

About 312 phase, the most focusing aim is Ti_3SiC_2, which has been systematically investigated by exploring the sintering temperatures, initial compositions, and initial particle sizes etc. Also, Ti_3AlC_2 was investigated using SPS or PDS. Additionally, the associated powdering method of prior mechanical activation (high energy milling) was applied as one rapid way to prepare nano-sized initial reactant powders.

(A) Ti_3SiC_2

PDS technique can be employed for rapid synthesis of Ti_3SiC_2 samples from elemental powders of Ti/Si/C at relatively low sintering temperature range of 1250-1300°C, which is about 100-200°C lower than those made by other methods [31]. To fabricate high purity Ti_3SiC_2, the composition of starting powder should be adjusted to be off-stoichiometric ratio from 3 : 1 : 2. The commercial powders of Ti (99.9%, 10 μm), Si (99.9%, 10 μm), and C (99.9%, 1 μm) were used and the molar ratios of Ti : Si : C were 3 : 1 : 2, 3 : 1.05 : 2, 3 : 1.10 : 2, 3 : 1.15 : 2, 5 : 2 : 3, and 3 : 1.5 : 2. All the powders were mixed in a Turbula mixer for 24 hours in Ar atmosphere. The mixture was compacted into a graphite mold (20 mm in diameter) and sintered in vacuum (10^{-4} Pa) in a temperature range of 1200-1500°C for 15-60 min by PDS. The heating rate was controlled in a range of 50-60°C/min and the applied constant pressure was 50 MPa. It was found that by using 5Ti/2Si/3C powder, the content of TiC could be decreased to 6.4 wt% in the samples sintered at 1300°C for 15 min. Additionally, Ti/Si/TiC mixture powders were prepared for the synthesis of Ti_3SiC_2 [32]. Commercial available Ti (99.9%, 10 μm), Si (99.9%, 10 μm), and TiC (2-5 μm) powders were employed. The molar ratios of Ti : Si : TiC were selected as 1 : 1 : 2 (M1) and 2 : 2 : 3 (M2). Before sintering, the powders were mixed in a Turbula shaker mixer in Ar atmosphere for 24 hours. During milling, some zirconia balls of 3 mm in diameter were put into the container to ensure better mixing. The mixture powders were filled in a cylindrical graphite mold with a diameter of 20 mm and then sintered by PDS. The chamber pressure was 10^{-3} Pa. During sintering, the heating rate was controlled in a range from 50-60°C/min and the applied pressure was maintained at 50 MPa. The sintering process was controlled at 1200-1400°C for 8 to 240 min. The results showed that the phases in all the samples consisted of Ti_3SiC_2 and a small amount of TiC, and the optimum sintering temperature was found to be in a small range of 1250-1300°C. For M1 sample, the lowest TiC content could be only decreased to about 3-4 wt%, whereas the content of TiC in M2 sample was always lower than that in M1. When M2 powder was sintered at 1300°C for 8 to 240 min, the TiC impurity was decreased to 1 wt%. The grain size of Ti_3SiC_2 increased from 5-10 μm to 80-100 μm with increasing holding time. The relative density of M2 sample was measured to be higher than 99%. Gao *et al.* [33] synthesized consolidated Ti_3SiC_2 using Ti/Si/2TiC by SPS and found that 2 wt% TiC existed in the sample in the sintering temperature range of 1250-1300°C.

Also, Sun *et al.* [34] synthesized Ti_3SiC_2 from four starting powder mixtures of 3Ti/Si/2C, 3Ti/SiC/C, Ti/Si/2TiC, and 2Ti/2Si/3TiC. The purity and particle size of the powders used were as follows: 99.9% and 10 μm for Ti, 99.9% and 10 μm for Si, 99% and 2-5 μm for TiC, as well as 99% and 1 μm for C. The four groups of powders were respectively mixed in argon with a Turbula mixer for 86.4 ks. The powder mixtures were sintered by PDS in a vacuum of 10^{-3} Pa. The sintering temperature and holding time were changed to investigate the respective effects and the pressure was a constant of 50 MPa. The experimental results demonstrated that when the starting material of 3Ti/Si/2C or 3Ti/SiC/C was used the impurity of TiC existed as higher than 30 wt%. In the materials sintered from Ti/Si/2TiC and 2Ti/2Si/3TiC an optimum sintering temperature existed at 1573 K. Especially for 2Ti/2Si/3TiC mixture, the sintered sample contained the second phase of TiC less than 1 wt%. The phase purity and density of sample showed very little dependence on the sintering time range from minutes to four hours. In another paper, Sun *et al.* [35] further investigated the Ti-Si-C system using five kinds of starting powders such as 5Ti/2Si/3C, 5Ti/2SiC/C, 2Ti/2Si/3TiC, 4Ti/2SiC/TiC, and Ti/$TiSi_2$/3TiC. Figure 10 shows the XRD patterns of samples sintered from the five groups of powders. The results also indicated that the 2Ti/2Si/3TiC was the best composition to prepare high purity Ti_3SiC_2 (99 wt%) at 1300°C for 15 min. The relative density of sample could be higher than 98% of theoretical value above 1275°C.

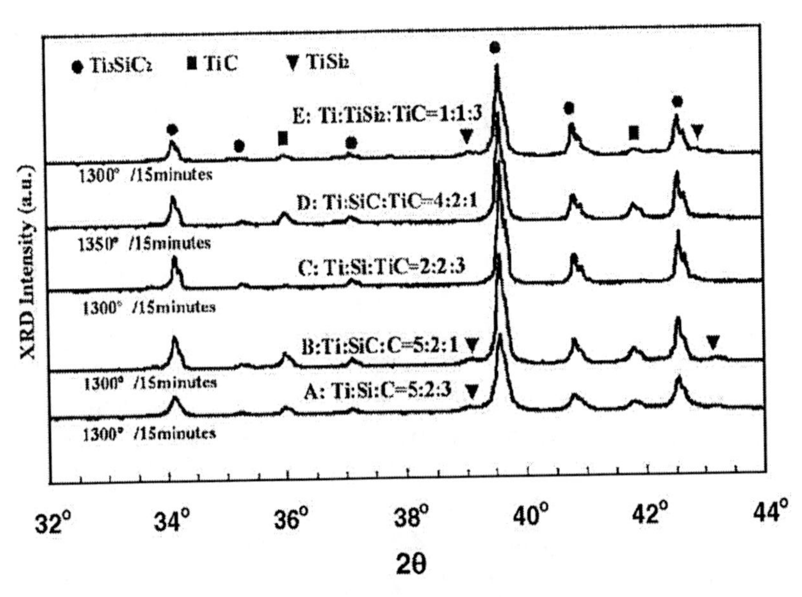

Figure 10. XRD patterns of the samples sintered at 1300 or 1350°C for 15 min from the five groups of powders. The molar ratios of constitutive powders are indicated in the figure [35].

To improve the purity of Ti_3SiC_2, the effect of Si content on the synthesis of Ti_3SiC_2 from Ti/xSi/2TiC (x = 1, 1.05, 1.1, and 1.15) was also investigated [36]. Ti (99.9%, 10 μm), Si (99.9%, 10 μm), and TiC (99%, 2-5 μm) powders were milled in a Tubular mixer in an Ar atmosphere for 24 h. During mixing, some zirconia balls with a diameter of 3 mm were put into the container to ensure the better mixing effect. The dried mixture powder was filled into a 20 mm diameter graphite die and sintered at 1200-1400°C by PDS. The chamber was evacuated to a pressure of 10^{-3} Pa before starting the sintering process. At the beginning, a rectangular pulse current with an intensity of 800 A and a pulse length of 30 ms was applied for 30 s, followed by a direct current superimposed with a pulse current wave. The current intensity was automatically controlled according to a temperature program. The heating rate was 50°C/min and the applied constant pressure was 50 MPa during sintering. It was found that the excess Si did not play a distinct role in improving the purity. Ti_3SiC_2 with a purity of about 97 wt% could be rapidly synthesized at 1250-1300°C.

For understanding the sintering mechanisms of 2Ti/2Si/3TiC for fabricating Ti_3SiC_2, the reaction path was investigated [37]. High purity Ti (99.9%, 10 μm), Si (99%, 2-3 μm), and TiC (99%, 2-5 μm) powders were mixed using a Tubular shaker mixer in argon atmosphere for 24 h and then compacted into a 20 mm diameter graphite die, and sintered in vacuum using PDS at 700-1300°C for 15 min with a constant pressure of 50 MPa. The results showed that Ti_5Si_3 formed as the intermediate phase during sintering. The reaction between Ti_5Si_3 and TiC as well as Si induced the formation of Ti_3SiC_2, and $TiSi_2$ appeared as the byproduct. At temperature above 1000°C, $TiSi_2$ reacted with TiC to form Ti_3SiC_2. Zou *et al.* [38,39] proved the assistant effect of Ti-Si liquid phase on the synthesis reaction of Ti_3SiC_2 using 2Ti/2Si/3TiC powder mixture. Starting powders were Ti (99.9%, 150 μm), Si (99.9%, 10 μm), and TiC (99%, 2-5 μm). These powders were mixed in a Turbula shaker mixer in Ar atmosphere for 24 h. The powder mixture was filled in a graphite mold (20 mm in diameter) and sintered in vacuum by PDS (PAS-V, Sodick Co. Ltd., Japan). The heating rate was controlled at 50°C/min, and the sintering temperature was selected in a range of 900-1500°C and held for 0-60 min. The axial pressure was 50 MPa. Ti_3SiC_2 was found to form through reaction among intermediate phases of Ti_5Si_3, $TiSi_2$, and the starting powders. Ti-Si liquid reaction above Ti-Ti_5Si_3 eutectic temperature (1330°C) appeared during sintering and was believed contributing to the product and densification of Ti_3SiC_2.

In further work, the powder mixture of TiH_2/SiC/C was found favorable for synthesizing single phase Ti_3SiC_2 [40]. Starting powders of TiH_2 (99%, 45 μm), SiC (99.9%, 2-3 μm), and C (99.7%, 5 μm) were weighed according to the molar ratios of 3 : 1 : 1, 5 : 2 : 1, and 2.8 : 1 : 0.8 and mixed in a Turbula shaker mixer in Ar atmosphere for 24 h. The mixture powder was filled in a graphite mold (20 mm in diameter) and sintered in vacuum by PDS (PAS-V, Sodick Co. Ltd., Japan). The heating rate was controlled at 50°C/min, and the sintering temperature was selected in a range of 800-1450°C and held for 0-20 min under a constant pressure of 50 MPa. It was found that dehydrogenation was accelerated by the synthesis reaction during sintering process and the formation of Ti_3SiC_2 was completed via the reactions among the intermediate phases of $Ti_5Si_3C_x$ and TiC. When using the molar ratio of 2.8 : 1 : 0.8, single phase dense Ti_3SiC_2 could be prepared at 1400°C for 20 min, as shown in Figure 11. The relative density of single phase sample was higher than 99%. In the microstructure, the coarse grains were 5-10 μm of width and beyond 20 μm of length.

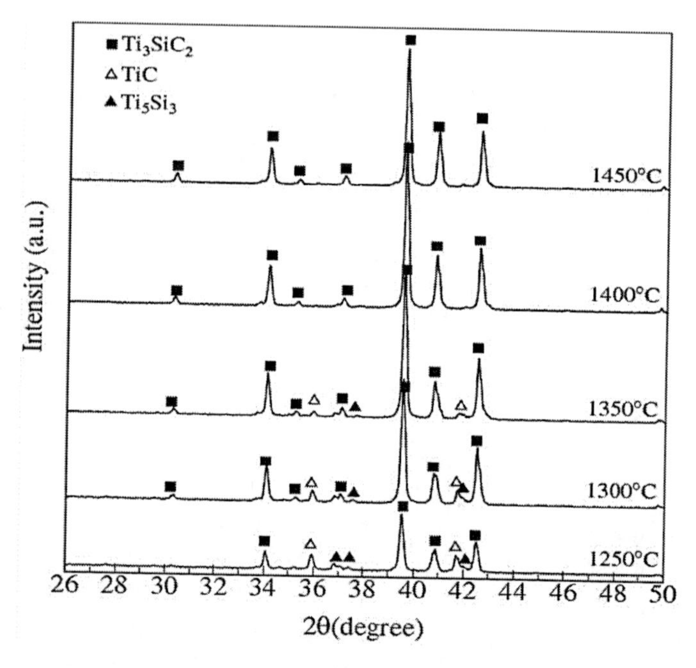

Figure 11. XRD patterns of samples sintered at various temperatures for 20 min using the mixture powder with a molar ratio of 2.8 : 1 : 0.8 [40].

According to aforementioned summary, it is seen that it is difficult to get pure Ti_3SiC_2 in Ti-Si-C system. Therefore, Al was added to explore the possibility for removing the residual TiC impurity in the samples. One example is that Sun *et al.* [41] revealed the effect of aluminum addition on the pure bulk Ti_3SiC_2 synthesis using Ti, SiC, C, and Al powder mixtures. Powders of Ti (99.9%, 10 μm), SiC (99%, 2-3 μm), C (99.7%, 5 μm), and Al (99.9%, 3 μm) in molar ratios of Ti : SiC : C : Al = 3 : 1 : 1 : x (x = 0, 0.05, 0.10, 0.15, and 0.20) were weighed and mixed by a Turbula shaker mixer in argon atmosphere for 24 h. The mixture was compacted into a graphite mold and sintered in vacuum using PDS technique (PAS-V, Sodick Co. Ltd., Japan) at 1200-1350°C for 15 min with a constant axial pressure of 50 MPa. Holding time was varied for certain temperatures to reveal the effects on the synthesis. It was found that a small amount of Al addition contributed to the pure Ti_3SiC_2 synthesis and decreasing the optimized sintering temperature. 100 wt% Ti_3SiC_2 could be synthesized from 3Ti/SiC/C/0.15Al powder mixture at temperature as low as 1200°C for 15 min.

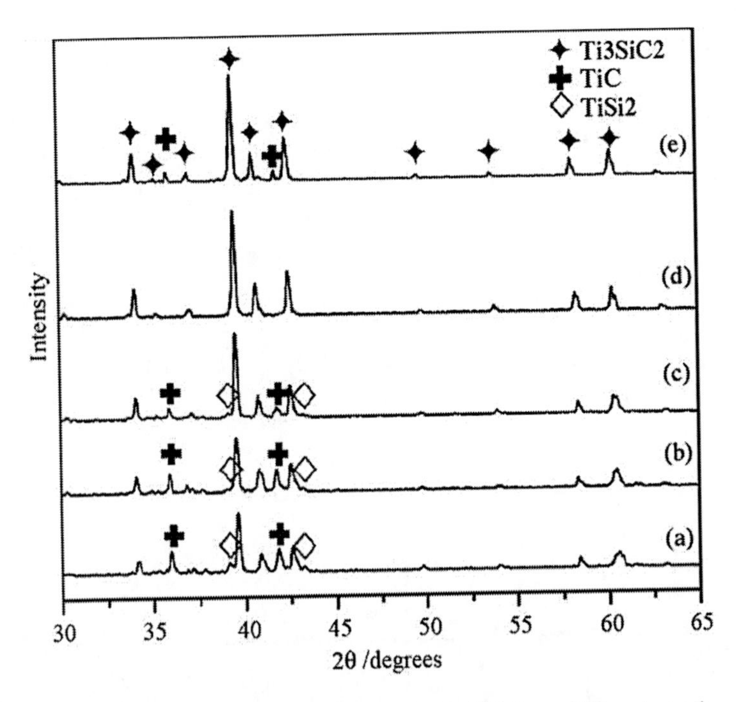

Figure 12. XRD patterns of samples fabricated by SPS at 1250°C from different starting compositions in molar ratios of (a) Ti:Si:TiC=1:1.2:2, (b) Ti:Si:TiC:Al=1:1.15:2:0.05, (c) Ti:Si:TiC:Al=1:1.1:2:0.1, (d) Ti:Si:TiC:Al=1:1:2:0.2, and (e) Ti:Si:TiC:Al=1:0.9:2:0.3 [42].

In another work, Zhou *et al.* [42,43] investigated the effect of Al additive into Ti, Si, and TiC mixture on the synthesis of pure Ti_3SiC_2. Titanium (99%, 10.6 μm), Si (99.5%, 9.5 μm), TiC (99.2%, 8.4 μm), and Al (99.8%, 12.8 μm) powders with the designed compositions of $Ti_3Si_{1.2-x}Al_xC_2$ (x = 0, 0.05, 0.1, 0.2, and 0.3) were mixed in ethanol for 24 h. After drying, the mixture was filled into a graphite die of 20 mm in diameter and then sintered in SPS system (Model SPS-1050, Sumitomo Coal Mining Co. Ltd., Japan). The samples were heated up to the requisite temperature at a rate of 80°C/min and the soaking time was 8 min under a pressure of 30 MPa. Figure 12 shows the XRD patterns of samples fabricated by SPS at

1250°C. When the aluminum addition was less than 0.1 mol, the main phase was Ti_3SiC_2 and a small amount of impurities of TiC and $TiSi_2$ were detected in the samples. For sample with the additive of 0.2 mol aluminum, pure Ti_3SiC_2 was obtained. However, with above 0.3 mol Al, TiC appeared again in the sample. It was revealed that the rapid formation of Ti_3SiC_2 was associated with the solid-liquid reaction between $TiSi_2$ liquid phase and TiC particles. It was also proposed that some Si atom sites in Ti_3SiC_2 have been substituted by Al to form a solid solution of $Ti_3Si(Al)C_2$ [44]. In addition, the optimized formation temperature of Ti_3SiC_2 was determined in a range of 1250-1300°C.

Furthermore, a method of using Ti, Si, and C element powders with the additive of Al to synthesize pure Ti_3SiC_2 by SPS was investigated, and the optimized composition was determined as 3Ti/Si/0.2Al/2C. Zhu *et al.* [45,46] found that fully dense and essentially single-phase polycrystalline Ti_3SiC_2 could be fabricated by SPS at 1250°C using this composition. Titanium (99%, 10.6 μm), Si (99.5%, 9.5 μm), carbon black (99%, 13.2 μm), and Al (99.8%, 12.8 μm) powders with the designed compositions were weighed and mixed in ethanol for 24 h. The powder mixture was sintered at 1250°C by SPS in vacuum with a heating rate of 80°C/min. The soaking time was 10 min and the pressure was 30 MPa. The obtained XRD results obviously indicated that no other phases but Ti_3SiC_2 existed in the sample with a starting composition of 3Ti/Si/0.2Al/2C, as shown in Figure 13. This result was also proved by Zhang *et al.* [47] who prepared high purity Ti_3SiC_2 at 1280°C for 6-36 min by SPS in vacuum (<6 Pa) with a heating rate of 100-200°C/min and a constant pressure of 50 MPa.

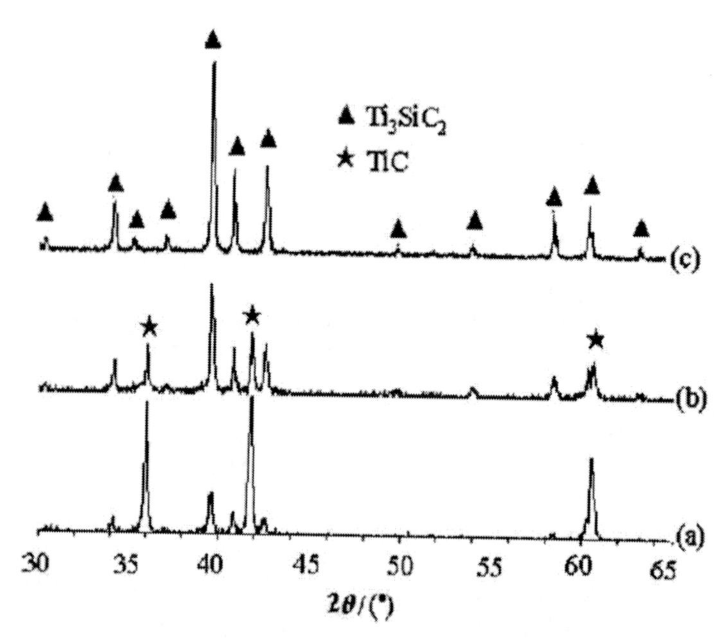

Figure 13. XRD patterns of samples with the starting compositions of (a) 3Ti/Si/2C, (b) 3Ti/1.2Si/2C, and (c) 3Ti/Si/0.2Al/2C [46].

Recently, in order to decrease the sintering temperature, Liang *et al.* [48,49] used mechanical alloying (MA) to enhance the reaction activity about mixed mixture. Ti (99.36%, 80 μm), Si (99.6%, 20 μm), carbon black (99%, 20 μm), and Al (99.6%, 50 μm) powders were weighed according to a molar ratio of 3Ti/Si/0.2Al/2C. The MA process was conducted in argon using a three-dimensional swinging, high energy ball mill with cylindrical jars and balls (10 mm in diameter) made of GCr15 steel. The powder ratio of ball to powder was 10 : 1 and the milling speed was 300 rpm. The 10 h-milled powders were put into a 10 mm graphite die and sintered by SPS with a heating rate of 100°C/min. The soaking time was 5 min and the applied pressure was 30 MPa. XRD result showed that TiC, Ti_3SiC_2, and $TiSi_2$ have already formed in the mixture and the TiC peaks were much stronger than $TiSi_2$ and Ti_3SiC_2 peaks. When sintering the powder at the temperature of 800-1100°C, it was found that the purity of Ti_3SiC_2 has reached 96.5 wt% at 850°C, as shown in Figure 14. At 1100°C, full dense Ti_3SiC_2 with a purity of 99.3 wt% and a relative density of 98.9% could be obtained. This investigation supported one way to prepare high purity Ti_3SiC_2 at much lower temperature than those reported before.

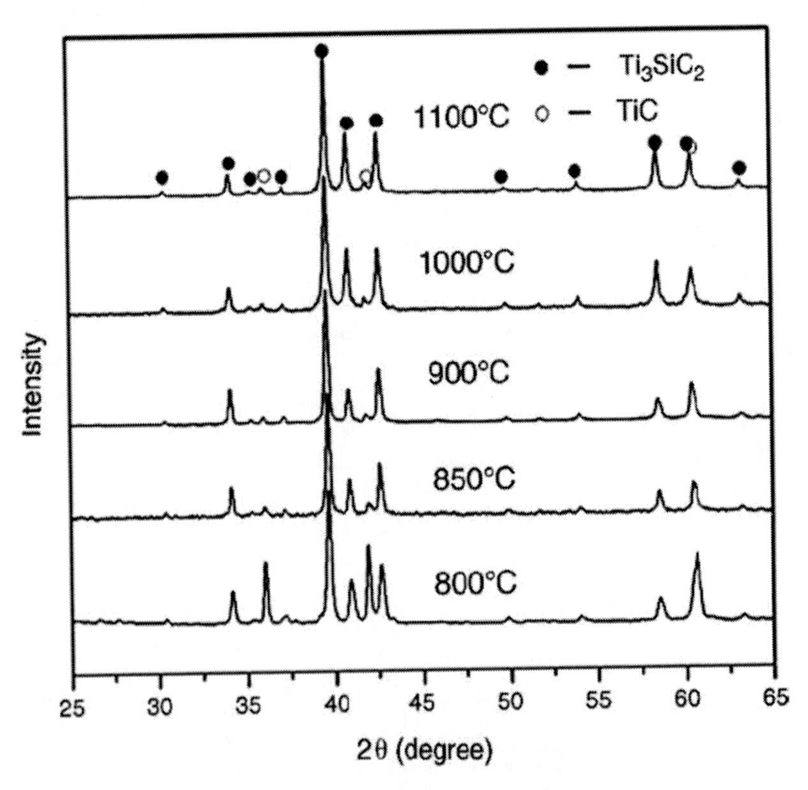

Figure 14. XRD patterns of compacts obtained by MA 10 h and then sintered at different temperatures using the starting mixture with the composition of 3Ti/Si/0.2Al/2C [48].

For synthesizing Ti_3SiC_2 composites by SPS or PDS, there are mainly two ways. Fist one is using *in-situ* reaction/pressing [50-59], and another is introducing second phase during powder milling and then sintering for densification [60-67]. Table 2 lists the relative initial compositions, powder preparation, sintering process parameters, microstructure, and optimized results of Ti_3SiC_2 composites. Obviously, it can be seen that all the composites could be densified at low temperatures, not higher than 1400°C.

Table 2. Initial compositions, powder preparation, sintering parameters, microstructure, and optimized results of Ti_3SiC_2 composites sintered by SPS or PDS

Samples	Initial compositions	Powder preparation	Sintering parameters	Microstructure	Optimized results
Ti_3SiC_2/TiC [50]	TiH_2/Si/1.5TiC $2TiH_2$/SiC/TiC $2.8TiH_2$/1.4SiC/TiC $3.5TiH_2$/1.75SiC/TiC $4TiH_2$/2SiC/TiC	TiC (99.9%, 1 μm), SiC (99%, 1 μm), Si (99.9%, 5 μm), C (99%, 5 μm), and TiH_2 (97.9%, 5 μm) powders. Milled with a mechanical pestle in isopropyl alcohol for 3 h.	Graphite molds of 30 and 50 mm diameter. PDSed, 1300°C, 50°C/min, 50 MPa, vacuum of 10^{-3} Pa.	TiC homogeneously distributed at the grain boundaries of Ti_3SiC_2.	(2.8-4)TiH_2/(1.4-2)SiC/TiC was the preferred composition.
Ti_3SiC_2/TiC [51]	Designed Ti, Si, C, and Al content	Ti (99%, 10 μm), Si (99%, 2 μm), C (99.9%, 1.6 μm), and Al (unknown) powders. Milled and dried.	SPSed, 1250-1350°C, 100-200°C/min, 70 MPa, holding 10 min, vacuum (less than 6 Pa).	Ti_3SiC_2-0, 10, 20, 30, and 40 vol.% TiC. Ti_3SiC_2 grains (2-10 μm) and TiC particles (about 1 μm). *In situ* addition of fine TiC particles to Ti_3SiC_2 matrix could evidently hinder the coarsening of Ti_3SiC_2 grains.	The addition of Al contributed to the effective control of theoretical TiC content.
Ti_3SiC_2/TiC [52]	Designed Ti, Si, and TiC content	Ti (99.9%, 45 μm), Si (99.9%, 45 μm), and TiC (99%, 1.72 μm) powders. Mixed by a Turbula shaker mixer for 24 h.	50 mm diameter graphite die. PDSed, 1250-1400°C, 50 MPa, holding 15 min.	Volume fractions of TiC in a range of 0-90%. The phase distribution of Ti_3SiC_2 and TiC was not uniform.	50 vol.% TiC composite has the maximum flexural strength.
Ti_3SiC_2/TiC/Ti_5Si_3 [53]	10Ti/4SiC	Ti (99%, 300 mesh) and SiC (99%, 0.5 μm) powders. Blended for 10 h with SiC media.	15 mm diameter graphite die. SPSed, 1260°C, 100-200°C/min, 60 MPa, holding 6 min.	Nanosized TiC particles homogeneously distributed in the Ti_5Si_3 matrix.	*In situ* formed TiC and Ti_3SiC_2 particles were very fine and uniformly distributed around Ti_5Si_3 grains.
Ti_3SiC_2/TiC/Ti_5Si_3 [54]	Ti : SiC = 5 : 2, 12 : 5, and 54 : 23	Ti (10 μm) and SiC (0.5 μm) powders. Milled and dried.	SPSed, 1260°C, 150-200°C/min, 60 MPa, holding 5 min, vacuum (less than 6 Pa).	0-50 vol.% Ti_3SiC_2. Fine grains were TiC, coarse grains were Ti_5Si_3, and Ti_3SiC_2 grains were plate-like.	The improvement of strength and toughness was attributed to the crack bridging of Ti_3SiC_2 grains.
Ti_3SiC_2/SiC/$TiSi_2$ [55]	C/6Si/4TiC	C (9 μm), Si (75 μm), and TiC (1.47 μm) powders. Blended in ethanol in Planet Ball Milling Machine for 12 h with agate media. Drying for 12 h at 70°C, and then sieving by a 200 mesh sieve.	SPSed, 1000-1260°C, 150-200°C/min, 60 MPa, holding 5 min, vacuum (less than 6 Pa).	The SiC particles distributed uniformly in the composites.	*In situ* formed SiC particles were very fine, as 200-300 nm.
Ti_3SiC_2/SiC [56]	0.61Ti/0.24Si/0.16C	Ti (99.47%, 45 μm), Si (99.9%, 75 μm), and C (99.99%, 50 μm) powders. Milled and dried.	15 mm diameter graphite die. SPSed, 1350°C, 100°C/min, 50 MPa, holding 6 min, vacuum (less than 6 Pa).	Ti_3SiC_2/20 vol.% SiC. TiC particles distributed in SiC phase.	The main phases were Ti_3SiC_2 and SiC with a small amount of TiC impurity.

Samples	Initial compositions	Powder preparation	Sintering parameters	Microstructure	Optimized results
Ti_3SiC_2/SiC [57]	Designed Ti, Si, and C content. Lower than 1 wt% Al additive.	Ti (99%, 10 μm), Si (99%, 2 μm), C (99.9%, 1.6 μm), and Al (99%, 45 μm) powders. Milled and dried.	SPSed, 900-1300°C, 100-200°C/min, 70 MPa, holding 6 min, vacuum (less than 6 Pa).	Ti_3SiC_2/20 vol.% SiC. Ti_3SiC_2 grains were 5 μm, and most of SiC grains were very fine as about 100 nm.	The composite showed improved mechanical properties than before.
Ti_3SiC_2/SiC [58]	Designed Ti, Si, and C content. Lower than 1 wt% Al additive.	Ti (99%, 10 μm), Si (99%, 2 μm), C (99.9%, 1.6 μm), and Al (99%, 45 μm) powders. Milled and dried.	SPSed, 1250-1350°C, 100-200°C/min, 70 MPa, holding 10 min, vacuum (less than 6 Pa).	0, 10, 20, 30, and 40 vol.% SiC. Fine SiC particles (about 100 nm) uniformly distributed in the Ti_3SiC_2 (2-10 μm) matrix.	The residual stress caused by thermal expansion mismatch was minimized.
Ti_3SiC_2/Al_2O_3 [59]	22.5 wt% Ti, 56.2 wt% TiC, 13.1 wt% Si, and 8.2 wt% Al_2O_3.	Ti (TSPT-350, 45 μm), Si (99.99%, 200 mesh), TiC (1.36 μm), and Al_2O_3 (AKP-30) powders. Milled and dried.	30 mm diameter graphite die. SPSed, 1200-1400°C, 100°C/min, 40 MPa, holding 5-15 min, vacuum.	Ti_3SiC_2/10 vol.% Al_2O_3. The agglomeration of Al_2O_3 appeared in the sintered composite at 1400°C.	When being sintered at 1300°C, the composite has the highest flexural strength of 600 MPa and fracture toughness of 7.4 $MPa{\cdot}m^{1/2}$.
Ti_3SiC_2/Al_2O_3 [60]	Designed Al_2O_3 and Ti_3SiC_2 content.	Al_2O_3 (99.9%, 0.2 μm), and Ti_3SiC_2 (99%, 10 μm) powders. Blended by ball milling for 48 h, dried by a rotary evaporator, and then sieved with 100 meshes.	20 mm diameter graphite die. SPSed, 1300°C, 30 MPa, vacuum (less than 6 Pa).	0-100 vol.% Ti_3SiC_2. The average alumina grain size tended to decrease with increasing Ti_3SiC_2 content.	More than 20 vol.% Ti_3SiC_2 was necessary to achieve a metallike composite.
Ti_3SiC_2/Al_2O_3 [61]	Designed Al_2O_3 and Ti_3SiC_2 content.	Al_2O_3 (99.9%, 0.2 μm), and Ti_3SiC_2 (99%, 10 μm) powders. Blended by ball milling for 48 h, dried by a rotary evaporator, and then sieved with 100 meshes.	20 mm diameter graphite die. SPSed, 1300°C, 30 MPa, vacuum (less than 6 Pa).	0-100 vol.% Ti_3SiC_2. The grain growth of alumina was clearly controlled by the presence of Ti_3SiC_2.	The composites containing 30-100 vol.% Ti_3SiC_2 could be machined by conventional Fe-Mo-W drills.
Ti_3SiC_2/HAP [62]	Designed HAP and Ti_3SiC_2 content.	HAP (20 nm), and Ti_3SiC_2 (99%) powders. Mixed by ball milling for 48 h, dried and sieved to pass 100 meshes.	20 mm diameter graphite die. SPSed, 1200°C, 200°C/min, 60 MPa, holding 5 min, vacuum (less than 6 Pa).	0-50 vol.% Ti_3SiC_2. The grain size of Ti_3SiC_2 and HAP were 3-10 and less than 1 μm, respectively.	The strength and fracture toughness of HAP were improved significantly by the addition of Ti_3SiC_2.
Ti_3SiC_2/HAP [63]	Designed HAP and Ti_3SiC_2 content.	HAP (20 nm), and Ti_3SiC_2 (99%) powders. Mixed by ball milling for 48 h, dried and sieved to pass 100 meshes.	20 mm diameter graphite die. SPSed, 1200°C, 200°C/min, 60 MPa, holding 4 min, vacuum (less than 6 Pa).	0-50 vol.% Ti_3SiC_2. No additional phase was found at the grain boundary.	The composites exhibited excellent machinability when the Ti_3SiC_2 content was higher than 20 vol.%.

Table 2. (Continued)

Samples	Initial compositions	Powder preparation	Sintering parameters	Microstructure	Optimized results
$Ti_3SiC_2/CaSiO_3$ [64]	Designed $CaSiO_3$ and Ti_3SiC_2 content.	$CaSiO_3$ and Ti_3SiC_2 powders. Milled for 24 h in alcohol, dried and sieved to pass 125 meshes.	20 mm diameter graphite die. SPSed, 1050-1150°C, 100-200°C/min, 50 MPa, holding 5 min, vacuum (less than 6 Pa).	10-30 volume fraction of Ti_3SiC_2. The grain size of Ti_3SiC_2 and $CaSiO_3$ were 10 and 1-2 μm in length, respectively.	When the volume fraction of Ti_3SiC_2 was below 30%, the composites showed good bioactivity with larger growth rate of HAP layer.
Ti_3SiC_2/3Y-TZP [65]	Designed 3Y-TZP and Ti_3SiC_2 content.	3Y-TZP (0.5 μm), and Ti_3SiC_2 (99%, 10 μm) powders. Ball milled for 48 h, dried and sieved by 100 meshes.	20 mm diameter graphite die. SPSed, 1300°C, 50 MPa, vacuum (less than 6 Pa).	0-50 vol.% Ti_3SiC_2. The laminated grain size of Ti_3SiC_2 was 3-10 μm, and the granular particle size of 3Y-TZP was less than 1 μm.	The fracture toughness of composites were improved greatly with increasing Ti_3SiC_2 content, up to 9.88 $MPa{\cdot}m^{1/2}$.
Ti_3SiC_2/3Y-TZP [66]	Designed 3Y-TZP and Ti_3SiC_2 content.	3Y-TZP (99.9%, 0.5 μm), and Ti_3SiC_2 (99%, 10 μm) powders. Ball milled for 48 h, dried and sieved by 100 meshes.	20 mm diameter graphite die. SPSed, 1300°C, 200-300°C/min, 50 MPa, holding 5 min, vacuum (less than 6 Pa).	0-50 vol.% 3Y-TZP. The laminated grain size of Ti_3SiC_2 was 5-30 μm, and the granular particle size of 3Y-TZP was less than 1 μm.	The maximum fracture toughness of 11.94 $MPa{\cdot}m^{1/2}$ was achieved for the composite with 30 vol.% 3Y-TZP.
FGM Ti_3SiC_2/Al_2O_3 [67]	Designed Al_2O_3 and Ti_3SiC_2 gradient content.	Al_2O_3 (99.9%, 0.2 μm), and Ti_3SiC_2 (99%, 20 μm) powders. Blended by ball milling for 48 h, dried by a rotary evaporator, and then sieved with 100 meshes. Stacked layer by layer.	24 mm diameter graphite die. SPSed, 1350°C, 25 MPa, vacuum (less than 6 Pa).	Gradient 0-100 vol.% Ti_3SiC_2. Ten layers. The structure of FGM was macroscopically heterogeneous due to the graded distribution of the chemical composition.	The electrical conductivity and knoop hardness showed the graded increase and decrease respectively with increasing Ti_3SiC_2 content.

(B) Ti_3AlC_2

The early work for synthesizing Ti_3AlC_2 by SPS was introduced at 2003 [68]. Ti (99.4%, 400 mesh), activated carbon (98%), and in-house synthesized Al_4C_3 (100 mesh) powders were used. Al_4C_3 was fabricated by sintering the mixture of Al (99.5%, 200 mesh) and activated carbon with a suitable stoichiometric composition at 1400°C for 1 h under Ar atmosphere. The mixture of Ti, Al_4C_3, and C powders with a molar ratio of Ti : Al : C = 3 : 1.2 : 2 was ball-milled in ethanol for 24 h. After being dried and sieved with 100 mesh screen, the mixture was placed into a 20 mm diameter graphite die and heated to 1300°C at a heating rate of 600°C/min by SPS (Sumitomo Coal Mining Co. Ltd., Japan). A pressure of 22 MPa was applied to assist the densification. The soaking time was selected as 1, 3, 5, and 7 min, respectively. It was found that when the soaking time increased from 1 to 5 min, the Ti_3AlC_2

peak intensities increased stably while those of TiC showed a revised trend. However, when the soaking time was longer than 7 min, almost pure TiC existed in the sample.

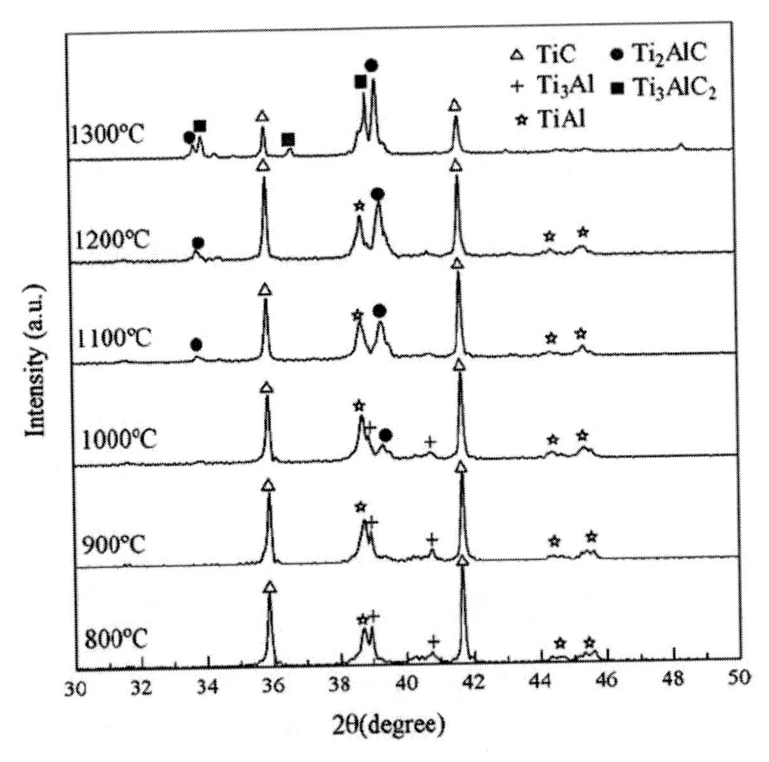

Figure 15. XRD patterns of the 2Ti/2Al/3TiC powder mixture and the samples heated to designed temperatures and immediately cooled down [70].

Almost single phase Ti_3AlC_2 was rapidly synthesized from Ti/Al/TiC powder mixture at 1300°C in vacuum for 15 min with a molar ratio of Ti : Al : TiC = 2 : 2 : 3 by using PDS technique [69]. Starting powders of Ti (99.9%, 10 μm), Al (99.9%, 10 μm), and TiC (2-5 μm) were used. The powders were mixed in a Turbula shaker mixer in Ar atmosphere for 24 hours. The mixture was filled in a graphite mold (50 mm in diameter) and sintered in a PDS facility (PAS-V, Sodick Co. Ltd., Japan). The mold temperature was monitored and controlled as the sintering temperature through an infrared camera. The heating rate was 50°C/min and a constant axial pressure of 50 MPa was applied. The size of plate-like grains in as-prepared Ti_3AlC_2 mainly distributed in a range of 5-15 μm. The stabilized Vickers hardness under a load of 19.6 N was 2.8 GPa and the room temperature flexural strength was 416±10 MPa. Additionally, for investigating the synthesis mechanism of Ti_3AlC_2, the mixture powder was heated to various intermediate temperatures of 800-1300°C and then immediately cooled down [70]. Figure 15 shows XRD patterns of as-prepared samples. When the sample was heated to 800°C, peaks of intermetallic compounds Ti_3Al and TiAl were detected firstly. With increasing temperature, the amount of TiAl increased while that of Ti_3Al decreased. When the temperature was up to 1000°C, Ti_2AlC appeared. At 1200°C, a little Ti_3AlC_2 was formed in the sample. Especially when heated to 1300°C, the diffraction peaks of Ti_3AlC_2 were clearly observed and those of TiAl disappeared. The relative possible reactions could be expressed as follows,

$$4Ti + 2Al = TiAl + Ti_3Al \tag{1}$$

$$Ti_3Al + 2Al = 3TiAl \tag{2}$$

$$TiAl + TiC = Ti_2AlC \tag{3}$$

$$Ti_2AlC + TiC = Ti_3AlC_2 \tag{4}$$

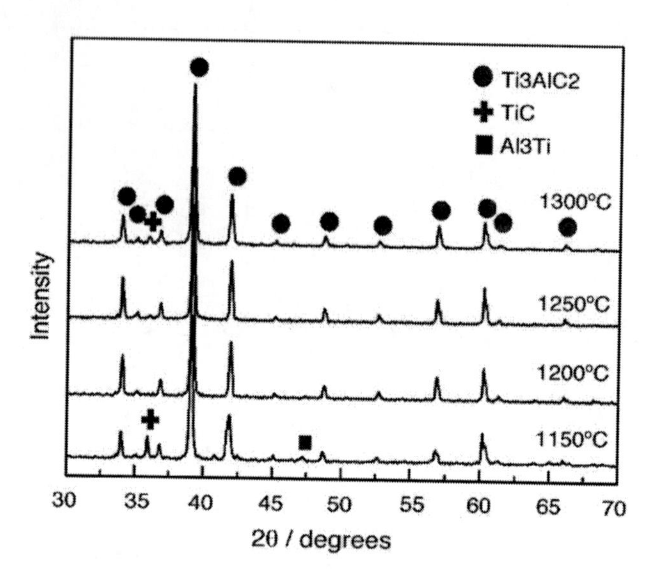

Figure 16. XRD patterns of samples with the starting composition of $Ti_3Al_{1.0}Si_{0.2}C$ sintered at 1150-1300°C [71].

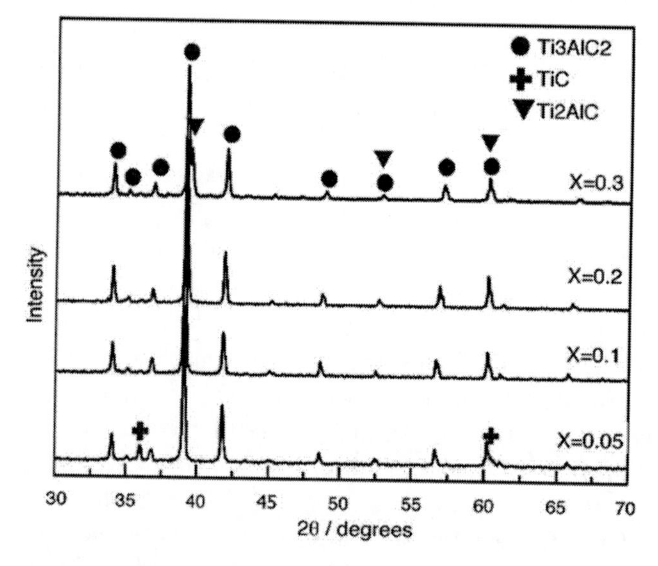

Figure 17. XRD patterns of samples with the starting composition of $Ti_3Al_{1.2\text{-}x}Si_xC$, where $x = 0.05$, 0.1, 0.2, and 0.3, sintered at 1250°C [71].

On the other hand, the effect of silicon on synthesis of Ti_3AlC_2 was investigated by SPS from Ti/Al/TiC powders [71]. The powders were titanium (99%, 10.6 μm), Si (99.5%, 9.5 μm), Al (99.8%, 12.8 μm), and TiC (99.2%, 8.4 μm). The mixture with designed composition was firstly mixed in ethanol for 24 h and then dried. The mixture was filled into a 20 mm diameter graphite die and finally SPSed (Dr. Sinter-1050, Izumi Technology Co. Ltd., Japan). The samples were heated at a rate of 80°C/min at 1150-1300°C. The soaking time was 8 min. Figure 16 shows XRD patterns of samples obtained by sintering $Ti_3Al_{1.0}Si_{0.2}C$ at different temperatures. When sintered at 1150°C, the content of Ti_3AlC_2 was very high, only very weak TiC and Al_3Ti peaks were identified. By increasing the sintering temperature up to 1200 and 1250°C, essentially single phase Ti_3AlC_2 was obtained. However, at 1300°C, TiC appeared again. It seemed that the optimized temperature was 1200-1250°C. Through modifying the Si content in initial compositions, the phase compositions were identified in sintered samples, as shown in Figure 17. For the sample with the addition of 0.05 mol Si, the product was highly pure, and only very weak peaks of TiC were found. With 0.1 and 0.2 mol Si addition, reaction products were pure Ti_3AlC_2. However, more Si content addition (0.3 mol) caused the appearance of Ti_2AlC. Therefore, the preferred Si content was 0.1-0.2 mol.

Also, there is one work about synthesizing Ti_3AlC_2 using TiH_2/Al/C powders by SPS [72]. Starting powders of TiH_2 (99%, 45 μm), Al (99.9%, 10 μm), and C (99%, 5 μm) were used. The powders in a molar ratio of TiH_2 : Al : C = 3 : 1.1 : 1.8 were mixed in a Turbula shaker mixer in Ar atmosphere for 24 h. The powder mixture was filled into a 20 mm diameter graphite die and sintered in vacuum by PDS (PAS-V, Sodick Co. Ltd., Japan). The heating rate was 50°C/min, and the sintering temperature was in a range of 900-1500°C and held for 0-60 min. A constant axial pressure of 50 MPa was applied. It was found that it was not possible to get pure Ti_3AlC_2. The impurities of Ti_2AlC and TiC always existed in the sintered samples. It was considered that incomplete dehydrogenation of TiH_2 postponed the formation of intermediate phases, which resulted in final products with more impurity phases.

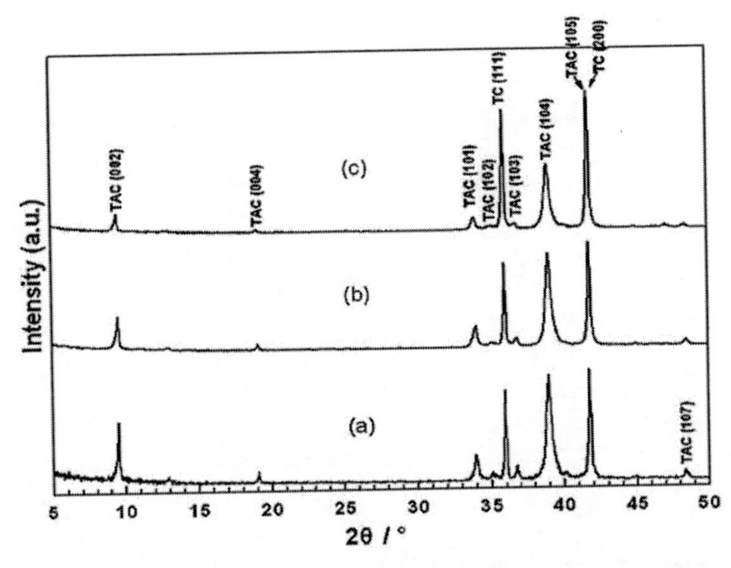

Figure 18. XRD patterns of the powders formed by 9.5 h mechanical alloying of different starting mixtures: (a) 3Ti/Al/2C, (b) 3Ti/1.1Al/2C, and (c) 3Ti/1.2Al/2C. TAC and TC stand for the Ti_3AlC_2 and TiC, respectively [73].

The combination of mechanical alloying and SPS contributes to the low temperature sintering of Ti_3AlC_2 [73,74]. Commercially available Ti (99.36%, 80 μm), Al (99.6%, 50 μm), and graphite (99%, 20 μm) powders were weighed according to a molar ratio of 3Ti/*x*Al/2C ($x = 1$, 1.1, and 1.2). The powder mixtures were put into steel jars and sealed in a glove box under argon protective atmosphere. The MA was conducted by swinging and high energy ball milling. The rotation speed was set at 400 rpm. The weight ratio of balls to powders was 10 : 1. After 9.5 h, the milled powder was sieved using 100 mesh sieve and put in a 10 mm graphite die and then sintered by SPS. The heating rate was controlled at 80°C/min and the sintering temperature was selected in a range of 900-1150°C. The soaking time was 5-20 min and the constant axial pressure was 35 MPa. Figure 18 shows XRD patterns of powders obtained by mechanical alloying. Obviously, a large amount of Ti_3AlC_2 has been formed during MA process. With increasing Al content in the initial composition, the relative peak intensities of Ti_3AlC_2 decreased in comparison with those of TiC, which indicated that excessive Al suppressed the formation of Ti_3AlC_2 during MA. The formation mechanism of Ti_3AlC_2 might be due to the self-propagating reaction. According to the XRD examination of SPSed 3Ti/Al/2C powder at 900-1150°C for 5 min, it was found that the optimized sintering temperature was 1050°C. By sintering the 3Ti/*x*Al/2C ($x = 1$, 1.1, and 1.2) powders at 1050°C for 5 min, it was revealed that the peak intensity of TiC was the lowest when the Al content was 1.1 molar, as shown in Figure 19. Furthermore, by increasing the annealing time from 5 to 20 min, the high purity Ti_3AlC_2 with the TiC content less than 1 wt% was obtained at holding time of 15-20 min. The size of grains could reach less than 0.5 μm in thickness and 10-20 μm in length.

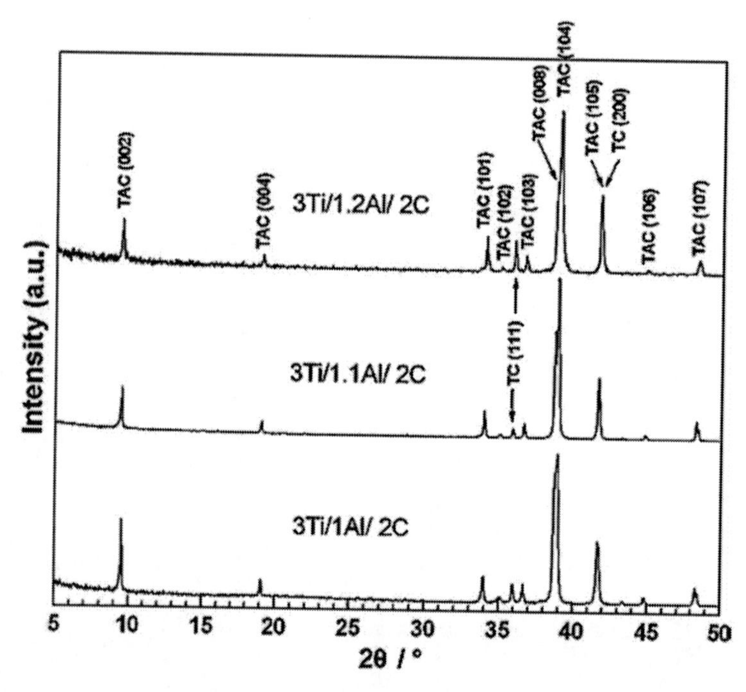

Figure 19. XRD patterns of samples sintered from MA powders with different Al content at 1050°C for 5 min [73].

About Ti_3AlC_2 composites synthesized by SPS, dense Ti_3AlC_2/TiB_2 was successfully prepared from B_4C/TiC/Ti/Al powders [75]. Ti (99.2%, 10.6 μm), B_4C (99.5%, 2.8 μm), TiC (99%, 2.6 μm), and Al (99.6%, 1.7 μm) powders were weighed according to the reactions of

$$Ti + 2TiC + Al = Ti_3AlC_2 \qquad (5)$$

$$3Ti + B_4C = TiC + 2TiB_2 \qquad (6)$$

The calculated volume content of TiB_2 was 5, 10, 20, and 30% in the composites, respectively. After ball milling in ethanol for 24 h, the powders were dried, sieved, and compacted uniaxially under 20 MPa in a graphite die pre-sprayed with a layer of BN. The green body was sintered by SPS (Dr. Sinter-1050, Izumi Technology Co. Ltd., Japan). The temperature was measured by means of an optical pyrometer focusing on a small hole in the die. The sintering parameters were 80°C/min, in vacuum of 0.5 Pa, and a pressure of 30 MPa. The soaking time was 8 min. According to the XRD analysis, there was no TiC_x impurity in Ti_3AlC_2/10-20 vol.% TiB_2 composites. When the TiB_2 content exceeded 20 vol.%, TiC appeared, which was probably ascribed to the incomplete reaction because of incorporation with too much TiB_2 content. The measured densities of all Ti_3AlC_2/TiB_2 composites were 98.4-99.2% of theoretical values. When the volume content of TiB_2 was above 10%, the density decreased a lot which was due to the agglomeration of TiB_2 in the microstructure. Additionally, Ti_2AlC-Ti_3AlC_2 composite was fabricated using Ti/Al/TiC mixture at 1200-1300°C [76] and Ti_2AlC-Ti_3AlC_2-Ti_3SiC_2 composite was synthesized using 2Ti/Al/0.2Si/C powders at 1100-1200°C by SPS [77].

5. *MAX* 413 CARBIDE

To date, there are eight discovered 413 phases, but only one 413 phase Nb_4AlC_3 has been investigated and fabricated by SPS [18]. Dense bulk Nb_4AlC_3 was *in-situ* fabricated by SPS using niobium (99.9%, 45 μm), aluminum (99.9%, 30 μm), and carbon black (99%, 20 nm) powders as initial materials. For investigating the reaction path, niobium, aluminum, and carbon black powers with a molar ratio of 4 : 1.5 : 2.7 were weighed using an electrical balance with an accuracy of 10^{-2} g. Excess aluminum and less carbon were selected because that aluminum would be lost at high temperature and carbon-defects exist in the Al-containing MAX phases. The powders were put into an agate jar and milled for 12 hours using ethanol as the dispersant. After milling, the mixed powders were dried in air and sieved using a 100 mesh sieve. The obtained mixture was put into a graphite die with a diameter of 20 mm. A layer of carbon sheet (~0.2 mm thickness) was put in the inner of die for lubrication. A layer of heat isolation carbon fiber was used to wrap the die for inhibiting the rapid heat diffusion. The mixture was firstly cold pressed as a compact green. Then the green together with the die was heated in a spark plasma sintering facility (100 kN SPS-1050, Syntex Inc., Japan). The sintering temperature was measured by an optical pyrometer focusing on a hole in the wall of die. From ambient temperature to 700°C, it took 5 min to heat the sample. Between 700 and 1400°C, a heating speed of 50°C/min was adopted. Above

1400°C, the heating speed was set as 10°C/min. The annealing temperatures were selected as 800, 1000, 1200, 1400, and 1600°C, respectively. The vacuum degree was 7-10 Pa. The holding time was 2 min. It was found that Nb_4AlC_3 was formed through the reaction between Nb_2AlC and NbC, and the decomposition of Nb_2AlC. Additionally, for investigating the effect of annealing temperature on the synthesis of single phase Nb_4AlC_3, the mixture powders with the molar ratio of Nb : Al : C = 4 : 1.5 : 2.7 were sintered at 1620, 1650, 1665, and 1680°C, respectively. It was found that the sintered sample decomposed into niobium carbide above 1665°C. Therefore, 1650°C was considered as the optimized annealing temperature. In order to determine the optimized composition, the mixed powders with the molar ratios of Nb : Al : C = 4 : 1.1 : 2.7, 4 : 1.3 : 2.7, 4 : 1.4 : 2.7, and 4 : 1.5 : 2.7 were sintered at 1650°C for 2 min, respectively. The phase compositions of sintered samples were examined by XRD. It was found that the molar ratio of Nb : Al : C = 4 : 1.5 : 2.7 was the optimized composition to synthesize Nb_4AlC_3 which did not contain niobium carbide. Additionally, the longer holding time of 4 min was used for preparing Nb_4AlC_3. However, the longer holding time of 4 min contributed to the decomposition of Nb_4AlC_3, i.e., the holding time of 2 min was better for the synthesis of Nb_4AlC_3. Therefore, dense bulk Nb_4AlC_3 was sintered at 1650°C for 2 min using the mixture powders with the molar ratio of Nb : Al : C = 4 : 1.5 : 2.7. The applied pressure was 30 MPa. Figure 20 displays XRD spectra of Nb_4AlC_3 sample sintered using the optimized parameters. It was seen that the primary phase is Nb_4AlC_3 and a few amount of Nb_2AlC and Al_3Nb existed in the sample. The impurities of Nb_2AlC and Al_3Nb were less than 6 wt%. The measured density was 6.71 g/cm^3, about 95% of the theoretical density (7.04 g/cm^3). In the sintered dense sample, the growth of grain did not show the preferable direction, i.e., textured microstructure. The mean grain size was determined as 21 μm in length and 9 μm in width. In the fracture surface, Nb_4AlC_3 grains exhibited the multiplex damage modes, such as transgranular fracture, intergranular fracture, kink bands, and delaminations.

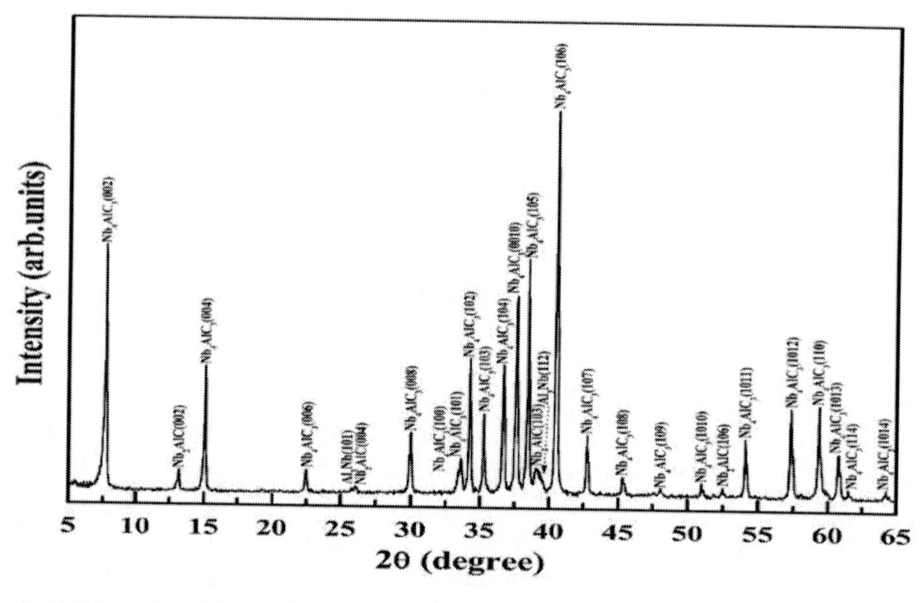

Figure 20. XRD pattern of dense Nb_4AlC_3 synthesized using the optimized parameters under a pressure of 30 MPa [18].

CONCLUSIONS

SPS or PDS could be used for successfully fabricating MAX phases and related composites at a lower temperature and in a shorter time. The synthesis parameters, such as initial composition, sintering temperature, heating rate, soaking time, as well as applied pressure, were systemically summarized. The most important three factors are initial composition, sintering temperature, and holding time, listed as follows: (1) For 211 phases, single phase Ti_2AlN, Ti_2AlC, and Cr_2AlC could be fabricated using the optimized process parameters of Ti/Al/TiN/1200°C/10 min, 2Ti/1.2Al/C/1100°C/8 min, and 2Cr/1.1Al/C/1400°C/5 min, respectively. (2) For 312 phases, pure Ti_3SiC_2 and Ti_3AlC_2 could be synthesized under the optimized conditions of 3Ti/Si/0.2Al/2C/1250°C/10 min and Ti/Al/2TiC/0.1-0.2Si/1200-1250°C/8 min, respectively. (3) For 413 phases, the optimized sintering parameters were 4Nb/1.5Al/2.7C/1650°C/2 min for preparing high purity Nb_4AlC_3.

ACKNOWLEDGMENTS

This work was partially supported by Grand-in-Aid for Scientific Research from the JSPS of Japan and also World Premier International Research Center (WPI) Initiative, MEXT, Japan.

REFERENCES

[1] Nowotny, H. *Prog. Solid State Chem.* 1970, 2, 27.
[2] Barsoum, M.W. *Prog. Solid State Chem.* 2000, 28, 201.
[3] Hu, C.F.; Li, F.Z.; He, L.F.; Liu, M.Y.; Zhang, J.; Wang, J.M.; Bao, Y.W.; Wang, J.Y.; Zhou, Y.C. *J. Am. Ceram. Soc.* 2008, 91, 2258.
[4] Hu. C.F.; He, L.F.; Liu, M.Y.; Wang, X.H.; Wang, J.Y.; Li, M.S.; Bao, Y.W.; Zhou, Y.C. *J. Am. Ceram. Soc.* 2008, 91, 4029.
[5] Low, I.M. *J. Eur. Ceram. Soc.* 1998, 18, 709.
[6] Barsoum M.W.; El-Raghy, T. *J. Am. Ceram. Soc.* 1996, 79, 1953.
[7] Barsoum, M.W.; Yoo, H.-I.; Polushina, I.K.; Rud', V.Yu.; Rud', Yu.V.; El-Raghy, T. *Phys. Rev. B* 2000, 62, 10194.
[8] Zhou Y.C.; Sun, Z.M. *Mater. Res. Innovat.* 1999, 2, 360.
[9] El-Raghy, T.; Zavaliangos, A.; Barsoum, M.W.; Kalidindi, S.R. *J. Am. Ceram. Soc.* 1997, 80, 513.
[10] Wang, J.M.; Wang, J.Y.; Zhou, Y.C.; Hu, C.F. *Acta Mater.* 2008, 56, 1511.
[11] Bao, Y.W.; Zhou, Y.C.; Zhang, H.B. *J. Mater. Sci.* 2007, 42, 4470.
[12] Hu, C.F.; Zhou, Y.C.; Bao, Y.W. *Ceram. Inter.* 2008, 34, 537.
[13] Jacques, S.; Di-Murro, H.; Berthet, M.-P.; Vincent, H. *Thin Solid Films* 2005, 478, 13.
[14] Tzenov, N.V.; Barsoum, M.W. *J. Am. Ceram. Soc.* 2000, 83, 825.
[15] Hu. C.F.; He, L.F.; Zhang, J.; Bao, Y.W.; Wang, J.Y.; Li, M.S.; Zhou, Y.C. *J. Eur. Ceram. Soc.* 2008, 28, 1679.

[16] Murugaiah, A.; Souchet, A.; El-Raghy, T.; Radovic, M.; Sundberg, M.; Barsoum, M.W. *J. Am. Ceram. Soc.* 2004, 87, 550.
[17] Wilhelmsson, O.; Eklund, P.; Högberg, H.; Hultman, L.; Jansson, U. *Acta Mater.* 2008, 56, 2563.
[18] Hu, C.F.; Sakka, Y.; Tanaka, H.; Nishimura, T.; Grasso, S. *J. Alloys Compd.* 2009, 487, 675.
[19] Orrù, R.; Licheri, R.; Locci, A.M.; Cincotti, A.; Cao, G. *Mater. Sci. Eng. R* 2009, 63, 127.
[20] Grasso, S.; Sakka, Y.; Maizza, G. *Sci. Technol. Adv. Mater.* 2009, 10, 053001.
[21] Munir, Z.A.; Tamburini, U.A.; Ohyanagi, M. *J. Mater. Sci.* 2006, 41, 763.
[22] www.spssyntex.net
[23] Shen, Z.J.; Nygren, M. *Chem. Rec.* 2005, 5, 173.
[24] www.ceramicindustry.com.
[25] Yan, M.; Mei, B.C.; Zhu, J.Q.; Tian, C.G.; Wang, P. *Ceram. Inter.* 2008, 34, 1439.
[26] Mei, B.C.; Zhou, W.B.; Zhu, J.Q.; Hong, X.L. *J. Mater. Sci.* 2004, 39, 1471.
[27] Zhou, W.B.; Mei, B.C.; Zhu, J.Q.; Hong, X.L. *Mater. Lett.* 2005, 59, 131.
[28] Tian, W.B.; Vanmeensel, K.; Wang, P.L.; Zhang, G.J.; Li, Y.X.; Vleugels, J.; Biest, O.V.D. *Mater. Lett.* 2007, 61, 4442.
[29] Tian, W.B.; Sun, Z.M.; Du, Y.L.; Hashimoto, H. *Mater. Lett.* 2008, 62, 3852.
[30] Tian, W.B.; Sun, Z.M.; Hashimoto, H.; Du, Y.L. *J. Alloys Compd.* 2009, 484, 130.
[31] Zhang, Z.F.; Sun, Z.M.; Hashimoto, H.; Abe, T. *J. Eur. Ceram. Soc.* 2002, 22, 2957.
[32] Zhang, Z.F.; Sun, Z.M.; Hashimoto, H. *Metal. Mater. Trans. A* 2002, 33, 3321.
[33] Gao, N.F.; Li, J.T.; Zhang, D.; Miyamoto, Y. *J. Eur. Ceram. Soc.* 2002, 22, 2365.
[34] Sun, Z.M.; Zhang, Z.F.; Hashimoto, H.; Abe, T. *Mater. Trans.* 2002, 43, 428.
[35] Sun, Z.M.; Hashimoto, H.; Zhang, Z.F.; Yang, S.L.; Tada, S. *Mater. Trans.* 2006, 47, 170.
[36] Zhang, Z.F.; Sun, Z.M.; Hashimoto, H.; Abe, T. *J. Alloys Compd.* 2003, 352, 283.
[37] Yang, S.L.; Sun, Z.M.; Hashimoto, H. *Mater. Res. Innovat.* 2003, 7, 225.
[38] Zou, Y.; Sun, Z.M.; Hashimoto, H.; Tada, S. *Mater. Trans.* 2006, 47, 1910.
[39] Zou, Y.; Sun, Z.M.; Tada, S.; Hashimoto, H. *J. Alloys Compd.* 2007, 441, 192.
[40] Zou, Y.; Sun, Z.M.; Tada, S.; Hashimoto, H. *Mater. Trans.* 2007, 48, 133.
[41] Sun, Z.M.; Yang, S.L.; Hashimoto, H. *J. Alloys Compd.* 2007, 439, 321.
[42] Zhou, W.B.; Mei, B.C.; Zhu, J.Q. *Mater. Lett.* 2005, 59, 1547.
[43] Zhou, W.B.; Mei, B.C.; Zhu, J.Q.; Hong, X.L. *J. Mater. Sci.* 2005, 40, 2099.
[44] Zhou, Y.C.; Chen, J.X.; Wang, J.Y. *Acta Mater.* 2006, 54, 1317.
[45] Zhu, J.Q.; Mei, B.C.; Xu, X.W.; Liu, J. *Scripta Mater.* 2003, 49, 693.
[46] Zhu, J.Q.; Mei, B.C. *J. Mater. Sci. Lett.* 2003, 22, 889.
[47] Zhang, J.F.; Wang, L.J.; Jiang, W.; Chen, L.D. *J. Alloys Compd.* 2007, 437, 203.
[48] Liang, B.Y.; Jin, S.Z.; Wang, M.Z. *J. Alloys Compd.* 2008, 460, 440.
[49] Liang, B.Y.; Jin, S.Z. *Adv. Appl. Ceram.* 2009, 108, 162.
[50] Konoplyuk, S.; Abe, T.; Uchimoto, T.; Takagi, T. *Mater. Lett.* 2005, 59, 2342.
[51] Zhang, J.F.; Wang, L.J.; Jiang, W.; Chen, L.D. *Mater. Sci. Eng. A* 2008, 487, 137.
[52] Tian, W.B.; Sun, Z.M.; Hashimoto, H.; Du, Y.L. *Mater. Sci. Eng. A* 2009, 526, 16.
[53] Wang, L.J.; Jiang, W.; Chen, L.D.; Bai, G.Z. *J. Mater. Res.* 2004, 19, 3004.
[54] Qin, C.; Wang, L.J.; Bai, S.Q.; Jiang, W.; Chen, L.D. *Key Eng. Mater.* 2007, 336-338, 1383.

[55] Qin, C.; Wang, L.J.; Jiang, W.; Bai, S.Q.; Chen, L.D. *Mater. Trans.* 2006, 47, 845.
[56] Zhang, J.F.; Wang, L.J.; Jiang, W.; Chen, L.D. *Key Eng. Mater.* 2007, 336-338, 1368.
[57] Zhang, J.F.; Wang, L.J.; Shi, L.; Jiang, W.; Chen, L.D. *Scripta Mater.* 2007, 56, 241.
[58] Zhang, J.F.; Wu, T.; Wang, L.J.; Jiang, W.; Chen, L.D. *Comp. Sci. Technol.* 2008, 68, 499.
[59] Wang, H.J.; Jin, Z.H.; Miyamoto, Y. *Ceram. Inter.* 2003, 29, 539.
[60] Luo, Y.M.; Li, S.Q.; Chen, J.; Wang, R.G.; Li, J.Q.; Pan, W. *J. Am. Ceram. Soc.* 2002, 85, 3099.
[61] Luo, Y.M.; Li, S.Q.; Pan, W.; Chen, J.; Wang, R.G. *J. Mater. Sci.* 2004, 39, 3137.
[62] Shi, S.L.; Pan, W.; Fang, M.H.; Fang, Z.Y. *J. Am. Ceram. Soc.* 2006, 89, 743.
[63] Shi, S.L.; Pan, W. *J. Am. Ceram. Soc.* 2007, 90, 3331.
[64] Zhao, S.J.; Wang, L.J.; Jiang, W.; Zhang, J.F.; Chen, L.D. *Mater. Trans.* 2008, 49, 2310.
[65] Pan, W.; Shi, S.L. *J. Eur. Ceram. Soc.* 2007, 27, 413.
[66] Shi, S.L.; Pan, W. *Mater. Sci. Eng. A* 2007, 447, 303.
[67] Luo, Y.M.; Pan, W.; Li, S.Q.; Wang, R.G.; Li, J.Q. *Mater. Sci. Eng. A* 2003, 345, 99.
[68] Zhou, A.G.; Wang, C.A.; Huang, Y. *Mater. Sci. Eng. A* 2003, 352, 333.
[69] Zou, Y.; Sun, Z.M.; Hashimoto, H.; Tada, S. *Mater. Trans.* 2007, 48, 2432.
[70] Zou, Y.; Sun, Z.M.; Hashimoto, H.; Tada, S. *J. Alloys Compd.* 2008, 456, 456.
[71] Zhou, W.B.; Mei, B.C.; Zhu, J.Q. *Ceram. Inter.* 2007, 33, 1399.
[72] Zou, Y.; Sun, Z.M.; Hashimoto, H.; Cheng, L. *J. Alloys Compd.* 2009, 468, 217.
[73] Yang, C.; Jin, S.Z.; Liang, B.Y.; Jia, S.S. *J. Eur. Ceram. Soc.* 2009, 29, 181.
[74] Yang, C.; Jin, S.Z.; Liang, B.Y.; Liu, G.J.; Duan, L.F.; Jia, S.S. *J. Alloys Compd.* 2009, 472, 79.
[75] Zhou, W.B.; Mei, B.C.; Zhu, J.Q. *Ceram. Inter.* 2009, 35, 3507.
[76] Zhou, W.B.; Mei, B.C.; Zhu, J.Q.; Hong, X.L. *J. Mater. Sci.* 2005, 40, 3559.
[77] Hong, X.L.; Mei, B.C.; Zhu, J.Q.; Zhou, W.B. *J. Mater. Sci.* 2005, 40, 2749.

In: MAX Phases: Microstructure, Properties and Applications ISBN 978-1-61324-182-0
Editors: It-Meng (Jim) Low and Yanchun Zhou

Chapter 2

SYNTHESIS OF TI-AL-C *MAX* PHASES BY ALUMINOTHERMIC REDUCTION PROCESS

Dominique Vrel**, *Ali Hendaoui and Mohamed Andasmas
LIMHP, Université Paris 13, CNRS; Institut Galilée,
Villetaneuse, France

ABSTRACT

Thermite reactions are inorganic oxydo-reduction reactions, based on the difference of affinity for oxygen with different metals. Generally speaking, these reactions are performed with aluminum or magnesium, whose affinity for oxygen is amongst the highest. After an overview of the many possibilities of thermite reactions from the literature of the last 10 years (2000-2010), we will detail some experiments comparing the direct synthesis of MAX phases, from the elements on the one hand, and from aluminothermic reduction of titanium dioxide on the other hand.

1. INTRODUCTION

Thermite reactions are inorganic oxydo-reduction reactions, based on the difference of affinity for oxygen of different metals. Generally speaking, these reactions are performed with aluminum or magnesium, whose affinity for oxygen is amongst the highest[1]. As Aluminum is probably the cheaper of the two, these reactions are often referred to as "aluminothermic reactions".

Because of the way these reactions propagates, they are most of the time considered as a sub-set of SHS reactions; however, they also present specific characteristics, including the possibility of synthesizing compounds with a low formation enthalpy, and the fact that, if

* LIMHP, Université Paris 13, CNRS; Institut Galilée, 99 avenue Jean-Baptiste Clément, 93430 Villetaneuse, France. * E-mail address: vrel@limhp.univ-paris13.fr.

1 Calcium, thorium, beryllium and lithium have a slightly larger formation free enthalpy ΔG, but beside the cost of these metals, they have safety hazards that make the use of aluminum and magnesium much more preferable.

phase separation is not performed by allowing the products to settle, if in the liquid phase, they always yield composite products.

The exothermicity of these reactions is directly linked to the difference of affinity for oxygen between the considered metals. When used with oxides with a low formation enthalpy, the heat release may be extreme, and make the reaction potentially dangerous, as temperatures as high as 4500°C may be reached.

As for the synthesis of MAX phases, thermite reactions are considered as an efficient way to produce low-cost compounds, e.g. by using titanium oxide instead of metallic titanium to produce Ti_2AlC, Ti_3AlC_2, Ti_3SiC_2, ... , according to an equation similar to the following:

$$6\ TiO_2 + 11\ Al + 3\ C \rightarrow 3\ Ti_2AlC + 4\ Al_2O_3 \quad (1)$$

2. Aluminothermy: Overview

2.1. Railroad Tracks Welding

The first historical use on a large scale of thermite reactions was, and still is, the welding of railroad tracks, since the end of the 19th century. This process is based on the reaction:

$$Fe_2O_3 + 2\ Al \rightarrow Al_2O_3 + 2\ Fe \quad (2)$$

This reaction is so exothermic that both of its products, alumina and iron, are in the liquid phase. Because of the density difference between the two products, it is quite easy to separate them before pouring the liquid iron between the two track pieces to weld. As it is, this process is quite cheap and easy to set up, but despite being used for over 100 years, there is still some research work performed, in order to better understand how damaged welds may appear [1] and how to perform non-destructive analysis to prevent them [2]. Modeling studies are also conducted, in order to determine the most efficient operating conditions; for example, it has been found that increasing slightly the gap between the rails from the standard 25mm may improve the final quality of the weld [3, 4].

2.2. Simulation of Nuclear Incidents

Nowadays, 'simple' thermite reactions are used for the production of large quantities, up to 100 kg, of molten material to simulate the melting of nuclear power plants core material, usually called 'corium', as it is a cheaper way to produce it than using ovens. A quite large number of papers have been published to predict wall failures due to thermal shock [5], or to steam explosion [6], using the usual iron oxide/aluminum thermite. Amongst the subjects studied, one can find the study of mixtures melting, including partial melting [7, 8], but also the thermo-mechanical behavior of the nuclear reactor walls in contact with such an extremely hot mixture, would it be steel [9] or concrete [10]. As expected, most of the work is aimed at reducing the impact of such accidents and to develop strategies to minimize their effect, including cooling technologies to prevent nuclear material leaks [11, 12].

Among the recently published works, a special mention should be delivered to Reference [13] for their use of an uranothermic mixture, in order to obtain a final composition as close as possible to the corium composition of nuclear power plants.

2.3. Nanothermites

At the other end of the volume range, nanothermites have recently received a large interest. Nanothermites are always involving intimately mixed reactants, e.g. mechanically activated by planetary ball milling, using mixtures such as MoO_3–Al, MoO_3–Mg, CuO–Al. All these reactions are extremely exothermic, and using mechanical activation makes them also very sensitive, and hence they may be triggered by very small amounts of energy. Indeed, some authors call this mechanical activation 'arrested reactive milling' [14, 15], to emphasize that a slightly increased milling time may induce mechano-synthesis of the mixture. If powders are concerned, dry ball milling generally yields micrometric nanostructured powders: grains with an average size of 20-1000μm are then composed of agglomerates of nanosized domains, down to 30-40 nm, and are produced by successive fracture-welding process. Each of these powder grain may then be considered as a 'perfectly' mixed reactor, with very small diffusion distances [16], and an associated improved chemical kinetic, altering the rate of energy release and the mechanism of combustion front propagation [17, 18]. Extreme front propagation velocities may be observed, from 600 to 1000 $m.s^{-1}$ [19, 16].

But nanothermites may also be prepared by the deposition of one reactant on the other, e.g. of an oxide on a metallic particle, using Atomic Layer Deposition (ALD) or a similar device [20, 21]. Some authors also perform a regular mixing of nanopowders of each reactant [19, 22] or even use a bimodal distribution (nano-micro) of one of them [23], while others developed new methods based on sol-gel chemistry [24, 25].

These processes make then possible the preparation of 'thermites on a chip' [26], as a very efficient energy source for what some authors call 'thermal batteries' [27], or the study of 'micro-reactors', where the propagation of these reactions may be studied in very small glass pipes or small metallic channels[14], down to the scale of 100μm [28].

This 'nanoreactors on a chip' technology may also be developed to generate gas pressure, using the extreme velocity of the reaction [18, 22]. On the other hand, the same reactions may also be triggered while dispersed in a gas phase, as an aerosol, and it has been shown that the flame propagation in such circumstances is much faster [15].

2.4. Materials Synthesis

When the aim of a study is the production of a specific material, thermites are either used

- to produce some additional heat to favor a specific synthesis reaction: experiments may be conducted either in the form of a 'chemical oven', where the thermite reaction is surrounding the main sample or with direct powder mixture;

- to significantly decrease the cost of a synthesis, the thermite reaction being used to reduce a metal which cost (as a pure metal powder) may be prohibitive;
- to synthesize specific composites, when the targeted material is a composite containing an oxide which is part of the targeted final composition.

As an example of the first case, the following reactions may be considered:

$$3CuO + 5Al \rightarrow 3AlCu + Al_2O_3 \quad (3)$$

as the formation enthalpy of the AlCu intermetallic is low, the thermite reaction promotes the reaction. . . , but the mixture is then so exothermic, that this reaction may then be mixed with a mixture of the elements to control the temperature reached during the reaction:

$$3xCuO + (3\text{-}3x)Cu + (3+2x)Al \rightarrow 3AlCu + xAl_2O_3 \quad (4)$$

The second case concerns directly the center of the subject presented here, following a reaction such as

$$6TiO_2 + 11Al + 3C \rightarrow 3Ti_2AlC + 4Al_2O_3 \quad (5)$$

This reaction however is not that highly exothermic, as the formation enthalpy of TiO_2 is quite high, and dilution is not required, except in the optimization of the process.

Finally, as an example of specific composites, one might consider the dilution of a Fe_2O_3- or a NiO-based aluminothermic reaction, such as:

$$3NiO + 2Al + xMgO \rightarrow 3Ni + (1\text{-}x)Al_2O_3 + xMgAl_2O_4 \quad (6)$$

$$3NiO + 2Al + xTiO_2 \rightarrow 3Ni + (1\text{-}x)Al_2O_3 + xTiAl_2O_5 \quad (7)$$

$$3NiO + 5Al + xMgO \rightarrow 3NiAl + (1\text{-}x)Al_2O_3 + xMgAl_2O_4 \quad (8)$$

where MgO, TiO_2 are adjusted to lower the exothermicity of the overall reaction.

However, within the literature, it may be quite difficult to find out the main reason of why the thermite reaction were used, as the effects are often combined; if segregation is sought for, one may think that the third reason is not the one but we will see below, about gravitational thermite, that this is not true. Moreover, beside this specific case, it is quite common to use these reactions to decrease the overall cost of the reaction *and* to increase the overall exothermicity, so any classification of these three 'types' of thermite reaction will always be somewhat artificial. Nevertheless, we propose hereunder such a classification with a list of the most significant examples.

2.4.1. Heat Production

As it is now clear from the above description of the thermite reactions, they may be used just in order to generate enough heat to trigger another reaction, which is in turn yielding the product we are aiming for. The thermite mixture is then either surrounding the sample, a configuration often referred to as the 'chemical oven', or mixed with the other reactants, in which case it might be more difficult to separate the products of the thermite reaction from the others.

As potential applications, it is sometimes exaggerated to speak of material *synthesis* here, since the heat source may be used just as a transformation of the considered product, such as the recrystallization of amorphous silicon [29], ZnO crystal growth [30], the melting of an alloy [31], or even to perform welding by providing intensive local energy [32].

In the case where the thermite mixture is surrounding the sample, it is only its exothermic aspect that is sought for, and thus there is no interaction between the sample and the thermite mixture. A typical example of a study involving such a configuration can be found in Reference [33] which aims at understanding the maximal temperature reached and the heat release rate of a double thermite, with the simultaneous reduction of titanium dioxide TiO_2 and ferrite Fe_2O_3 by aluminum, by varying the composition of the thermite mixture. Such a study therefore has a direct application on chemical ovens, e.g. to synthesize materials for which the exothermicity of their formation is not high enough to obtain self-sustaining reactions, as for the formation of aluminum oxinitrides [34].

If a thermite mixture is mixed in with the other reactants, it may simply be in order to change the temperature at which the reaction happens, e.g. to change the final microstructure of the product. Such reactions have been studied with nickel and aluminum mixtures to produce nickel aluminides, either NiAl [35] or Ni_3Al [36] by changing progressively part of the nickel by its monoxide. The temperature reached during the reaction is by this mean varied, even through the products which are sought for are still the same.

In some other cases, some authors have been studying reactions where thermite mixtures would be used only to synthesize additives to the main product: Reference [37] studies the synthesis of nickel silicide Ni_5Si_3 , with a doping of the final product with chromium, which is itself produced by a thermite reaction; in Reference [38], the study aims at the synthesis of titanium carbide, for which nickel is known to be a good sintering aid; whereas the titanium carbide is produced from the elements, nickel is introduced through a thermite reaction, which will also increase the global temperature of the final sample and thus may improve the final density of the sample.

Finally, some authors use thermite reactions in order to synthesize products whose synthesis is not exothermic enough to be self-sustained. Such is the case in Reference [39] for the synthesis of Ti_2PTe_2, or in Reference [40] for the synthesis of Cu-$MoSi_2$ composites, as copper would normally quench the molybdenum disilicide synthesis reaction. This has been performed using a double thermite reaction, reducing copper monoxide and molybdenum trioxide with aluminum, to increase the final temperature. From the authors' observations, it seems that copper oxide is reduced more easily than molybdenum oxide, and thus the ignition of the aluminum/copper oxide mixture triggers the ignition of the aluminum/molybdenum trioxide mixture, and the molybdenum produced will finally react with silicon; we therefore have a succession of three different exothermic reactions.

2.4.2. Cost Effectiveness

This application of thermite reactions may be considered as a subset of the previous one, the main difference being that instead of using only the heat of the reaction, one of the products from the thermite reaction is used within the 'main' reaction. For example, if a compound of titanium is sought for, one may choose to use titanium dioxide TiO_2 instead of titanium, as the oxide is much cheaper than the metal. This will mainly be the case of MAX phase synthesis, but to produce the pure product some leaching has to be performed in order to remove the supplemental oxide byproduct, whether it is alumina or magnesia. Some of the examples cited above, e.g. References [37, 38] are not very far from this description.

As the first industrial use of thermite reactions are with the production of molten iron, we should start this section by the description of the formation of special steels. Reference [41] describes a complex thermite, using aluminum to reduce simultaneously Fe_2O_3, B_2O_3 and TiO_2 to produce TiB_2-reinforced steel. A similar study has been performed with TiC-reinforced steel [42], or ZrC-reinforced steels, using carbon and $ZrSiO_4$ [43]. The former reference is also interesting as it provides a review on this kind of processing, and as it gives comparative results with other processing methods. In all these experiments, the steel and alumina are separated by natural phase segregation of the products, as alumina has a significantly lighter density.

Amongst the different materials that could be synthesized through such a procedure, the synthesis of borides should be emphasized. In such a case, the metallic element as a well as boron could be reduced either by aluminum or by a magnesium. In Reference [44], the synthesis of titanium and zirconium borides has been studied, and in Reference [45] a similar synthesis and has been performed by improving the kinetics of the reaction using mecanosynthesis, with nanocrystalline final materials. In these cases, magnesium oxide has been leached by dilute acid, in order to get the pure boride products.

But one of the most spectacular application of such a reaction can be found with space applications. Reference [46] presents preliminary results of reactions using regolith[2] as the source of metallic oxides, the final aim of such a study being the possibility of building sturdy ceramic/metal composites for either the Moon or Mars colonization. In such a case, the reducing agents only, which are usually light metals, have to be brought in space, and thus most of the weight of the final structure comes from materials found on site.

2.4.3. Specific Synthesis

Here is the most interesting part for thermite-based synthesis for ceramics or composites, as thermite reactions are here used at their best, using them as a source of heat, as a source of cheap reactants, and where the products are actually the one or amongst the ones which are sought for.

As in the previous paragraph, we may start with an application close to railroad welding. Reference [47] describes the production of steels in the FeCrNi composition range. The overall reaction is thus similar to References [41, 43, 42] cited above, except that here phase segregation between steel and alumina is clearly avoided by a control of the final temperature, using a mixture of metals and their oxides, and alumina is considered as a valuable additive to improve the mechanical properties of the final product, such as hardness and friction behavior.

---

2 Regolith is a layer of loose, heterogeneous material covering solid rock. It covers the surface of the Moon with a thickness ranging from 4 to 15 m. Although it seems that this layer is thinner on Mars, it is well known that the reddish color of the planet is mainly due to the intensive presence of iron oxide, with which thermite reactions are very efficient. Moreover, the presence on the planet of dust storms despite the fact that the atmosphere is so thin would be an indication that the average particle size is very small and would therefore have a very high reactivity.

While metals and alloys are being considered, we should mention the formation of nanometric tungsten powders, by mechanically activated magnesio-thermite reactions [48]. The extreme heat of the reaction using tungsten trioxide WO_3 as a source of tungsten tends to disperse the final products as fine powders, not giving the system time for grain growth.

But naturally, such an approach has been used for the production of ceramics and intermetallics. Concerning ceramics, Reference [49] describes the synthesis of titanium nitride/alumina functionally graded materials (FGM), which is a perfect example where thermites are used for their products and not for their exothermicity, as most of the heat brought to the system is controlled by an oven. From this point of view, this is a counter example of the section above describing the use of thermite reaction only (or mainly) for the heat they produce.

Among ceramics, we would find once more the production of borides, with the control of the composition and/or the exothermicity of the reaction by using mixtures of elemental and oxide powders. Reference [50] thus describes the synthesis of Al_2O_3/TiB_2 and Al_2O_3/NbB_2 composites, where both the source of metal and of boron may be oxides (TiO_2, Nb_2O_5, B_2O_3); due to the comparative thermodynamical stabilities of titanium dioxide, niobium oxide and alumina, titanium diboride and niobium diboride, increasing the thermitic part of the overall reaction tends to decrease the temperature for titanium oxide, and to increase it for niobium oxide. A similar system has been studied in Reference [51] using $Al_4B_2O_9$ as a source of boron. But other ceramics may be produced by such a process. Among them, we may mention the synthesis of $Al_2O_3/TiC/Al$ composites [52], the chemical pathway on a similar system having been studied by a copper block quenching method [53].

Finally, intermetallic systems have also been studied, especially iron aluminide [54] and nickel aluminide [55, 56, 57] with or without phase segregation.

A) Coatings

It may seem from what has been said before that phase segregation is either undesirable for the product, if a natural composite is sought for, or desirable only if one product has to be separated from the other only to be eliminated. There is however a third case, where a natural phase segregation is favored, in order to take full advantage of a gradient, and to promote the synthesis of FGMs. First, a brief mention of reactions taking place with bulk materials must be made.

Indeed, aluminothermic reactions could be used to create inhomogeneous materials. Once more, the most standard type of reaction will imply aluminum and iron or nickel oxides. In Reference [58], an aluminothermic reaction is started at the surface of an aluminum substrate either with Fe_3O_4 or with FeO; similar experiments have been performed in References [59, 60, 61], especially to promote the resistance to abrasion of aluminum substrates by the synthesis of a thick alumina layer.

Reaction with nickel oxide has also been considered in Reference [62], but for the alumina layer to be deposited on nickel substrates, and it has also been demonstrated that a doping of the alumina layer with zirconia is possible. Similar aim on a steel substrate has also been studied in Reference [63]. However, because of the high thermal conductivity of bulk materials, such reactions are rarely self-sustaining, except if the thickness of the layer is sufficient, and thus, as in this last reference, techniques to promote the reaction, such as scanning the surface with a laser may be used. Because the exothermicity is thus not an issue any more, dilution with silica and/or nickel may be performed.

B) Gravitational Thermites

Because phase segregation is naturally promoted by gravity and density differences between the phases, the influence of gravity on aluminothermic reduction process has been studied by various authors. Beside experiments performed in low gravity environments for space applications [64, 65], experiments are either conducted in centrifuge machines or inside rotating pipes. It is generally believed that gravity has a predominant role on phase segregation, but it has been proven [65] that this is not always true: in some cases, and at least for limited gravitational ranges, some authors have shown a significant influence of other parameters such as thermal conductivity [66]. But naturally, while these parameters may have an influence on *whether* the phases might segregate, if segregation happens, gravity will then play a key role to determine *where* the different phases might migrate.

The simplest case of gravitational thermite concerns the aluminothermic reduction of iron oxide Fe_2O_3 inside a rotating steel pipe. Due to enhanced segregation from increase gravity, the iron generated tends to move closer to the inside surface of the steel pipe, when alumina goes at the top of this layer. Consequently, this system yields steel pipes where the inner surface is covered with alumina, with therefore a much improved abrasion resistance, much needed for the transportation of mineral charged liquids. While the main problem of such pipes remains in the welding of the pipes, many studies have been performed in the recent years to improve the quality of these layers, by diluting the aluminothermic mixture with iron [67] or by changing the final composition of the layer by adding silica [68, 67] or zirconia [69, 70]. Other pipe compositions have also been studied, although uses are sparser, and the thermite mixture has then to be adapted to the chemical composition of the pipe: CuO/Al thermite have thus be used in Cu pipes [71], using additives such as $Na_2B_4O_7$, SiO_2, ZrO_2 to improve compressive and shear mechanical properties.

But to investigate extensively the role of increased gravity, it remains necessary to compare not only the products but also the behavior of the reaction under such circumstances. Such a study is performed in Reference [72] for aluminothermic reduction of iron and chromium oxides (Fe_2O_3, Cr_2O_3), using high speed video recording under normal and rotating conditions. An application of such a study would be the direct synthesis of Fe-Cr alloys, as found in Reference [73], in which the influence of additives such as $Na_2B_4O_7$, Ni, Al, and Ti+B_4C, yielding byproducts such as AlFe, TiC, and TiB_2 is also investigated. The welding of the ceramic layer on the steel pipe may then be performed not through a metallic, but through an intermetallic layer. Reference [74] describes such a study, where a NiO-based thermite is performed to synthesize not metallic nickel, but nickel aluminide; this reaction may be controlled by adding supplemental nickel and aluminum, in order to reduce the overall exothermicity of the reaction, but also to alter the respective proportions of the intermetallic and ceramic layers. One paper [75] even describes the synthesis of NiAl from the elements, where a thermite reaction is added to increase the temperature: all in all, this is then a very peculiar case of chemical oven under high gravity.

To complete this picture of thermite reactions under high gravity, we may mention Reference [76], where two different reactions are simultaneously performed, one of them being a thermite reaction, the aim being to ensure a good adhesion of the products by forcing one of the products to infiltrate the porosity of the other.

3. MAX PHASES PRODUCED BY SHS

We will not here describe fully the many possibilities of SHS to synthesize MAX phases, as this topic is by itself the subject of another chapter. However, we will briefly present some results, obtained in conditions similar to the ones where aluminothermic reactions are involved.

Samples were prepared from commercial powders of aluminum, titanium and graphite, which were carefully weighed to obtain the compositions of the required stoichiometry. Once weighed, the powders were thoroughly mixed using a Turbula® mixer, co-milled using a Fritsh® Pulverisette 6 planetary mill at 200RPM for 2h with a Ball-to-Powder Ratio (BPR) of 10:1 (i.e. the weight of the balls is 10 times the weight of the powders) and were then pressed.

The resulting samples were then set in the reaction chamber, described in References [77, 78]. Ignition of the powder mixtures was then carried out by a graphite plate heated up to a temperature of about 2000 K by a high intensity electrical current (up to 12 V – 200 A), and the reaction was then analyzed by infrared thermography using an AVIO® TVS 2000 ST IR–camera.

X-Ray Diffraction (XRD) analysis has been performed using an XRG-3000 diffractometer from INEL®. This diffractometer works in the asymmetric Bragg-Brentano geometry, with a curved detector with a 90° aperture; with 50 cm curvature radius and 8192 digital channels, this detector provides a precision close to 0.01°. The radiation wavelength used is the Cu-$k_{\alpha 1}$ radiation, 1.54056 Å, monochromated by a germanium monocrystal; out of the two slits placed just before the sample, the one parallel to the linear beam is opened for only 13 μm in order to eliminate the Cu-$k_{\alpha 2}$ radiation.

3.1. Influence of the Cooling Rate

In order to have comparative and regular densities, the samples were isostatically pressed at a pressure chosen at 80 MPa. Because aluminum is a very soft metal this pressure is sufficient to get a sample with mechanical properties good enough to manipulate it.

Due to an inhomogeneity in the temperature field and/or in the emissivity at the surface of the sample during reaction, we suspected that the cooling rate could influence strongly the reaction kinetics. A detailed analysis by x-ray diffraction, presented in Reference [79], made us believe that the purity of the final product could be improved by reducing the maximum temperature and/or by reducing the time the sample spent at high temperature, which naturally brought us to try to increase the cooling rate of our sample.

With such an aim, it may seem paradoxical to use mechanical activation, as it is known that such a treatment has a tendency to *increase* the maximal temperature. However, the increased stability of the reaction front allowed us to increase the heat losses beyond the quenching threshold, i.e. beyond the point where a self-propagating reaction could not propagate in unmilled powders.

Three samples with different diameters of 10, 14 and 18 mm were then pressed; they were maintained using thin tungsten wires, making radiation and natural convection the dominant heat loss phenomena, and heat conduction between the sample and the sample holder could be neglected.

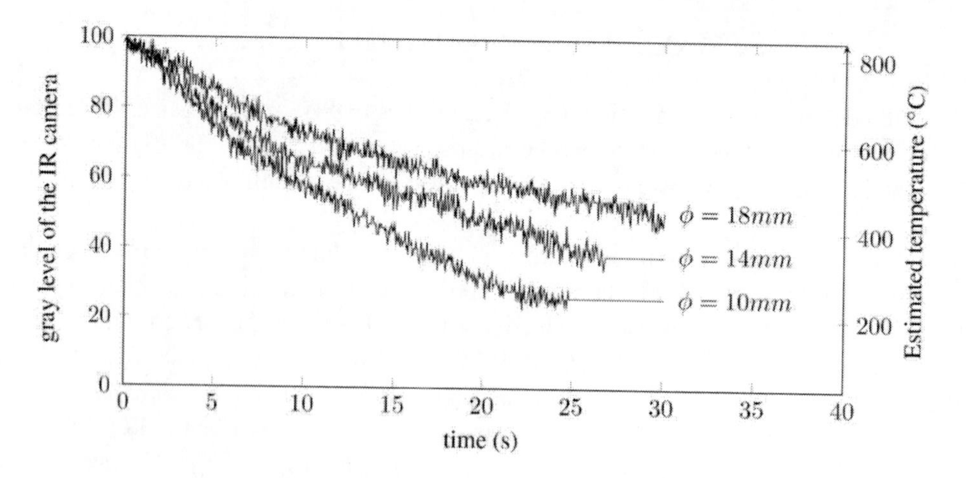

Figure 1. Apparent temperature of 3 samples during cooling after SHS from Ti-Al-C with a 2:1:1 stoichiometry. The gray levels are the physical precision of the IR camera; the estimated temperature is determined by setting the camera emissivity parameter to 1.

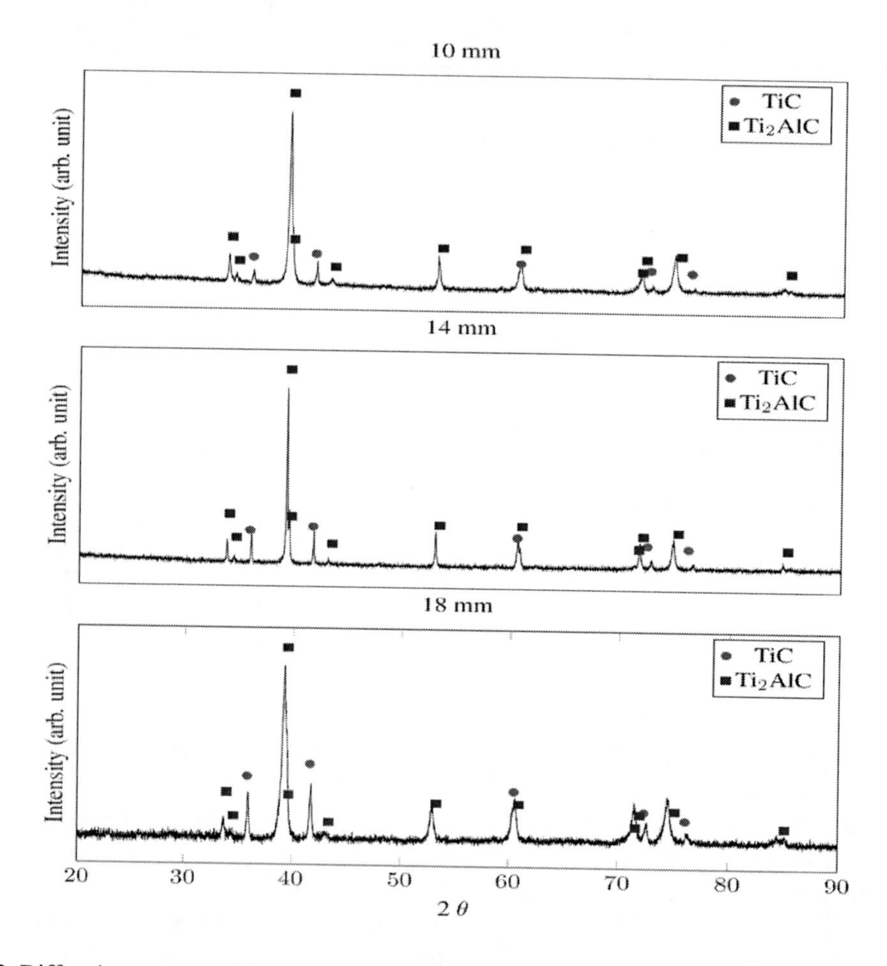

Figure 2. Diffraction patterns of the three samples. The relative intensity of the TiC peaks increase with the samples diameter.

Figure 1 presents the cooling rate of these three samples, as seen at their surface by the infrared camera. It should be here noted that, because the emissivity parameter of the camera was set arbitrary to 1, and is necessarily smaller, the temperature of the sample is always underestimated by the camera. However, the emissivity of our samples is not well defined, as it must evolve during the reaction, due to temperature, physical state and chemical composition. Considering for example a sample with an initial temperature of 840°C, the cooling down to a temperature of 500°C would take 10s for the ϕ=10mm diameter sample, 15s for the ϕ=14mm diameter sample and 25s for the ϕ=18mm diameter sample. Beside this, the propagation of the reaction front is very similar for all samples, considering the maximal temperature and the heating rate.

The resulting compositions of these samples as determined by X-ray diffraction are presented in Figure 2, on which it can clearly be seen that a decrease in the samples diameter yields immediately a decrease into the TiC impurity. Estimating the TiC content from the comparative area of the main peak of each phase yields a respective TiC content of 6.2% TiC for the f=10 mm sample; 13.7% for f=14mm; and 17,1% for f=18mm. Although this is not meant to be an accurate measure of the TiC content, it proves very clearly the key role of the cooling rate in the impurity content.

From these numbers, it would seem that a sample with a diameter of about 5 to 7 mm would present a pure Ti_2AlC phase. Unfortunately, the method cannot be applied for so small a sample, for we didn't succeed in retrieving an unbroken sample after isostatic pressing with a diameter smaller than 10 mm. Nevertheless, the same method could be applied with other geometries, e.g. powder beds.

3.2. Influence of the Stoichiometry

In order to improve our control on the temperature history of the samples, we decided to change the stoichiometry of the initial powder mixture. As the final composition of the samples observed in the previous section showed the presence of TiC, we therefore chose to increase slightly the aluminum content to improve the chances of the formation of a ternary compound. The reaction equation was then:

$$2Ti + (1+n/2)\,Al + C \rightarrow Ti_2Al_{1+n/2}C \qquad (9)$$

with $n = 0 \rightarrow 3$.

As the mechanical behavior of the powder mixture must evolve with the composition, we may expect an increase of the density with the aluminum content if we keep the same compacting pressure, this metal being the softest of the three elements. To avoid this, we chose to uniaxially press the sample, fixing not the compaction pressure but the sample final volume to obtain a relative density of 60% of the theoretical value (which naturally changes with the composition).

Figure 3 presents the evolution of phase composition after reaction as a function of the initial aluminum content. Surprisingly, the Ti_2AlC content decreases sharply even for n=1, dropping from 83 to 49%; however, during the same time, the TiC amount also drops from 17 to 4.5%, which is exactly why we developed this dilution method. The rest of the sample is made of Ti_3AlC_2, which is also a MAX phase and therefore may be considered as a positive result, the total amount of MAX phases being 95.5%. Nevertheless, this is quite surprising, since this later MAX phase has a theoretical Al content smaller than Ti_2AlC. From this, it may be deduced that the measured composition by X-ray diffraction may be interesting as long as crystal structures are concerned, but should be taken cautiously, as evidently the stoichiometry of the observed compounds are far from being the theoretical ones.

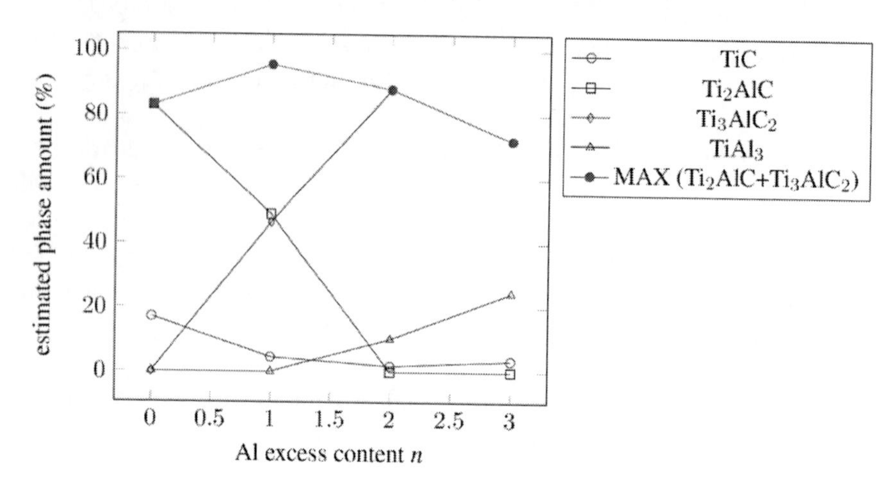

Figure 3. Phase compositions of the samples after SHS as a function of the initial composition of the powder mixture (equation 9). Percentage of each phase is determined by the relative area of its main peak.

If the content in aluminum is further increase (n=2), the TiC amount in our samples further decreases to a value of 1.8%. In the mean time, Ti_2AlC completely disappears from the sample composition, and, if Ti_3AlC_2 increases, it only reaches a value of 88%, and the total amount of MAX phases in our sample is thus decreasing: indeed, a new impurity appears in the sample, $TiAl_3$. Going then from n=2 to n=3 does not improve the situation: TiC increases back, and $TiAl_3$ reaches a value close to 25%.

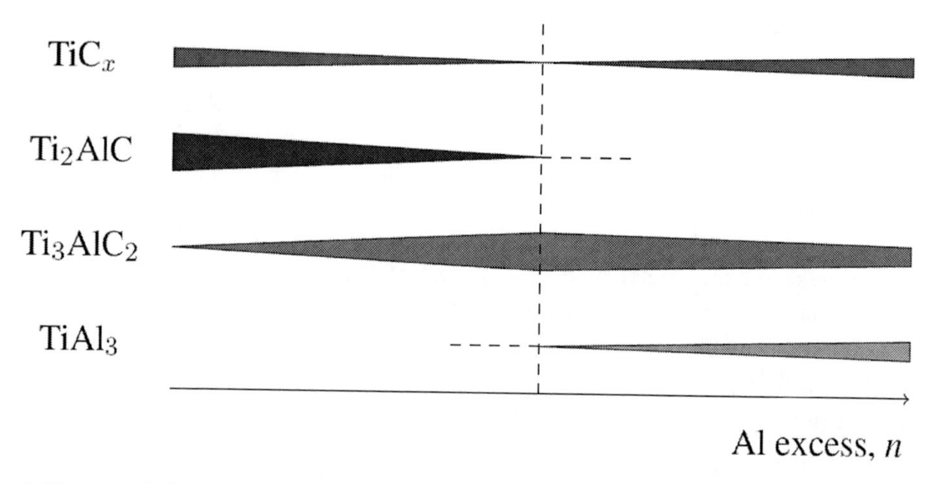

Figure 4. Phase evolution as a function of the initial composition of the powder mixture.

Figure 4 summarizes the phase evolution of the system, as a function of Al content in the initial mixture. An optimal value for n may be found between 1 and 2, where the appearance of $TiAl_3$ is not yet detectable, and where the TiC amount reaches a minimum value. As for MAX phases compositions, they would be in favor of Ti_3AlC_2, but may still contain some Ti_2AlC.

4. MAX PHASES PRODUCED BY THERMITE REACTIONS

Using aluminothermic reactions to produce MAX phases in the Ti-Al-C system has for main objective the cost effectiveness of the reaction, using cheaper TiO_2 as a source of titanium, according to the reaction:

$$6TiO_2 + 11Al + 3C \rightarrow 3Ti_2AlC + 4Al_2O_3 \quad (10)$$

Following the study we performed on the synthesis of MAX phases from the elements, we will use non-stoichiometric mixtures, increasing the Al content, according to the following equation:

$$6TiO_2 + (11+n)Al + 3C \rightarrow 3Ti_2Al_{1+n/3}C + 4Al_2O_3 \quad (11)$$

with $n = 0 \rightarrow 5$. Indeed, we increased the variation range of n, because its influence is slightly smaller in equation 11 than it was in equation 9, n being divided by 3 instead of 2.

4.1. Results

4.1.1. Velocity of the Reaction Front

It is well known that when diluting an SHS reaction, the maximum temperature reached during the reaction decreases, but the velocity of the reaction front also decreases, and may induce instabilities such as pulsating reaction fronts or spin propagation where hot spots propagate around the sample. In our case however, the problem is slightly more complex, as the very notion of dilution could be questioned, as we do not introduce in our sample an inert mass. From what we observed in the previous section, we do change the final composition of our products, and we may think that the byproducts that we observe such as $TiAl_3$ are less thermodynamically stable, and even that the exothermicity of the synthesis of the MAX phases is reduced as the final stoichiometry is non-perfect. But all in all, the term 'dilution' is still slightly questionable.

To assert the validity of such a designation, and to quantify the thermal effect of the variation of the stoichiometry, using an infrared thermographic camera, we measured the reaction front velocity. Results are presented in Figure 5. Within all the studied composition range, the reaction front was always propagating at a constant velocity, and no instability appeared. The front velocity decreased regularly from 2.0 $mm.s^{-1}$ to 1.4 $mm.s^{-1}$, proving the dilution effect of the stoichiometry variation.

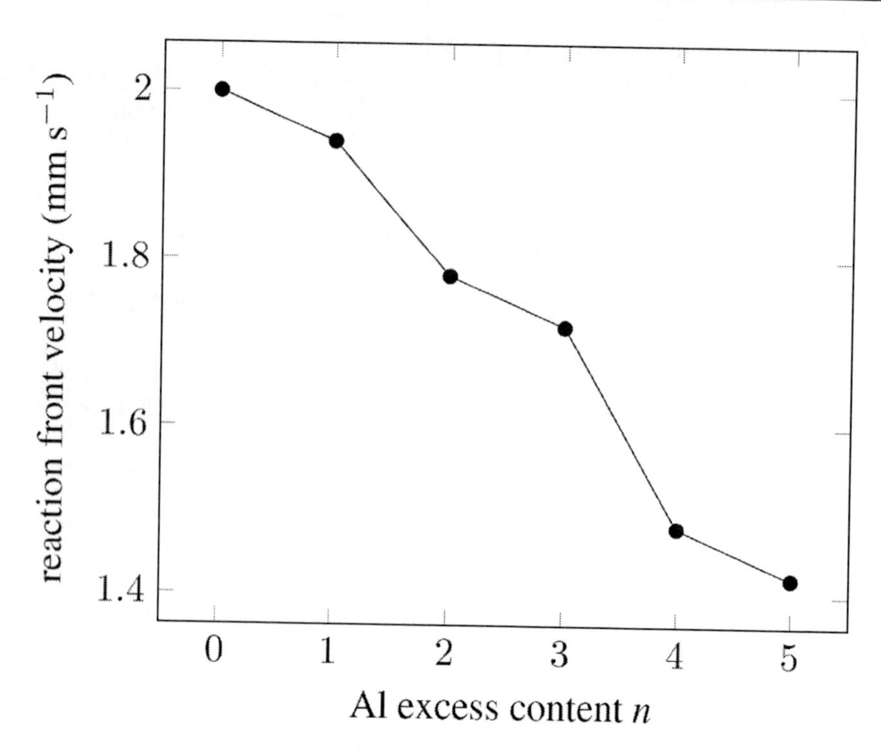

Figure 5. Velocity of the reaction front as a function of the initial composition of the sample (equation 11), observed by infrared thermography. Initial density is fixed at 60% of its theoretical value.

4.1.2. X-Ray Diffraction

Figure 6 presents the X-ray diffraction results for *n* values ranging from 0 to 3. At first, when no additional aluminum is added, the only observable products are alumina Al_2O_3, titanium carbide TiC and the Ti_2AlC MAX phase. As a result, we immediately can point out that the aluminothermic part of the reaction is complete: no TiO_2 is detectable under any of its known phases, and no metallic aluminum can be found. This first result is also valid for the other *n* values. The TiC value for this first diffraction pattern yields a value of 13.8%, which is smaller than the 17% found when synthesizing the MAX phase from the elements, and is thus already a good omen for the use of aluminothermic reaction in such syntheses.

A first increase in the *n* value to 1 does not bring any dramatic difference to the diffraction pattern except for two points: the apparent intensity of the TiC diffraction peaks is smaller for an estimated amount of 9.1%, and the main peak at $2\theta \approx 39°$ is now a doublet, the main one representing Ti_2AlC, the other Ti_3AlC_2.

At the next step, $n=3$, the optimal situation is reached, with very small TiC peaks, for an estimated amount of 4.9%. Beside alumina, the only other detectable phases are the two MAX phases of the Ti–Al–C system, Ti_2AlC and Ti_3AlC_2, with respective estimated amounts of 15.7% and 79.2%.

Unfortunately, increasing further the *n* value does not improve the situation. Ti_2AlC is now undetectable; the only MAX phase being then Ti_3AlC_2, which amount only reaches the value of 74.1%: indeed, as previously observed with the synthesis from the elements, the TiC amounts are now increasing again, and $TiAl_3$ is a second impurity present in our final product.

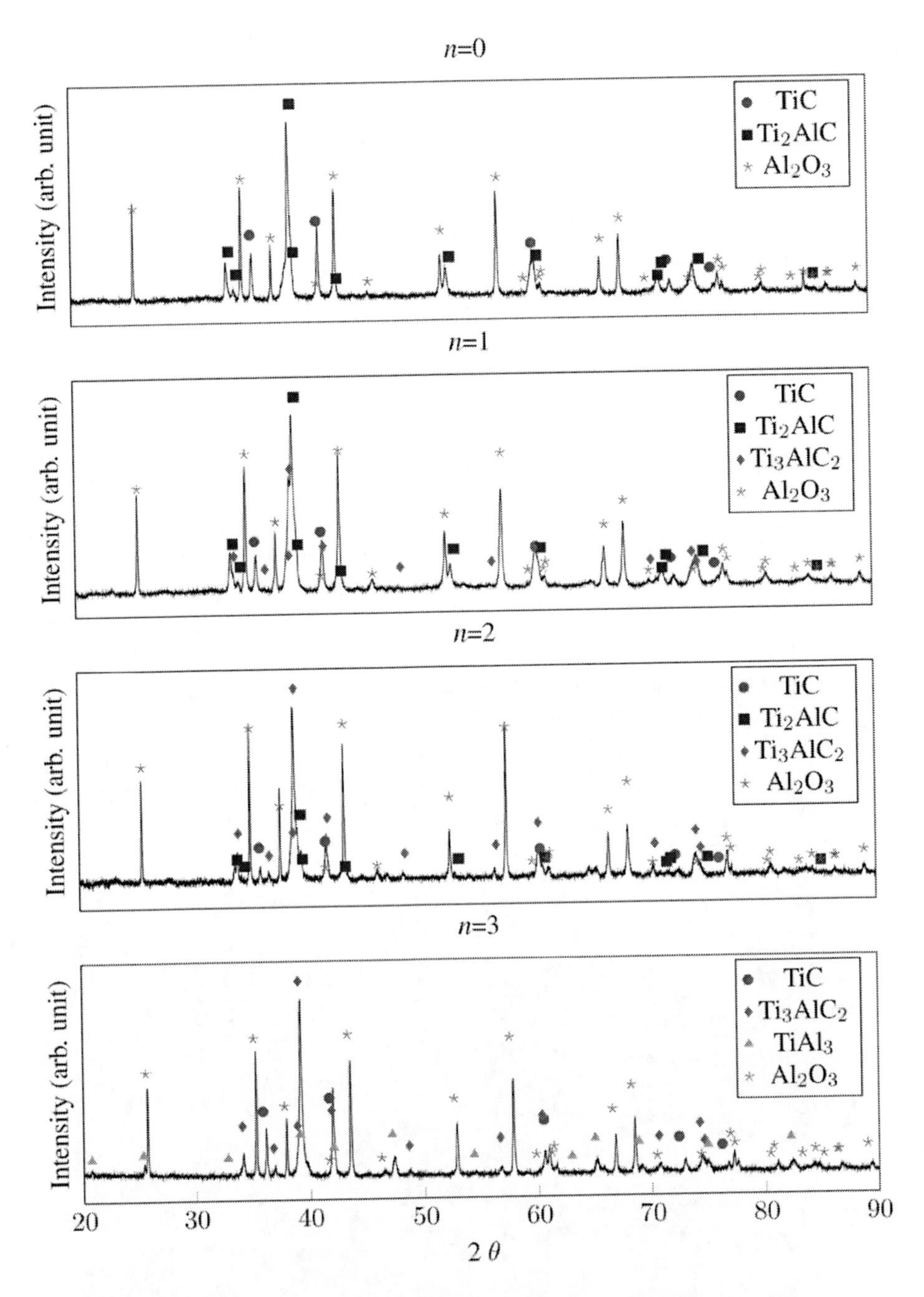

Figure 6. Diffraction patterns as a function of the initial stoichiometry (equation 11).

Figure 7 summarizes all these data, together with the results obtained for n=4 and n=5, where we can observe that the changes observed for n=3 are evolving with the same tendency: no Ti_2AlC can be found, the Ti_3AlC_2 decreases progressively, while TiC and $TiAl_3$ increase and reach values above 25%.

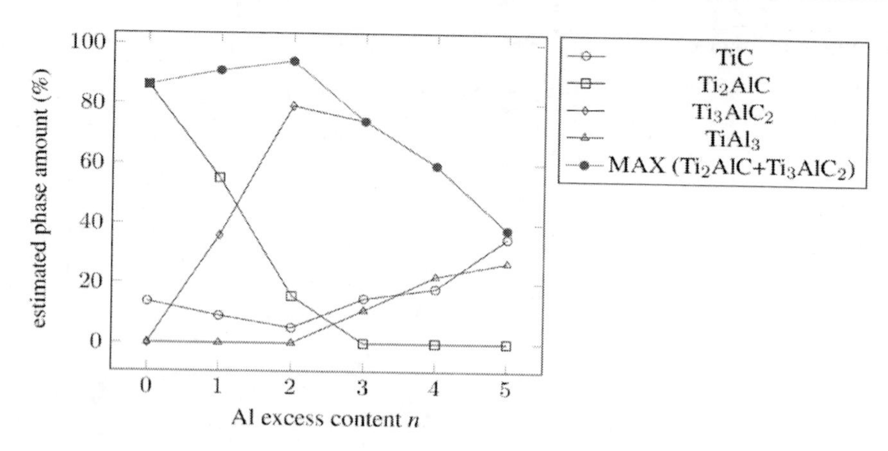

Figure 7. Phase compositions of the samples after SHS as a function of the initial composition of the powder mixture (equation 11). Percentage of each phase is determined by the relative area of its main peak.

Once more, an optimum can be found, here for $2 \leq n \leq 3$, for which a minimum amount of TiC is synthesized; this optimum corresponds to a mixture of MAX phases in which, surprisingly, the Al-poor MAX phase is predominant. The same caution concerning the stoichiometry of the products should therefore be considered.

4.1.3. SEM Micrographs

Figure 8 presents the microstructure of the sample produced with the stoichiometric mixture (n=0). Small rounded particles are clearly visible and are made of titanium carbide, TiC. Most of the observed surface is made of smooth particles, identified as alumina, Al_2O_3, but some platelets are also visible and have been identified as Ti_2AlC. All identifications have been performed using EDX.

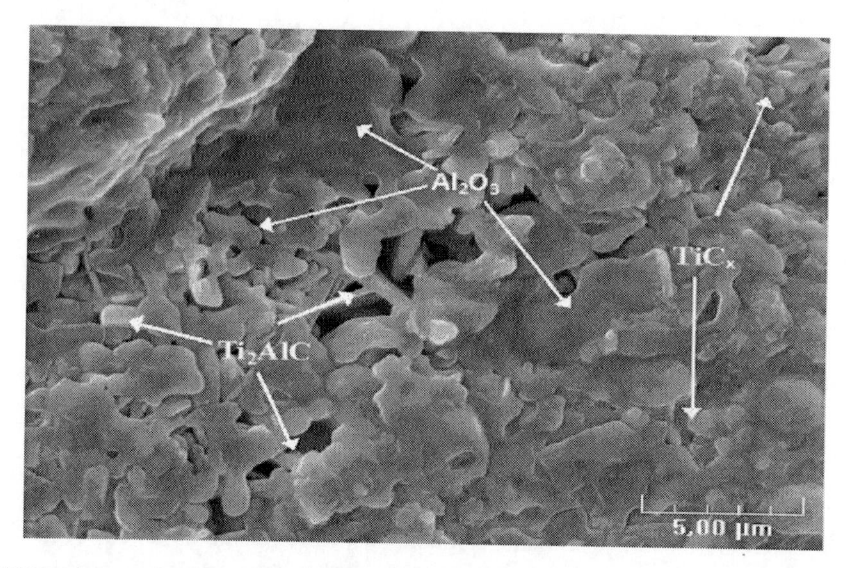

Figure 8. SEM micrograph of the stoichiometric mixture (n=0).

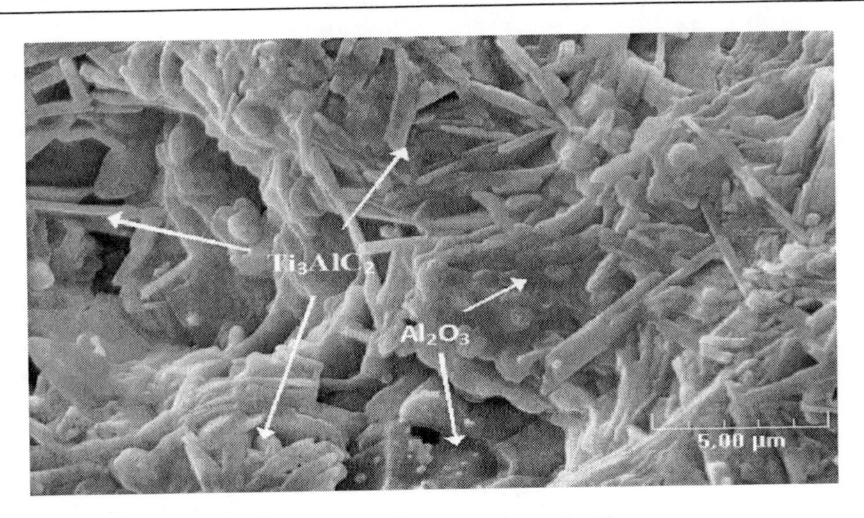

Figure 9. SEM micrograph of the mixture with added aluminum (n=2).

Figure 9 presents finally the microstructure of the sample whose X-ray diffraction pattern presented the best composition, for n=2. On this sample, still using EDX, no TiC particles have been detected. However, smooth alumina particles are still easy to distinguish, and most of the sample surface is now clearly made of platelet grains, which are now made of Ti_3AlC_2.

CONCLUSIONS

All the samples synthesized during the study concerning the stoichiometry were prepared with similar conditions, with an initial relative density of 60%, and therefore the influence of changing the geometry of the samples has not been studied. In the mean time, final products present porosity similar to the initial conditions. A way of improving further the highest purity we obtained would be to use once more samples with a small diameter, or thin powder beds. However, this control on the geometry could be increased by performing compaction of the final products, e.g. by uniaxial pressing while the sample is still hot. Such a process would reduce the final porosity and thus increase the thermal conductivity of the sample allowing a better thermal flow throughout the sample and a higher cooling rate, especially because the conductive heat flows between the sample and the mold would be much higher than the convective heat flows we have during air cooling.

Because one of our byproduct was titanium carbide, we sought increasing the aluminum content in the system would be a good idea to increase the Ti_2AlC content in the final product. The system has however proved to behave differently, and, increasing the Al content in the system, the Al-poor MAX phase has been favored. Although this is still an interesting result as long as a specific MAX composition is not sought for, other ways must be explored. One of the most promising ways would be to simply use self-dilution, using the products of a reaction, i.e. containing Ti_2AlC as a main phase, to dilute the reactive mixture. The effect of such a dilution would therefore be purely thermal, as no significant chemical reaction would take place with such a compound, and, Ti_2AlC could probably be obtained without Ti_3AlC_2.

Although magnesium is significantly more expensive than aluminum, magnesothermite reactions to produce MAX phases should also be studied, and purification steps should also be performed: indeed, simple techniques have been developed to remove the oxide with dilute HCl.

As for cost effectiveness, we calculated the cost of our products, considering the price list of a well known laboratory chemicals provider, which naturally is much more expensive than what could be found at an industrial volume range, but yields good comparative results. Using aluminothermic reaction described in equation 11 with n=2, the cost would be 106€.kg^{-1}, considering only the mass of the MAX phases contained in the sample. Naturally, if it is not the MAX phases that are sought for, but the Al_2O_3/MAX composite, the price would be even further reduced, and would reach the value of 57€.kg^{-1}. On the contrary, if alumina has to be removed using hydrochloric acid, the cost of such a process would have to be added to the total. Nevertheless, the same synthesis from the elements, using the same catalog, would be of 220€, and the use of aluminothermic reaction therefore allows a price cut of about 52%. In the mean time, purities are very comparable, with 94.9% purity with aluminothermic reactions (disregarding Al_2O_3), and 95.5% when the reaction is performed from the elements.

REFERENCES

[1] H. Shitara, Y. Terashita, M. Tatsumi, and Y. Fukada, "Nondestructive testing and evaluation methods for rail welds in Japan," *Quarterly Report of RTRI*, vol. 44, no. 2, pp. 53–58, 2003.

[2] Y. Terashita and M. Tatsumi, "Analysis of damaged rail weld," *Quarterly Report of RTRI*, vol. 44, no. 2, pp. 59–64, 2003.

[3] Y. Chen, F. Lawrence, C. Barkan, and J. Dantzig, "Heat transfer modeling of rail thermite welding," Journal of Rail and Rapid Transit, Part F, *Proceedings of the Institution of Mechanical Engineers*, vol. 220, no. F3, pp. 207–217, 2006.

[4] Y. Chen, F. Lawrence, C. Barkan, and J. Dantzig, "Weld defect formation in rail thermite welds," Journal of Rail and Rapid Transit, Part F, *Proceedings of the Institution of Mechanical Engineers*, vol. 220, no. F4, pp. 373–384, 2006.

[5] T. Sawada, "Computational modeling and verification for hexcan wall failure under simulated core disruptive accident condition," *Annals of Nuclear Energy*, vol. 28, pp. 1457–1468, Sept. 2001.

[6] W. Cherdron and A. Kaiser, "The eco steam explosion experiments," *Atw Internationale Zeitschrift für Kernenergie*, vol. 51, pp. 518–525, Aug.-Sept. 2006.

[7] J. Foit, "Large-scale ecokats experiments: Spreading of oxide melt on ceramic and concrete surfaces," *Nuclear Engineering and Design*, vol. 236, no. 24, pp. 2567–2573, 2006.

[8] K. Mwamba, P. Piluso, D. Eyries, and C. Journeau, "Self-propagating high-temperature synthesis of a nuclear reactor core melt for safety experiments," *International Journal of Self-Propagating High-Temperature Synthesis*, vol. 15, no. 4, pp. 284–296, 2006.

[9] R. Krieg, J. Devos, C. Caroli, G. Solomos, P. Ennis, and D. Kalkhof, "On the prediction of the reactor vessel integrity under severe accident loadings (RPVSA)," *Nuclear Engineering and Design*, vol. 209, no. 1-3, pp. 117–125, 2001.

[10] M. Nie, M. Fischer, and W. Koller, "Status of interpretation of selected transient mcci experiments conducted in the frame of the coresa randd project," in *Proceedings of the*

Annual Meeting on Nuclear Technology, pp. 185–188, 2002. Stuttgart, Germany. 14-16 May 2002.

[11] K.-H. Kang, R.-J. Park, K.-M. Koo, S.-B. Kim, and H.-D. Kim, "Experimental investigations on in-vessel corium retention through inherent gap cooling mechanisms," *Journal of Nuclear Science and Technology*, vol. 43, no. 12, pp. 1490–1500, 2006.

[12] R. Wittmaack, "Simulation of free-surface flows with heat transfer and phase transitions and application to corium spreading in the EPR," *Nuclear Technology*, vol. 137, no. 3, pp. 194–212, 2002.

[13] P. Piluso, K. Mwamba, and C. Journeau, "Uranothermic reaction as an efficient shs process to synthesize severe accident nuclear materials," *International Journal of Self-Propagating High-Temperature Synthesis*, vol. 18, no. 4, pp. 241–251, 2009.

[14] M. Schoenitz, T. S. Ward, and E. L. Dreizin, "Fully dense nano-composite energetic powders prepared by arrested reactive milling," *Proceedings of the Combustion Institute*, vol. 30, no. 2, pp. 2071–2078, 2004.

[15] S. Umbrajkar, S. Seshadri, M. Schoenitz, V. Hoffmann, and E. Dreizin, "Aluminum-rich Al-MoO_3 nanocomposite powders prepared by arrested reactive milling," *Journal of Propulsion and Power*, vol. 24, pp. 192–198, March-April 2008.

[16] S. M. Begley and M. Q. Brewster, "Radiative properties of MoO_3 and Al nanopowders from light-scattering measurements," *Transactions of the ASME Journal of Heat Transfer*, vol. 129, pp. 624–33, May 2007. (Mech. Sci. and Eng. Dept., Univ. of Illinois, Urbana, IL, USA).

[17] J. A. Puszynski, C. J. Bulian, and J. J. Swiatkiewicz, "The effect of nanopowder attributes on reaction mechanism and ignition sensitivity of nanothermites," in *Multifunctional Energetic Materials Symposium*, vol. 896, pp. 147–58, Materials Research Society, 2006.

[18] M. Comet, D. Spitzer, and J.-P. Moeglin, "Nanothermites for space and defense applications.," in *Proceedings of the SPIE - The International Society for Optical Engineering*, vol. 7314, p. (8 pp.), 2009.

[19] B. S. Bockmon, M. L. Pantoya, S. F. Son, B. W. Asay, and J. T. Mang, "Combustion velocities and propagation mechanisms of metastable interstitial composites," *Journal of Applied Physics*, vol. 98, pp. 64903–1–7, Sept. 2005.

[20] S. M. George, J. D. Ferguson, K. J. Buechler, and A. W. Weimer, "SnO_2 atomic layer deposition on ZrO_2 and Al nanoparticles: pathway to enhanced thermite materials," *Powder Technology*, vol. 156, pp. 154–163, Aug. 2005.

[21] Wang-Yi, Jiang-Wei, Cheng-Zhi-Peng, Zhang-Xian-Feng, An-Chong-Wei, Song-Xiao-Lan, and Li-Feng-Sheng, "Thermal performance investigation of core-shell cu/al micron-nano composites with WO_3," *Acta Physico Chimica Sinica*, vol. 23, pp. 1753–1759, Nov. 2007.

[22] K. Martirosyan, L. Wang, A. Vicent, and D. Luss, "Synthesis and performance of bismuth trioxide nanoparticles for high energy gas generator use," *Nanotechnology*, vol. 20, no. 40, p. 405609, 2009.

[23] K. Moore, M. L. Pantoya, and S. F. Son, "Combustion behaviors resulting from bimodal aluminum size distributions in thermites," *Journal of Propulsion and Power*, vol. 23, pp. 181–185, Jan.-Feb. 2007.

[24] B. J. Clapsaddle, Z. Lihua, A. E. Gash, J. H. Satcher-Jr., K. J. Shea, M. L. Pantoya, and R. L. Simpson, "Synthesis and characterization of mixed metal oxide nanocomposite

energetic materials," in *Synthesis, Characterization and Properties of Energetic/ Reactive Nanomaterials Symposium, vol. Mater. Res. Soc. Symposium Proceedings* Vol. 800, pp. 91–96, 2004.

[25] A. E. Gash, J. H. Satcher-Jr., R. L. Simpson, and B. J. Clapsaddle, "Nanostructured energetic materials with sol-gel methods," in *Synthesis, Characterization and Properties of Energetic/ Reactive Nanomaterials Symposium, vol. Mater. Res. Soc. Symposium Proceedings* Vol. 800, pp. 55–66, 2004.

[26] S. Apperson, S. Bhattacharya, Y. Gao, S. Subramanian, S. Hasan, M. Hossain, R. V. Shende, P. Redner, D. Kapoor, S. Nicolichand, K. Gangopadhyay, and S. Gangopadhyay, "On-chip initiation and burn rate measurements of thermite energetic reactions," in *Multifunctional Energetic Materials Symposium*, vol. 896, pp. 81–86, Materials Research Society, 2006.

[27] G. C. S. Freitas, F. C. Peixoto, and A. S. Vianna-Jnr, "Simulation of a thermal battery using phoenicsreg.," *Journal of Power Sources*, vol. 179, pp. 424–429, Apr. 2008.

[28] S. F. Son, B. W. Asay, R. A. Y. T. J. Foley, M. H. Wu, and G. A. Risha, "Combustion of nanoscale Al/MoO_3 thermite in microchannels," *Journal of Propulsion and Power*, vol. 23, no. 4, pp. 715–721, 2007.

[29] M. Hossain, S. Subramanian, S. Bhattacharya, Yuanfang-Gao, S. Apperson, R. Shende, S. Guha, M. Arif, Mengjun-Bai, K. Gangopadhyay, and S. Gangopadhyay, "Crystallization of amorphous silicon by self-propagation of nanoengineered thermites," *Journal of Applied Physics*, vol. 101, pp. 054509–1–6, March 2007.

[30] K. Minato, D. Nezaki, T. Okamoto, and M. Takata, "Growth conditions and luminescence of ZnO crystals grown by electric current heating with thermite reaction," *Key Engineering Materials*, vol. 248, pp. 95–98, 2003. Electroceramics in Japan VI. 22nd Electronics Division Meeting of the Ceramic Society of Japan. Kawasaki, Japan. 24-25 Oct. 2002.

[31] Xia-Dongdong and Wu-Xiaodong, "Melting high titanium ferroalloy by thermite method," *Shanghai Metals*, vol. 30, pp. 28–31, March 2008.

[32] K. Blobaum, M. Reiss, L. J. Plitzko, and T. Weihs, "Deposition and characterization of a self-propagating CuO_x/Al thermite reaction in a multilayer foil geometry," *Journal of Applied Physics*, vol. 94, pp. 2915–2922, Sept. 2003.

[33] V. Kobyakov, L. Mashkinov, and M. Sichinava, "Heat release in combustion of burning composite thermite mixtures on the base of $Fe_2O_3/TiO_2/Al$ system," *High Temperature*, vol. 47, no. 1, pp. 119 – 122, 2009.

[34] [34] H. Joo, J. Yun, K. Hwang, and B. Jun, "Synthesis of AlON refractory raw materials from aluminum dross using thermite reaction process." *Materials Science Forum*, vol. 510-511, pp. 866–869, 2006.

[35] A. Igtatev, M. Shiriaeva, D. Kovalev, V. Ponomarev, V. Sanin, and V. Yukhvid, "Dynamics of phase and chemical transformations in the combustion wave of the thermite composition NiO/Ni/Al," *International Journal of Self-Propagating High-Temperature Synthesis*, vol. 14, no. 1, pp. 41–53, 2005.

[36] T. Ohmi, J. Yanoma, and M. Kudoh, "Melting behavior of reaction products during thermite-type combustion synthesis of ni-al intermetallic compounds," *Journal of the Japan Institute of Metals*, 2001.

[37] B. Qinling, L. Peiqing, L. Weimin, X. Qunji, and D. Yutian, "Microstructure and properties of Ni_3Si alloyed with Cr fabricated by self-propagating high-temperature

synthesis casting route," *Metallurgical and Materials Transactions A* (Physical Metallurgy and Materials Science), vol. 36A, pp. 1301–1307, May 2005.

[38] M. Ali-Rachedi, W. Ramdane, D. Vrel, A. Benaldjia, P. Langlois, and M. Guerioune, "The role of sintering additives on synthesis of cermets by auto-combustion," *Powder Technology*, vol. 197, pp. 303–308, January 2010.

[39] F. Philipp, P. Schmidt, E. Milke, M. Binnewies, and S. Hoffmann, "Synthesis of the titanium phosphide telluride Ti_2Pte_2: a thermochemical approach," *Journal of Solid State Chemistry*, vol. 181, pp. 758–767, April 2008.

[40] H. Qiaodan, L. Peng, Q. Da, and Y. Youwei, "Self-propagation high-temperature synthesis and casting of Cu-$MoSi_2$ composite," *Journal of Alloys and Compounds*, vol. 464, pp. 157–161, 22 Sept. 2008.

[41] K. Das and T. Bandyopadhyay, "Synthesis and characterization of tib2-reinforced iron-based composites.," *Journal of Materials Processing Technology*, vol. 172, pp. 70–76, 20 Feb. 2006.

[42] K. Das, T. Bandyopadhyay, and S. Das, "A review on the various synthesis routes of tic reinforced ferrous based composites," *Journal of Materials Science*, vol. 37, pp. 3881–3892, 15 Sept. 2002.

[43] K. Das and T. Bandyopadhyay, "Synthesis and characterization of zirconium carbide-reinforced iron-based composite," *Materials Science and Engineering A* (Structural Materials: Properties, Microstructure and Processing), vol. A379, pp. 83–91, 15 Aug. 2004.

[44] S. Gunchoo and C. Good-Sun, "Synthesis and sintering of titanium and zirconium diborides," *Journal of the Korean Institute of Metals and Materials*, vol. 40, no. 9, pp. 976–983, 2002.

[45] R. Ricceri and P. Matteazzi, "A fast and low-cost room temperature process for TiB_2 formation by mechanosynthesis," *Materials Science and Engineering A* (Structural Materials: Properties, Microstructure and Processing), vol. A379, pp. 341–346, 15 Aug. 2004.

[46] E. J. Faierson, K. V. Logan, B. K. Stewart, and M. P. Hunt, "Demonstration of concept for fabrication of lunar physical assets utilizing lunar regolith simulant and a geothermite reaction," *Acta Astronautica*, vol. 67, pp. 38–45, July-August 2010.

[47] N. Travitzky, P. Kumar, K. Sandhage, R. Janssen, and N. Claussen, "Rapid synthesis of Al_2O_3 reinforced Fe-Cr-Ni composites," *Materials Science and Engineering A* (Structural Materials: Properties, Microstructure and Processing), vol. A344, pp. 245–252, 15 March 2003.

[48] R. Ricceri and P. Matteazzi, "A study of formation of nanometric W by room temperature mechanosynthesis," *Journal of Alloys and Compounds*, vol. 358, pp. 71–75, 25 Aug. 2003.

[49] J. Nie, Y. Li, H. Yan, Y. Liang, and Y. Li, "TiN/Al_2O_3 functionally graded material fabricated by in-situ reaction," *Key Engineering Materials*, vol. 368-372, pp. 1835–1837, 2007.

[50] C. Yeh, R. Li, and Y. Shen, "Formation of Ti_3SiC_2-Al_2O_3 in situ composites by SHS involving thermite reactions.," *Journal of Alloys and Compounds*, vol. 478, no. 1, pp. 699 – 704, 2009.

[51] R. Plovnick and E. Richards, "New combustion synthesis route to TiB_2-Al_2O_3," *Materials Research Bulletin*, vol. 36, pp. 1487–1493, May-June 2001.

[52] T. Xia, T. Liu, W. Zhao, B. Ma, and T. Wang, "Self-propagating high-temperature synthesis of Al_2O_3-TiC-Al composites by aluminothermic reactions," *Journal of Materials Science*, vol. 36, pp. 5581–5584, Dec. 2001.

[53] C. Cho and D. Kim, "Microstructure evolution and isothermal compaction in TiO_2-Al-C combustion reaction," *Journal of Materials Synthesis and Processing*, vol. 10, pp. 127–134, May 2002.

[54] T. Talaka, T. Grigorieva, P. Vitiaz, N. Lyakhov, A. Letsko, and A. Barinova, "Structure peculiarities of nanocomposite powder $Fe_{40}Al/Al_2O_3$ produced by MASHS," *Materials Science Forum*, vol. 534-536, pp. 1421–1424, 2007.

[55] T. Ohmi, M. Kudoh, K. Matsuura, and M. Iguchi, "Reaction mechanism for thermite combustion synthesis of Ni3Al," *International Journal of Self-Propagating High-Temperature Synthesis*, vol. 14, no. 2, pp. 99–110, 2005.

[56] D. Vrel, P. Langlois, E. M. Heian, N. Karnatak, S. Dubois, and M.-F.Beaufort, "Reaction kinetics and phase segregation in the 3NiO + 2Al □ 3Ni + Al_2O_3 thermite system," *International Journal of Self-Propagating High-Temperature Synthesis*, vol. 12, no. 4, pp. 261–270, 2003.

[57] D. Vrel, A. Hendaoui, P. Langlois, S. Dubois, V. Gauthier, and B. Cochepin, "SHS reactions in the NiO-Al system: Influence of the stoichiometry," *International Journal of Self-Propagating High-Temperature Synthesis*, vol. 16, no. 2, pp. 62–69, 2007.

[58] S. Nayak, W. Hsin, E. Kenik, I. Anderson, and N. Dahotre, "Observation of exothermic reaction during laser-assisted iron oxide coating on aluminum alloy," *Materials Science and Engineering A* (Structural Materials: Properties, Microstructure and Processing), 2005.

[59] H. Kaijin, L. Xin, X. Changsheng, and T. Yue, "Microstructure and wear behaviour of laser-induced thermite reaction Al_2O_3 ceramic coating on AA7075 aluminum alloy," *Journal of Materials Science and Technology*, vol. 23, no. 2, pp. 201–206, 2007.

[60] T. Yue, K. Huang, and H. Man, "Laser cladding of Al_2O_3 coating on aluminium alloy by thermite reactions," *Surface and Coatings Technology*, vol. 194, pp. 232–237, 1 May 2005.

[61] T. Yue, K. Huang, and H. Man, "In situ laser cladding of Al_2O_3 bearing coatings on aluminum alloy 7075 for improvement of wear resistance," *Surface Engineering*, vol. 23, pp. 142–146, March 2007.

[62] T. Yue, H. Yang, T. Li, and K. Huang, "The synthesis of graded thermal barrier coatings on nickel substrates by laser induced thermite reactions," *Materials Transactions*, vol. 50, no. 1, pp. 219-221, 2009.

[63] L. Yanxiang, Y. Jiankui, and L. Yuan, "Synthesis and cladding of Al_2O_3 ceramic coatings on steel substrates by a laser controlled thermite reaction," *Surface and Coatings Technology*, vol. 172, pp. 57–64, 15 July 2003.

[64] E. Miyazaki and O. Odawara, "Shs technology for in-situ resource utilization in space," *International Journal of Self-Propagating High-Temperature Synthesis*, vol. 12, no. 4, pp. 323–332, 2003.

[65] C. Lau, A. Mukasyan, and A. Varma, "Reaction and phase separation mechanisms during synthesis of alloys by thermite type combustion reactions," *Journal of Materials Research*, vol. 18, no. 1, pp. 121–128, 2003.

[66] I. S. Gordopolova, T. P. Ivleva, K. G. Shkadinsky, and V. I. Yukhvid, "Formation of the composition structure under gravity-induced phase separation and heat transfer in the

system of high-temperature melt-metal substrate. ii. the influence of phase thermal conductivity on structure formation," *International Journal of Self-Propagating High-Temperature Synthesis*, vol. 10, no. 2, pp. 177–191, 2001.

[67] S. Wang, W. Gao, T. Ma, K. Liang, and X. Zhang, "The monolithic glass-ceramic lined steel elbow made by SHS gravitational-thermite process," *Key Engineering Materials*, vol. 280-283, no. 1, pp. 1631–1634, 2005.

[68] J. Lee, M. Le, and H. Chung, "Physical properties of ceramic layer prepared by SHS in centrifugal field," *Materials Transactions*, vol. 48, pp. 2960–2963, Nov 2007.

[69] L. Zhang, Z. Zhao, Y. Song, W. Wang, and H. Liu, "Al_2O_3/ $ZrO_2(Y_2O_3)$ prepared by combustion synthesis under high gravity," *International Journal of Modern Physics B*, vol. 23, pp. 1148–1153, 20 March 2009.

[70] Z. Hao, Z. Zhongmin, Z. Long, S. Yalin, and P. Chuanzeng, "Composition, microstructure and properties of Al_2O_3/ZrO_2 composites prepared by combustion synthesis under high gravity," *Special Casting and Nonferrous Alloys*, vol. 29, no. 5, pp. 455–458, 2009.

[71] Z. Du, H. Fu, H. Fu, and Q. Xiao, "A study of ceramic-lined compound copper pipe produced by SHS-centrifugal casting," *Materials Letters*, vol. 59, pp. 1853–1858, June 2005.

[72] O. Menekse, J. Wood, and D. Riley, "Investigation of Fe_2O_3-Al and Cr_2O_3-Al reactions using a high speed video camera," *Materials Science and Technology*, vol. 22, pp. 199–205, 1 Feb. 2006.

[73] Q. Meng, S. Chen, J. Zhao, H. Zhang, H. Zhang, and Z. Munir, "Microstructure and mechanical properties of multilayer-lined composite pipes prepared by SHS centrifugal-thermite process," *Materials Science and Engineering A* Structural Materials: Properties, Microstructure and Processing, vol. 456, pp. 332–6, 15 May 2007.

[74] T. Ohmi, Y. Murota, K. Kirihara, and M. Kudoh, "Centrifugal casting of Ni-Al intermetallic compounds produced by thermite-type combustion synthesis," *Journal of the Japan Institute of Metals*, vol. 65, pp. 458–463, May 2001.

[75] R. Seshadri, "Centrifugal force assisted combustion synthesis of intermetallics," *Materials and Manufacturing Processes*, vol. 17, no. 4, pp. 501–518, 2002.

[76] V. Sanin, V. Yukhvid, and A. Merzhanov, "The influence of high-temperature melt infiltration under centrifugal forces on SHS processes in gasless systems," *International Journal of Self-Propagating High-Temperature Synthesis*, vol. 11, no. 1, pp. 31–44, 2002.

[77] D. Vrel, N. Girodon-Boulandet, S. Paris, J.-F. Mazué, E. Couqueberg, M. Gailhanou, D. Thiaudière, E. Gaffet, and F. Bernard, "A new experimental setup for the time resolved x-ray diffraction study of self-propagating high-temperature synthesis," *Review of Scientific Instruments*, vol. 73, no. 2, pp. 422–428, 2003.

[78] F. Bernard, S. Paris, D. Vrel, M. Gailhanou, J.-C. Gachon, and E. Gaffet, "Time-resolved XRD experiments adapted to SHS reactions: past and future," *International Journal of Self-Propagating High-Temperature Synthesis*, vol. 12, no. 2, pp. 181–190, 2002.

[79] A. Hendaoui, M. Andasmas, A. Amara, A. Benaldjia, P. Langlois, and D. Vrel, "SHS of high-purity MAX compounds in the TiAlC system," *International Journal of Self-Propagating High-Temperature Synthesis*, vol. 17, no. 2, pp. 129–135, 2008.

In: MAX Phases: Microstructure, Properties and Applications ISBN 978-1-61324-182-0
Editors: It-Meng (Jim) Low and Yanchun Zhou

Chapter 3

REACTIVE INFILTRATION PROCESSING OF Ti_3AlC_2 AND Ti_3SiC_2-BASED COMPOSITES

Xiaowei Yin[1], Nahum Travitzky[2] and Peter Greil[2]

[1]Northwestern Polytechnical University, National Key Laboratory of Thermostructure Composite Materials, Xi'an, Shaanxi, China

[2]Department of Materials Science, Glass and Ceramics, University of Erlangen-Nuremberg, Martensstrasse 5, Erlangen, Germany

ABSTRACT

Owing to the unique nanolaminate crystal structure, the *MAX* phases with *A* = Al and Si offer superior mechanical properties, which make them a potential reinforcement for brittle ceramic matrix materials. This chapter will focus on the processing of Ti_3AlC_2 and Ti_3SiC_2-based composites by reactive melt infiltration technique. Capillary driven infiltration of a reactive melt allows near-net shape manufacturing of *MAX* phase reinforced composites with high flexibility in component geometry at low production costs. Reaction infiltration formation of *MAX* phase in ceramic matrix composites such as fiber reinforced ceramic matrix composites and interpenetrating phase composites may extend the application fields of advanced ceramic composite materials.

1. INTRODUCTION

Intermetallics and ceramics have high strength, high hardness, good wear resistance, and high temperature resistance at low weight, making these materials attractive thermostructure materials for applications in the industries of aeronautic, aerospace, military, and nuclear. In recent years, it has been extensively demonstrated that the combination of *MAX* phase with intermetallics and ceramics could improve their mechanical properties considerably [1-7]. Up to now, *MAX*-phase-based materials (e.g., Ti_3SiC_2 and Ti_3AlC_2) were prepared mainly by hot-pressing at elevated temperatures which limited shaping capability to rather simple component geometries. In order to achieve near-net and complicate shape component manufacturing pressureless melt infiltration techniques are of particular interest. In the present chapter, three-dimensional printing, chemical vapor infiltration and reactive melt infiltration were explored to realize a near net-shape fabrication process with great flexibility in component shaping. The microstructure and mechanical properties of $TiAl_3$-based and SiC-based composites containing the *MAX* phases Ti_3AlC_2 and Ti_3SiC_2 fabricated by the near net-shape fabrication process will be analyzed [8-12].

2. Near Net-Shape Fabrication Process

2.1. Three-Dimensional Printing

Additive Manufacturing allows the fabrication of a three-dimensional part of arbitrary shape directly from a computer-aided design (CAD) model by a fast and highly automated manufacturing process [13, 14]. Among an increasing number of Additive Manufacturing techniques, Layered Objective Manufacturing (LOM), Three Dimensional Printing (3DP™), Selective Laser Sintering (SLS) and Fused Deposition Modeling (FDM) gain interest for ceramic prototyping. Common to all of these techniques is to decompose a 3D-CAD model into 2D layers (slicing) and layer-by-layer building of the part. Therefore, these techniques are often referred to as solid freeform fabrication or layered manufacturing. In contrast to *"subtractive"* processes (e.g., drilling, milling, grinding) the *"additive"* techniques allow fabrication of products with complex internal structure that cannot be manufactured by other approaches. Furthermore, Additive Manufacturing can significantly shorten fabrication times with small personnel expenditure and reduce product costs when applied properly.

3D printing (3DP™) involves the deposition of powdered material in layers and the selective binding of the powder by a modulated ''ink-jet'' printing of a binder material. Two basically different approaches of 3DP™ can be distinguished, a direct and an indirect one: direct means that the powder in the form of a colloidal suspension (e.g. ink) passes through the print head to form the freestanding shape [15], while indirect means that the liquid binder is printed onto preplaced powder [16-18]. Depending on the preprocessing of the powder bed (e.g., powder composition, particle size, and its distribution) and the printing process (e.g., thickness of an individual powder layer and the rheological parameters of the binder), geometrical accuracy, porosity, and surface roughness of the printed object may vary in a wide range [16-23]. Owing to the low strength of the porous printed object, a separate consolidation (sintering) step before infiltration might be necessary to achieve volume and shape stability upon infiltration and reaction. Unfortunately, a large shrinkage of 20 % or more may occur during sintering of a porous preform [19]. Thus, increasing the shape stability and strength of the printed object without inducing a dimensional change is a major challenge for development of 3D printing process for ceramic and ceramic composite materials.

2.2. Chemical Vapor Infiltration

Chemical vapor infiltration (CVI) is widely used to fabricate the matrix in continuous fibers reinforced carbon and ceramic matrix composites [24-25]. During CVI reactive gases are flowing through the porous fiber preform in a reactor chamber and decomposition or reaction of the gas phase results in the deposition of solid matter on the inner surface of the porous preform. CVI is controlled by vapor phase diffusion, which may be constrained by chemical (multi-component gas phase) and physical (pore geometry) limitations [26]. In order to improve the infiltration depth, CVI is usually conducted at low temperature (800-1100°C) and reduced pressure (1-10kPa) to achieve high diffusion rates of the reaction gases. Therefore, the CVI reactors are operated at low temperature and low pressure, to maintain very low deposition rates and high transportation rates [27, 28]. The internal surface area of porous preform decreases continuously due to the progressive densification, which influences the interactions of gas-phase and surface kinetics in the micropores. When performed under isothermal and isobaric conditions, the deposition reaction results in internal reactant depletion, and thus may lead to a different deposit thickness obtained in the core and on the surface of the porous preform.

CVI method has various advantages: (1) the process can be conducted at low pressure and low temperature, no external pressure is required, so fiber reinforced ceramic matrix composites with high performance can be fabricated owing to the small residual stress and less damage of the fiber; (2) the phase composition of the matrix can be easily designed by adjusting the kinds, concentration, and deposition sequence of the reaction gases; (3) components with complex shape and high volume content of fibers can be fabricated by a near-net shape process. However, CVI method may suffer from some technical disadvantages: (1) the processing time is long; (2) component thickness is limited and (3) a high density e.g. low porosity is difficult to achieve.

2.3. Reactive Melt Infiltration

Metal melt infiltration into a porous solid preform is one of the preferred methods to fabricate ceramic-metal and ceramic matrix composites. A molten metal can penetrate into the pore channels driven either by an external force (squeeze casting) or by the action of capillary pressure created when the liquid wets the solid [29]. Reaction between porous ceramic preform and the molten metal may considerably enhance the wettability and, thus lead to reactive melt infiltration (RMI). Compared to the pressure-assisted infiltration techniques such as squeeze casting or gas-pressure infiltration, RMI offers a high flexibility in component geometry and low production costs [30-34].

For reactive infiltration to be successful, infiltration rate of the melt into the porous preform should be faster compared to the reaction rate in order to avoid pore clogging by the reaction product near the surface and incomplete filling of the preform. The wetting behavior of the melt on the pore surface in the preform is among the main factors affecting kinetics of infiltration. In reactive melt infiltration of initially wettable systems e.g. wetting angle $\theta <$ 90° as for example Al-melt (1000°C) on Si_3N_4 ($\theta \sim 0°$) or SiC ($\theta \sim 30°$), application of an external pressure is not required and spontaneous self-infiltration takes place by capillary suction. For many metal-ceramic combinations, however, the wettability is not sufficient as for example Al-melt (830°C) on Al_2O_3 ($\theta \sim 140°$). Therefore, an external pressure has to be applied to promote infiltration. The threshold pressure necessary to move the molten metal along the pores depends on the surface tension of molten metal, wetting angle and pore diameter. It has been observed, however, that chemical reactions between the molten metal and the porous preform may enhance infiltration and thus lower the required pressure. Moreover, from active brazing it is well known, that reactive transition metals like Ti or Cr improve wetting of liquid metals on ceramic materials (e.g. Cu–Ti alloys on SiC).

As the reaction between the molten metal and the porous ceramic preform proceeds, a continuous layer of the reaction product is formed at the melt/preform interface. As a result, the channels through which the molten metal can travel become narrow thereby reducing the flow. If the reaction kinetics is fast and pores are very fine, complete pore closure and flow cessation may occur (pore clogging). This is the case, when reaction products prevent further infiltration of molten Al into Reaction-Bonded Silicon Nitride (RBSN) with very fine porosity (pore diameter $< 0.05\mu m$) [35]. Thus, it is very important to select the pore size and distribution of porous preforms before RMI process. While the Additive Manufacturing techniques may produce porous preforms with a high degree of freedom in geometry and shape, capillary-driven infiltration of the metal melt followed by a subsequent reaction into the final composite material finally may result in a dense microstructure of the component. Net-shape manufacturing is possible by controlling dimensional changes associated with a displacive reaction between the infiltrated melt and the ceramic preform phase (e.g. volume shrinkage is compensated by volume expansion).

In the following sections, the process, microstructure and properties of Ti_3AlC_2 toughened $TiAl_3$-Al_2O_3 composites and Ti_3SiC_2 toughened C/C-SiC composites fabricated by the combination of three-dimensional printing and chemical vapour infiltration with reactive melt infiltration will be discussed.

3. Ti_3AlC_2 TOUGHENED $TiAl_3$-Al_2O_3 COMPOSITES

3.1. 3D Printing of Preforms

$TiAl_3$ is a potential thermo-structure material due to its advanced characteristics such as low density (3.3 g/cm^3), high melting temperature (1303 °C), and good oxidation resistance. However, some weaknesses, such as low fracture toughness (2 MPa·m$^{1/2}$) and shaping difficultly limit its potential applications [36-37]. Al_2O_3 can be used as the dispersive toughening phase in a composite because of its high hardness (18 GPa) and high modulus (Young's modulus 386 GPa, shear modulus 175 GPa). Fracture toughness of Al_2O_3 toughened $TiAl_3$ composite (Al_2O_3-$TiAl_3$) can reach 5.0-8.6 MPa·m$^{1/2}$ [38-39]. Nanolaminate *MAX*-phase Ti_3AlC_2 is of particular interest to act as highly effective reinforcement since it combines the toughness and strain tolerance of a metal with the high elastic modulus, strength and wear resistance of a ceramic [40-42]. The unique nanolaminate structure of alternating layers of edge sharing Ti_6C octahedra and close-packed Al layers gives rise for a pronounced anisotropic deformation and fracture behavior governed by the distinct differences of bonding energy in the ceramic (Ti_6C) and the metal (Al) layers in Ti_3AlC_2. An increase of fracture toughness is attributed to the microscopic delamination along the weak Al-plane which triggers extended crack deflection and crack bridging mechanisms during crack growth in the *MAX* phase reinforced composites. Thus, Ti_3AlC_2 achieved an elastic modulus, compressive strength and bending strength of 289 GPa, 785 MPa and 375 MPa, respectively [40-41].

Ti_3AlC_2 ceramics are machinable and have a high compressive strength of 800 MPa up to 700°C [42]. Particularly, its fracture toughness could reach 7.2 MPa·m$^{1/2}$ [41], which is much higher than those of $TiAl_3$ and Al_2O_3. Compared with $TiAl_3$ and Al_2O_3, Ti_3AlC_2 exhibits superior thermal shock resistance [43]. Therefore, dispersing Ti_3AlC_2 in an Al_2O_3-$TiAl_3$ matrix composite is expected to result in improved toughness and thermal shock resistance of the composite. In this section, a combination process of indirect 3DP™ and pressureless reactive melt infiltration was used to fabricate three phase Ti_3AlC_2-Al_2O_3-$TiAl_3$ composites.

In order to form Ti_3AlC_2 toughened $TiAl_3$-Al_2O_3 composites the various reactions between the preform phase (TiO_2 + TiC) and the infiltrating Al-melt have to be analyzed. Carbon produced from the decomposition of dextrin binder at 800 °C was used to reduce TiO_2 into Ti_2O_3 at 1400 °C which was reported to offer better wetting behavior for an Al-melt than TiO_2 [44]:

$$2TiO_2(s) + C(s) = Ti_2O_3(s) + CO(g)\uparrow \quad (1)$$

Once Al-melt has spontaneously infiltrated into the porous preform at temperatures exceeding 1000 °C, Al reduces the oxide phase to form a multiphase reaction product:

$$2TiC(s) + Ti_2O_3(s) + 6Al(l) \rightarrow Ti_3AlC_2(s) + TiAl_3(s) + Al_2O_3(s) \quad (2)$$

At the reaction temperature of 1200 °C, the change of Gibbs free energy calculated for reaction (2) (FactSage 5.4.1 database, Montreal, Quebec, Canada) is -342kJ/mol indicating that a three phase composite microstructure Ti_3AlC_2 - $TiAl_3$ - Al_2O_3 is likely to form.

The 3DP™ ceramic preform was prepared using powder blends of TiC (mean particle size 1.2 μm) and TiO_2 (30 nm). Dextrin powder ($(C_6H_{10}O_5)_n$, n = 10 – 200) with a mean particle size of 115 μm was added to the powder blend, which served as a binder when coming in contact with the injected water based printer solution. Thermal analysis in Ar atmosphere revealed that the decomposition of dextrin started above 250 °C leaving a total residue of amorphous carbon of 25 wt.% (from the initial dextrin fraction) at 800 °C [8]. A typical powder blend applied for printing consisted of 63 wt% TiC, 31 wt% TiO_2 and 6 wt% dextrin. All the ceramic and the binder materials used in the printing experiments were reagent grade.

Three dimensional printing was carried out by a 3D printer. A water based printing solution was injected through a bubble jet print head. In the region where the water solution was injected, swelling and gelling of the dextrin binder triggered bonding of the powder particles whereas the neighbouring powder remained unbound (Figure1 (a)). The thickness of individual print layer was 90 μm. A build-up speed of 20 mm/h was applied. After drying in air at room temperature for 48 h, the printed parts were removed and cleaned from the unbound powder bed. The printed bodies were subsequently pyrolyzed in flowing N_2 at 800 °C for 2 h, and then sintered in flowing Ar at 1400 °C for 0.5 h. During pyrolysis, dextrin decomposed into amorphous carbon, which at 1400 °C caused reduction of TiO_2 into Ti_2O_3 at least on the powder particle surface.

a)

b)

Figure 1. Photos of (a) the printed parts buried in the powder bed and (b) computer-aided design model of a gearwheel and the corresponding printed part after reactive melt infiltration [9].

Figure 2 shows a representative fracture surface of a printed object after annealing at 1400 °C. The preform microstructure is characterized by aggregates which gave rise for a bimodal pore structure with large inter-aggregate pores of approximately 30 μm and small intra-aggregate pores of 0.07 μm. After consolidation of the preform by sintering at 1400 °C, grain growth of the nanoscale TiO_2 particle caused a pronounced shift of the small pore size from 0.07 to 0.7 μm. The pores were fully interconnected and were found to provide excellent conditions for capillary driven spontaneous infiltration of the Al melt. The open porosity and the bulk density of the sintered preform were 63 vol% and 1.61 g/cm^3, respectively.

Figure 2. Microstructure of the sintered preform: (a) Low magnification and (b) High magnification.

3.2. Reactive Melt Infiltration and Composite Formation

Al-melt infiltration started from 1050 °C and the temperature was dwelled at 1200 °C for 1 h. Subsequently the temperature was raised to 1300 °C or to 1400 °C for 1.5 h. The infiltration depth of the metal melt into the porous preform, h, was estimated by Darcy's equation as a function of time, t, and effective pressure difference, ΔP, according to:

$$h = \left[\frac{2K\Delta P}{\mu \varepsilon_p} t \right]^{1/2} \quad (3)$$

where μ denotes the viscosity of the melt, ε_p the pore volume fraction, and K is the permeability of the porous powder-packed preform. K depends on the particle radius, r, by (Carmen–Kozeny equation)

$$K = \frac{\varepsilon_p{}^3 r^2}{37.5(1-\varepsilon_p)^2} \tag{4}$$

Combining Eqs. (3) and (4) and replacing the effective pressure difference ΔP by the capillary pressure ΔP_c:

$$\Delta P_c = \frac{3\lambda}{r}\left(\frac{1-\varepsilon_p}{\varepsilon_p}\right)\gamma\cos\theta \tag{5}$$

finally yields

$$h = \left[t\left(\frac{\varepsilon_p}{1-\varepsilon_p}\right)\frac{r\lambda\gamma}{6.25\mu}\cos\theta\right]^{1/2} \tag{6}$$

Taking reasonable values for $\varepsilon_p = 0.51$ (inter-agglomerate) and 0.14 (intra-agglomerate), particle shape factor $\lambda = 0.5$ [45], viscosity of the Al-melt at 1000 °C $\mu = 0.885$ mPa·s, surface tension of Al-melt at 1000 °C $\gamma = 0.815$ J/m^2 [46], the infiltration depth h was calculated as a function of time for a wetting angle $\theta = 50°$ [47]. Assuming that the wetting angle θ and the surface tension γ remain constant for the large inter-agglomerate pores as well as for the small intra-agglomerate pores, the calculation results suggest that reasonable infiltration depths could be achieved within minutes ($h \rightarrow 2$ cm for intra- agglomerate and $h \rightarrow 20$ cm for inter-agglomerate pores within 1 min, respectively). Once the Al-melt has completely penetrated into the titania/titanium carbide aggregates, further reaction may then occur at the metal melt – ceramic interface. Due to the lower molar volume of the solid reaction product (Al_2O_3) compared to the solid phase of the preform (TiO_2), continuous access of the melt through the reaction layer facilitates the complete reaction within reasonable processing time.

The microstructure of the composite after reactive melt infiltration with Al contains $TiAl_3$-rich and $TiAl_3$-free regions. $TiAl_3$-rich regions were composed of 35 vol% Ti_3AlC_2, 55 vol% $TiAl_3$, 5 vol% Al_2O_3, 3 vol% TiC and 2 vol% Al, whereas $TiAl_3$- free regions have 35 vol% Ti_3AlC_2, 35 vol% Al, 20 vol% Al_2O_3 and 10 vol% TiC. The content of $TiAl_3$-rich regions in the composite ranged from 40 vol% to 50 vol%. At temperatures exceeding the melting point of $TiAl_3$, 1303 °C, the melt phase accelerated grain growth. The grain size of the Ti_3AlC_2 reaction phase attained 5 μm in length and 2 μm in thickness at reaction temperature of 1300 °C (Figure 3). Exaggerated grain growth at the higher reaction temperature of 1400 °C resulted in larger Ti_3AlC_2 platelet size of 50 μm in length and 5 μm in thickness. Total shrinkage of the final composite compared with the CAD model dimensions applied for 3D printing was less than 3.2% with only a small anisotropy between the *in-plane* (e.g. the printing layer) and *out-of-plane* directions (e.g. perpendicular to the printed layers), as shown in Figure1 (b) [9].

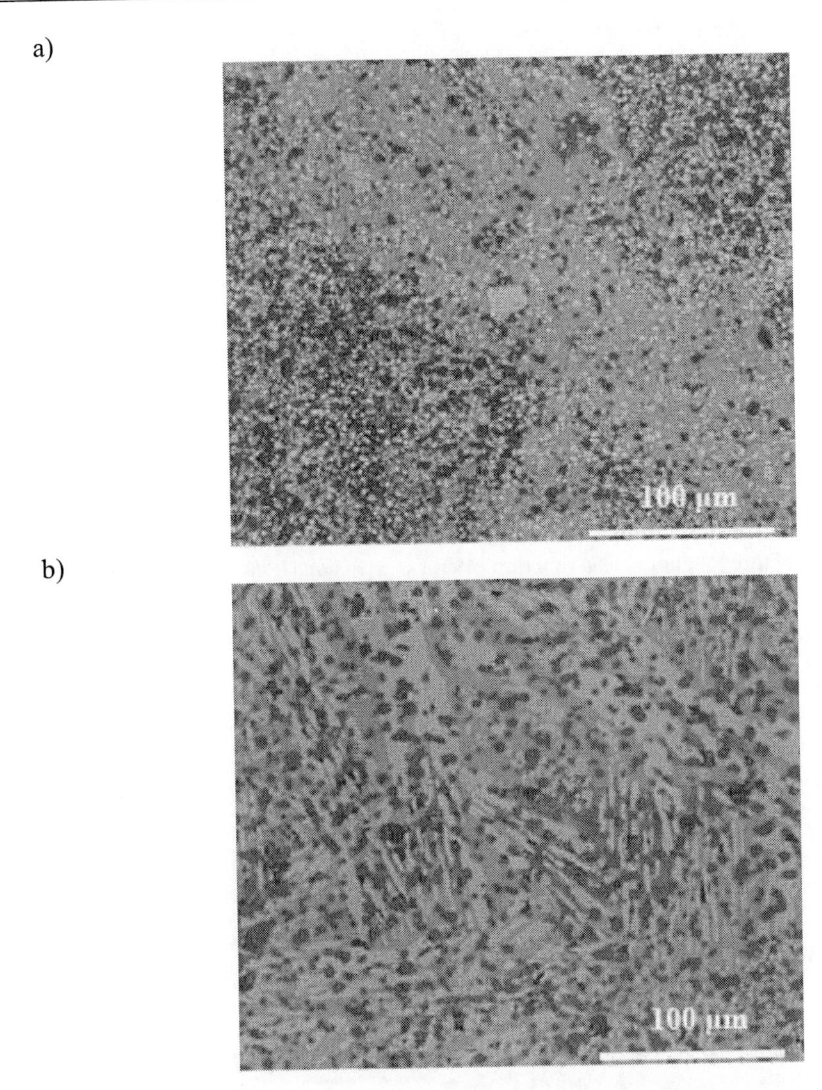

Figure 3. Backscattered electron (BSE) images of (a) the polished cross section of Ti_3AlC_2-$TiAl_3$-Al_2O_3 composite fabricated at 1300 °C and (b) the polished cross-section of composite fabricated at 1400 °C.

3.3. Mechanical Properties

The Ti_3AlC_2-$TiAl_3$-Al_2O_3 composites attained a four-point bending strength of 320 ± 40 MPa, a strain-to-failure of 0.36 % and a Young's modulus of 184 ± 24 GPa. Fracture toughness measured on the specimen reacted at 1300 °C resulted in K_{IC} = 8.1±1.7 MPa·m$^{1/2}$ which was increased to K_{IC} = 9.7 ± 0.8 MPa·m$^{1/2}$ for the composite reacted at 1400 °C [10]. Figure 4 shows the fracture resistance-curves (R-curves) of both composites. While the specimen reacted at 1300 °C having a small grain size only shows a moderate increase of crack resistance with increasing crack length, the coarse grained specimen reacted at 1400 °C exhibits an unusual increase of apparent toughness values.

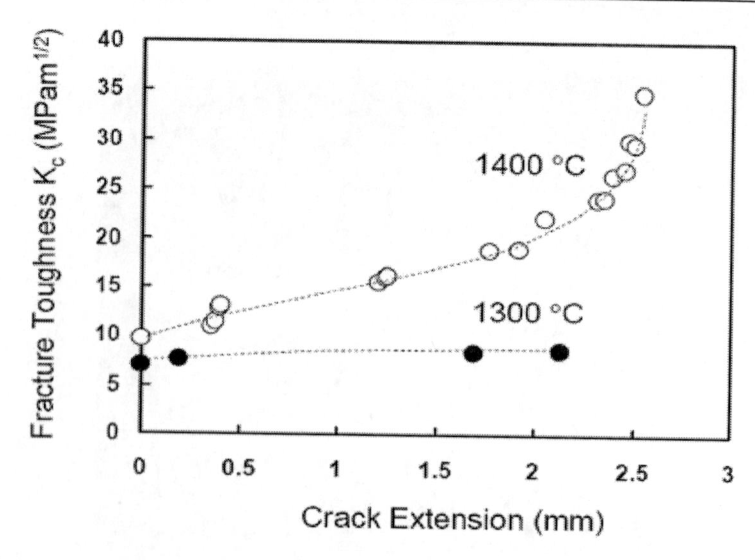

Figure 4. Variation of fracture toughness as a function of crack extension (R curves) [10].

Figure 5. BSE image of the Ti_3AlC_2-$TiAl_3$-Al_2O_3 composite fabricated at 1400°C after R-curve testing, showing the bridging of the main crack.

In brittle materials, the effect of a rising resistance to crack propagation with increasing crack extension (R-curve) has been associated with both the growth of a bridging zone in the wake of the crack and a damage zone ahead of the crack tip [48]. Indeed, a pronounced bridging effect was observed in the coarse-grained specimen prepared at 1400 °C. Fracture analyses revealed also extensive crack deflection, crack bridging (Figure 5), crack branching, grain delamination and pull-out to occur in the composite. X-ray photoelectron spectroscopy showed that the bonding between the aluminum layer and the Ti_6C-octahedra along the (0001) basal plane in Ti_3AlC_2 is characterized by low bonding energies [49]. The energies are 281.0 - 281.5 eV, for C 1s (which is below the lowest values measured for carbides), 454.0 - 454.7 eV for Ti 2p (comparable to that of Ti metal and TiC), while the energy of the Al 2p, 71.9 eV, is lower by ca. 0.8 eV than that for Al metal. Those observations suggest that crack growth and delamination in the nanolaminate Ti_3AlC_2 grains is likely to occur along the weakly bonded Al-planes. In addition to regular slip, mechanisms for ambient temperature plastic deformation in Ti_3AlC_2 are thought to involve the readjustment of local stress fields from kink band (boundaries) formation, buckling, and delamination of individual grains. Specifically, delamination along the weaker basal planes leads to the creation of microlaminae within a single grain and consequently, the deformation and distortion of such laminate provides a potential contribution to toughening [50].

4. Ti_3SiC_2 Toughened C/C-SiC Composites

4.1. Fiber Reinforced SiC Composites

Carbon fiber reinforced carbon/silicon carbide binary matrix composites (C/C-SiC) attract more and more attentions owing to their excellent oxidation resistance, high strength and thermal shock resistance particularly at high temperatures [51-55]. The fabrication methods of porous C/C-SiC composites mainly include Chemical Vapor Infiltration (CVI) and Polymer Infiltration and Pyrolysis (PIP). The main advantages of CVI and PIP are superior mechanical properties of the fabricated composites and ability to near-net shape fabrication. The disadvantages are high residual porosity, long processing time and high cost. C/C-SiC composites fabricated by both processes typically contain a porosity of 10 ~ 15 vol%. The pores in the C/C-SiC composites may blunt the matrix cracks and the interconnected pores may act as diffusion channel for oxidizing gases, leading to the oxidation of the C-fibers. Reduction of residual porosity in C/C-SiC may be achieved by liquid silicon infiltration (LSI), which may shorten the fabrication time and, therefore, reduce production cost. The main disadvantages of LSI, however, may be seen in a high content of residual free Si and the reaction of Si melt with C-fibers which deteriorates the mechanical properties of the composite. A combination of LSI and CVI or PIP was successfully applied for the fabrication of C/C-SiC composites characterized by low porosity and improved properties [51]. Dense C/C-SiC composite is distinguished by superior oxidation resistance, but it has low fracture toughness and high content of residual silicon, which may limit applications at temperatures above 1300 °C. Therefore, improving the toughness of dense fiber reinforced C/C-SiC composites is of great interest. Increase of toughness of dense C/C-SiC composites mainly depends on triggering effective crack bridging and fiber pull-out during crack propagation. Forming a weak interface between the strong C-fiber reinforcement and the matrix is a key factor for activating effective crack bridging and pull-out mechanisms. Commonly, thin layers of amorphous pyro-carbon or boron nitride interphase are formed by coating the fibers prior to matrix generation. The same kind of weak interface was applied in SiC/graphite and SiC/BN layered composites [56-57]. Alternatively, nano-layered *MAX* phases seem to be of particular interest for tailoring the fiber/matrix interface bonding. The nano-laminate crystal structure of Ti_3SiC_2 *MAX*- phase exhibits a high toughness of 16 $MPa{\cdot}m^{1/2}$ [4, 58], good machinability and a melting temperature as high as 3000 °C. Its high stability and compatibility in contact to SiC makes Ti_3SiC_2 a candidate for reinforcement of SiC based composites with improved mechanical properties as demonstrated earlier [59-61].

4.2. Thermodynamic Aspects

Reaction infiltration formation of the matrix in the carbon fiber reinforced ceramic composite was analyzed by thermodynamic calculations (FactSage 5.4.1 database, Montreal, Quebec, Canada). Infiltration of liquid Si into a porous TiC powder preform which forms the matrix between C-fibers results in the formation of Ti_3SiC_2 and SiC according to :

$$3TiC + 2Si \rightarrow Ti_3SiC_2 + SiC \quad (7)$$

Taking the standard free energies of Ti_3SiC_2 formation as a function of temperature from [62] ($\Delta G_f^0 = -547145 + 24.845T$), a change of free reaction energy of ΔG_r = -75.0 kJ/mol was calculated for reaction (7) at 1500 °C. Concurrently, liquid Si may react with the carbon fibers:

$$Si + C \rightarrow SiC \quad (8)$$

which at 1500 °C yields a slightly lower value of ΔG_r = -56.9 kJ/mol. Since the concentration of infiltrated Si may change upon infiltration and reaction process, a variety of possible reactions were calculated for the reaction temperature of 1500 °C. Thus, with increasing Si content titanium silicides may form:

$$TiC + 3Si \rightarrow TiSi_2 + SiC \qquad \Delta G_r = -12.2 \text{ kJ/mol} \quad (9)$$

$$5TiC + 8Si \rightarrow Ti_5Si_3 + 5SiC \qquad \Delta G_r = -55.3 \text{ kJ/mol} \quad (10)$$

Once the silicides have formed a large change of reaction energy will favour the reaction with TiC to yield Ti_3SiC_2

$$7TiC + 2TiSi_2 \rightarrow 3Ti_3SiC_2 + SiC \qquad \Delta G_r = -200.4 \text{ kJ/mol} \quad (11)$$

$$2C + 4TiC + Ti_5Si_3 \rightarrow 3Ti_3SiC_2 \qquad \Delta G_r = -283.5 \text{ kJ/mol} \quad (12)$$

$$Ti_5Si_3 + 10TiC + 2Si \rightarrow 5Ti_3SiC_2 \qquad \Delta G_r = -319.6 \text{ kJ/mol} \quad (13)$$

The highest release of exothermic reaction energy, however, was calculated for the case that TiC reacts with silicon and carbon directly into Ti_3SiC_2

$$9TiC + 8Si + 2C \rightarrow 3Ti_3SiC_2 + 5SiC \qquad \Delta G_r = -338.7 \text{ kJ/mol} \quad (14)$$

Neglecting to a first approximation kinetic effect on the possible reaction path, reaction (14) should be thermodynamically favored in the presence of free C. In order to remove residual silicon by the in-situ formation of Ti_3SiC_2, the molar ratio of TiC and Si should be higher than 1 according to both reaction (7) and reaction (14).

4.3. Ti_3SiC_2 Toughened C/C-SiC Composite Manufacturing

A three-dimensional needle-punched C/C preform was fabricated by CVI, which has a density of 1.43 g/cm^3 to 1.51 g/cm^3 [63]. This porous C/C preform was infiltrated with an aqueous slurry of TiC powder with a mean particle size of 1 μm. Carboxymethyl cellulose (CMC, $C_6H_7O_2(OH)_2OCH_2COONa$) was dissolved into distilled water in a ratio of 5 g/l. TiC powder was mixed with the above water solution in a ratio of 750 g/l. The slurry was ball-milled for 12 - 48 hours. The C/C specimens were then infiltrated with the slurry and subsequently freeze-dried. The infiltration step was repeated up to five times to increase TiC loading. The density of the infiltrated composite of 1.67 g/cm^3 corresponds to a TiC volume fraction of 4 % and an open porosity of 20 %. Known from the pore size distributions of the C/C preform and the TiC infiltrated preform (Figure 6), partial filling of the large inter-fiber bundle pores with a pore size maximum at approximately 32 μm generated a second peak of significantly smaller inter-particle pores centring at 1.1 μm. The slurry infiltrated composites were named as SI-composites.

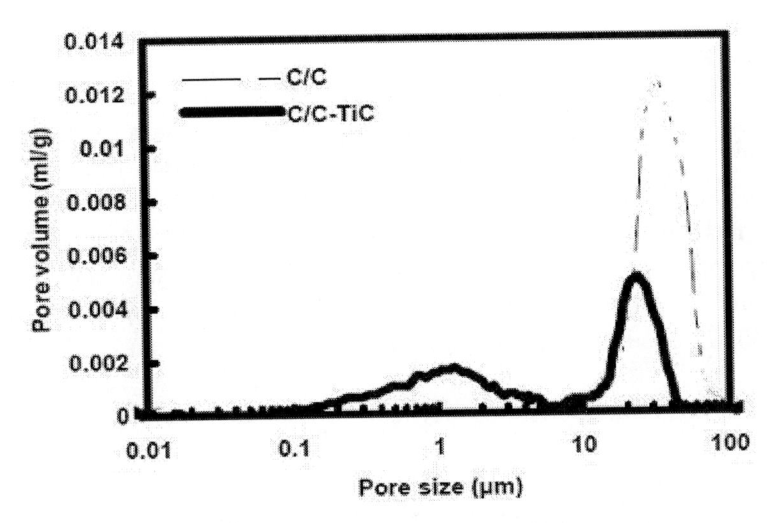

Figure 6. Pore-size distributions of C/C composite and the composite with TiC particles [12].

Subsequently, the SI-composites were infiltrated with liquid Si at 1500 °C in a vacuum furnace. At 1500 °C the contact angle of liquid Si on TiC is about 32° [64] and about 35° on vitreous carbon [65]. As expected, liquid Si was spontaneously infiltrated into the SI-composites and the reactions of liquid Si with TiC and carbon led to the *in-situ* formation of Ti_3SiC_2 and SiC as confirmed by XRD (Figure 7). A large amount of Ti_3SiC_2 was detected, which may exist in the original inter-fiber bundle pores of the C/C preforms. No peak of residual Si was observed. The composite is composed of approximately 73 vol% carbon (fiber and matrix), 22 vol% SiC, and 5 vol% Ti_3SiC_2. The as-received materials were named as SI-LSI composites [12].

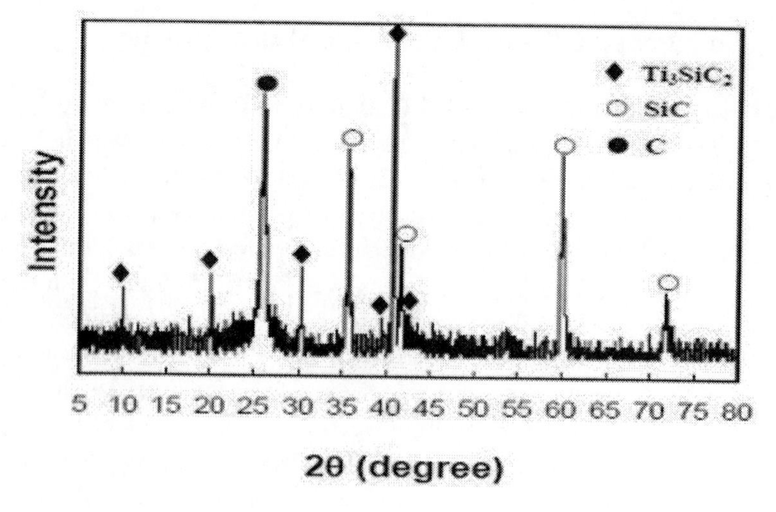

Figure 7. Micro-area XRD pattern of C/C-SiC-Ti_3SiC_2 composite.

4.4. Mechanical Properties

During liquid Si infiltration, some C-fibres unavoidably reacted with Si, which is the main reason that LSI C/C-SiC composites usually had decreased fracture toughness. As shown in Table 1, the needle-punched C/C composite attained a fracture toughness of 8 MPa·$m^{1/2}$ and flexural strength of 125 MPa. After LSI, the C/C-SiC composite revealed a low open porosity and improved flexural strength of 178 MPa at the same level of fracture toughness. After formation of Ti_3SiC_2, the C/C-SiC-Ti_3SiC_2 composite attained a flexural strength of 200 MPa, and fracture toughness of 9 MPa·$m^{1/2}$.

Table 1 Comparison on properties of three kinds of composites [12]

Materials	ρ (g/cm^3)	P_o (vol. %)	σ_f (MPa)	K_{IC} (MPa·$m^{1/2}$)
C/C	1.5	24	125 (±5)	8 (±1.2)
LSI C/C-SiC	1.9	11	178(±20)	7.9 (±0.3)
LSI C/C-SiC-Ti_3SiC_2	2.2	8	200 (±15)	9 (±0.7)

Figure 8 (a) and (b) show typical fracture surface morphologies of the C/C-SiC-Ti_3SiC_2 composite. The inter-fiber bundle pores in the initial C/C preform were filled by *in-situ* formed Ti_3SiC_2 and SiC, and carbon fibers were extensively pulled-out during loading. Ti_3SiC_2 grains distributed in the SiC matrix exhibited typical features of complex deformation modes (discussed previously). It is interesting to note, that compared with LSI C/C-SiC, the carbon fibers in LSI C/C-SiC-Ti_3SiC_2 composite were less seriously attacked by liquid Si, presumably due to the reaction of Si with TiC instead of C. Since the amount of residual Si was below the detection limit, the mechanical properties of LSI C/C-SiC-Ti_3SiC_2 composite were improved significantly, and thus, the promising high-temperature mechanical properties are also expected.

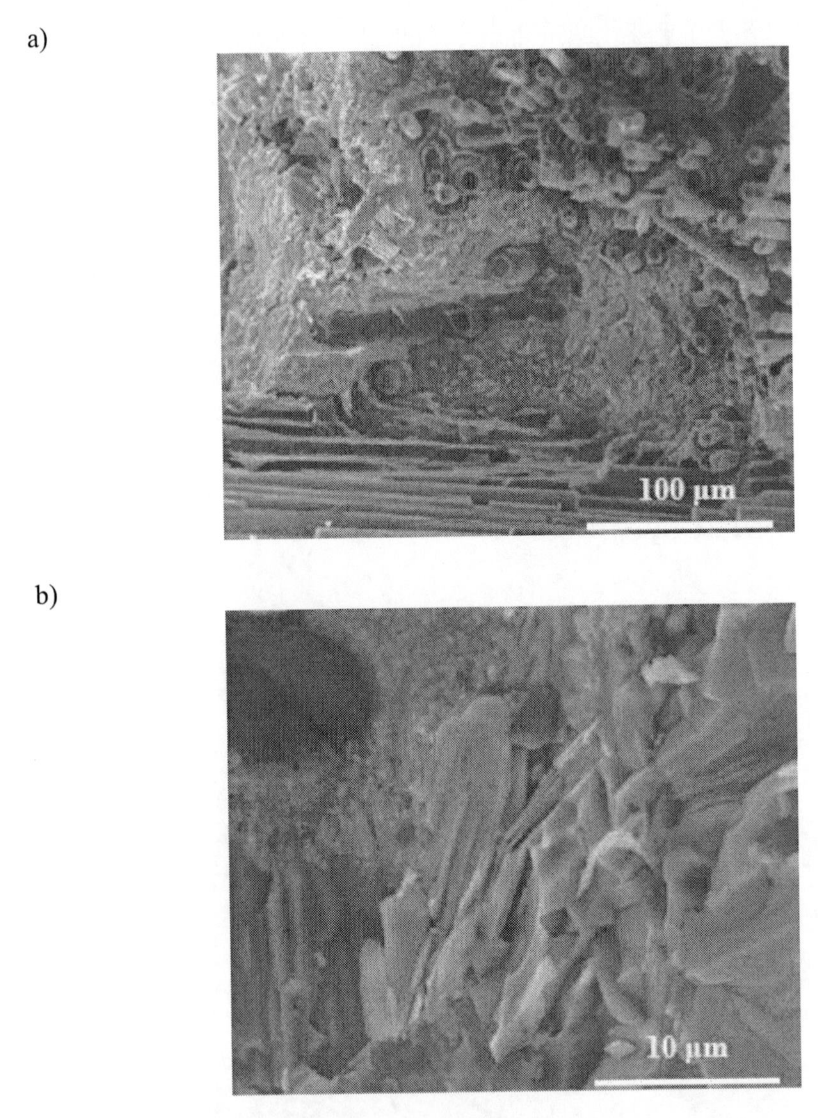

Figure 8. Fracture-surface morphologies of C/C-SiC-Ti_3SiC_2 composite: (a) Low magnification showing the fiber-pullout and (b) High magnification showing the kinking of Ti_3SiC_2.

4.5. Friction Behavior

C/C-SiC composites fabricated by LSI offer excellent friction properties and low sensitivity to service environments which have brought them to new applications as high performance brake materials in automotive, aircraft and elevator industries. Though residual Si in brake materials may be beneficial for increased friction, long term stability of the composites may suffer from free Si content. Ti_3SiC_2 ceramics have been proved to be promising tribological materials [66-68]. The wear resistance increases with increasing grain size for sliding and abrasive wear tests. Under dry conditions at room temperature a friction coefficient of Ti_3SiC_2/Ti_3SiC_2 pair of 1.16-1.43 was observed, but that of Ti_3SiC_2 plate/ diamond pin pair was 0.05-0.1, which was even lower than that of graphite plate/diamond pin pair [67]. The difference may depend on whether lubricious oxide tribo-films, TiC_xO_y, form or not [68]. Tribological properties of the Ti_3SiC_2 toughened C/C-SiC composites were tested under a braking pressure 0.8 MPa at inertia of 0.235 kg·m^2, which was compared with those of C/C-SiC composites.

Figure 9 shows the wear rates of the samples including both rotating and stationary disks. Both disks of the two materials showed increasing tendencies in weight and linear wear rates with increasing initial braking velocity. However, wear rates of Ti_3SiC_2 toughened C/C-SiC composites were much lower than those of C/C-SiC at higher velocities and the differences increased significantly with increasing velocity.

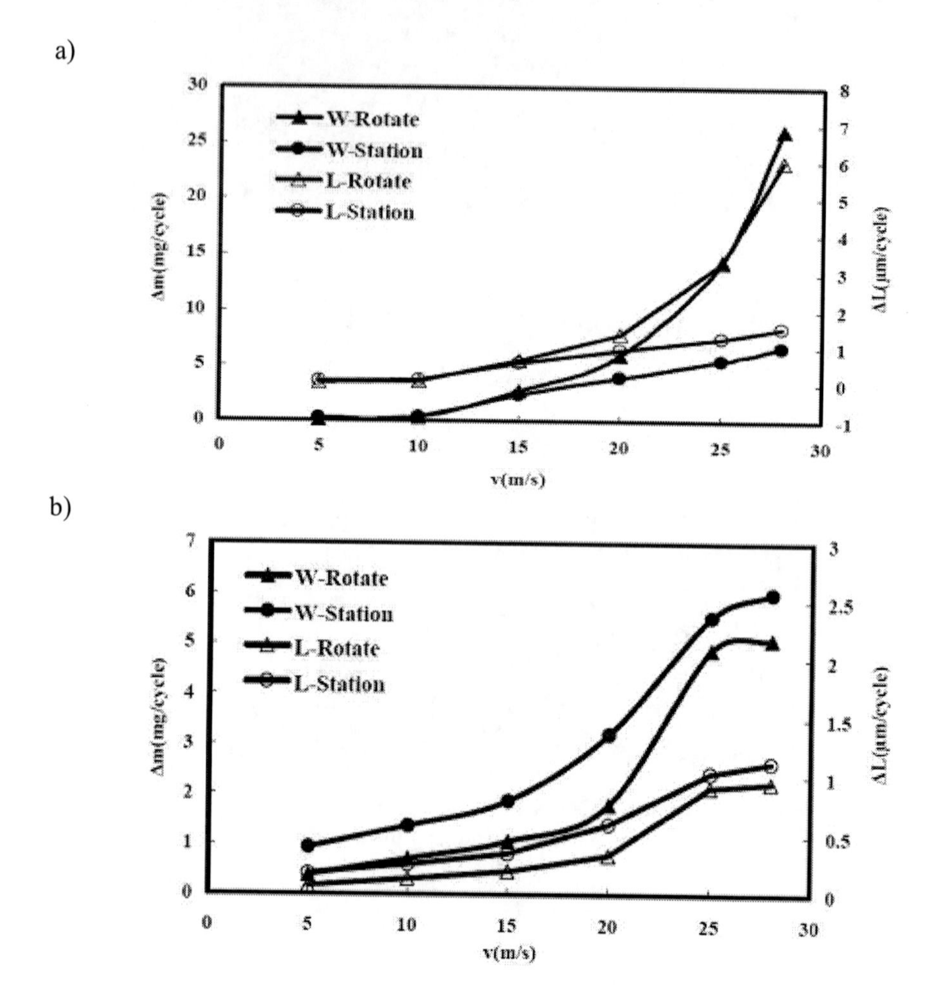

Figure 9. Relationship between wear rate and braking velocity of (a) C/C-SiC and (b) C/C-SiC-Ti_3SiC_2 composites (Δm: weight wear rate; ΔL: linear wear rate) at pressure of 0.8 MPa.

In the present work, both materials have the same reinforcement geometry and the content of carbon fibers, and the similar content of C and SiC matrix. The considerable difference was that C/C-SiC contained 8 vol% residual silicon and C/C-SiC-Ti_3SiC_2 had 5 vol% Ti_3SiC_2. As a result, both friction and wear properties of the latter were improved by the formation of Ti_3SiC_2 matrix. After braking at 28 m/s, the debris ejected from the disks were collected and characterized by SEM microscopy. While the debris of C/C-SiC composite was mainly composed of fine particles, the debris of C/C-SiC-Ti_3SiC_2 was film-like.

It was assumed, that the generation of debris is a process of crack nucleation and propagation and that oxidation controls the crack propagation [69]. During braking, the debris on the rubbing surfaces may undergo oxidation. It is important to point out that metals, such as Ti, Zn, W, Ta, Al, Ni, etc. and their alloys and several carbides and nitrides such as TiC, TiN, TiCN and Ti_3SiC_2 exhibit good tribological properties at elevated temperatures owing to the formation of lubricious oxides by tribo-oxidation [70-72]. In Ref. [72], tribo-films, which were mainly comprised of amorphous oxides of the *M* and *A* elements were formed on the contact surfaces of *MAX* phase and Al_2O_3. In contrast, the debris of C/C-SiC composite is not self-lubricated and the debris particles removed from the surface are unable to form a continuous wear layer. During braking, the particles freely move between the contacting surfaces, causing abrasion damage to both rubbing surfaces. Therefore, wear rate of C/C-SiC composite was much higher than that of composite containing Ti_3SiC_2.

Conclusions

MAX phases are promising reinforcement for both interpenetrating phase composites and fiber reinforced ceramic matrix composites. Capillary driven infiltration of a reactive melt allows near-net shape manufacturing of *MAX* phase reinforced composites with high flexibility in component geometry at low production costs.

Extending the application fields of advanced ceramic matrix composite materials requires the relations between content of *MAX* phase and properties of composite to be established. For the applications of fiber reinforced ceramic matrix composites requiring superior mechanical properties, the approach to prevent fiber (e.g., Hi-Nicalon SiC fiber) from thermal degradation and to protect fiber from reacting with melt during RMI should be further investigated.

References

[1] J. Daniel Whittenberger; Ranjan Ray; Serene C. Farmer. *Interm.* 1994, 2, 167-178.
[2] Bingchu Mei; Yoshinari Miyamoto. *Mater. Chem. Phys.* 2002, 75, 291-295.
[3] I. M. Low. *Mater. Lett.* 2004, 58, 927-932.
[4] M.W. Barsoum. Prog. *Solid State Chem.* 2000, 28, 201-281.
[5] X. H. Wang; Y. C. Zhou. *J. Mater. Sci. Technol.* 2010, 26, 385-416.
[6] Shibo Li; Jianxin Xie; Litong Zhang; Laifei Cheng. *Mater. Lett.* 2003, 57, 3048-3056.
[7] J. X. Chen; Y. C. Zhou. *Scrip. Mater.* 2004, 50, 897-901.
[8] X. Yin; N. Travitzky; R. Melcher; P. Greil. Intern. *J. Mater. Res.* 2006, 97, 492-498.
[9] Xiaowei Yin; Nahum Travitzky; Peter Greil. *Int. J. Appl. Ceram. Technol.* 2007, 4, 184–190.
[10] X. Yin; N. Travitzky; P. Greil. *J. Am. Ceram. Soc.* 2007, 90, 2128-2134.
[11] Shanshan He; Xiaowei Yin; Litong Zhang; Xiangming Li; Laifei Cheng. *Trans. Nonferrous Met. Soc. China* 2009, 19, 1215-1221.

[12] Xiaowei Yin; Shanshan He; Litong Zhang; Shangwu Fan; Laifei Cheng; Guanglai Tian; Tong Li. *Mater. Sci. Eng. A* 2010, 527, 835-841.

[13] C. K. Chua; K. F. Leong. Rapid Prototyping: Principles and Applications in Manufacturing; Wiley, New York, (1997).

[14] D. King; T. Tansey. *J. Mater. Process. Techn.* 2002, 121, 313-317.

[15] E. M. Sachs; M. Cima; P. Williams; D. Brancazio; J. Cornie. *Trans. ASME. Ser. B.* 1992, 114, 481-488.

[16] J. Windle; J. Derby. *J. Mater. Sci. Lett.* 1999, 18, 87–90.

[17] R. Melcher; S. Martins; N. Travitzky; P. Greil. *Mater. Lett.* 2006, 60, 572–575.

[18] C. R. Rambo; N. Travitzky; K. Zimmermann; P. Greil. *Mater. Lett.* 2005, 59, 1028–1031.

[19] M. Turker; D. Godlinski; F. Petzoldt. *Mater. Character.* 2008, 59, 1728-1735.

[20] W. Sun; D. J. Dcosta; F. Lin; T. EI-Raghy. *J. Mater. Process. Techn.* 2002, 127, 343-351.

[21] Kathy Lu; William T. Reynolds. *Powder Techn.* 2008, 187, 11-18.

[22] Jooho Moon; Amador C. Caballero; Leszek Hozer; Yet-Ming Chiang; Michael J. Cima. *Mater. Sci. Eng. A* 2001, 298, 110–119.

[23] P. Patirupanusara; W. Suwanpreuk; T. Rubkumintara; J. Suwanpratee. *J. Mater. Process. Techn.* 2008, 207, 40-45.

[24] R. Naslain; F. Langlais. *Mater. Sci. Res.* 1986, 20, 145-164.

[25] Gerard L. Vignoles; Francis Langlais; Cédric Descamps; Arnaud Mouchon; Hélène Le Poche; Nicolas Reuge; Nathalie Bertrand. *Surf. Coat. Techn.* 2004, 188-189, 241-249.

[26] J. Y. Ofori; S. V. Sotirchos. *Ind. Eng. Chem. Res.*, 1996, 35, 1275–1287.

[27] W. V. Kotlenski. In: Chemistry and Physics of Carbon; P. L. Walker Jr.; P. A. Thrower; Marcel Dekker; Ed.; New York, 1973; Vol. 9, pp 173–262.

[28] N. Birakayala; E. A. Evans. *Carbon* 2002, 40, 675–683.

[29] R. Voytovych; V. Bougiouri; N.R. Calderon; J. Narciso; N. Eustathopoulos. *Acta Materia.* 2008, 56, 2237–2246.

[30] Peter Greil. *Mater. Chem. Phys.* 1999, 61, 64-68.

[31] N. A. Travitzky; A. Shlayen. *Mater. Sci. Eng. A* 1998, 244, 154-160.

[32] N. A. Travitzky; E. Y. Gutmanas; N. Claussen. *Mater. Lett.* 1997, 33, 47-50.

[33] P. Kumar; N. A. Travitzky; P. Beyer; K. H. Sandhage; R. Janssen; N. Claussen. *Scrip. Materia.* 2001, 44, 751-757.

[34] Xiaowei Yin; Laifei Cheng; Litong Zhang; Yongdong Xu; Chang You. *Mater. Sci. Eng. A* 2000, 290, 89-94.

[35] Nahum A. Travitzky; Nils Claussen. *J. Europ. Ceram. Soc.* 1992, 9, 61-65.

[36] Y. V. Milman; D. B. Miracle; S. I. Chugunova; I. V. Voskoboinik; N. P. Korzhova; T. N. Legk-aya; Y. N. Podrezov. *Interm.* 2001, 9, 839-845.

[37] T. Wang; J. S. Zhang. *Interm.* 2006, 99, 20-25.

[38] F. Wagner; D. E. Garcia; A. Krupp; N. Claussen. *J. Europ. Ceram. Soc.* 1999, 19, 2449-2453.

[39] H. X. Peng; D. Z. Wang; L. Geng; C. K. Yao. *Scrip. Mater.* 1997, 37, 199-204.

[40] N. V. Tzenov; M. W. Barsoum. *J. Am. Ceram. Soc.* 2000, 83, 825-832.

[41] X. H. Wang; Y. C. Zhou. *Acta Mater.* 2002, 50, 3143-3151.

[42] Y. Khoptiar; I. Gotman; E. Y. Gutmanas. *J. Am. Ceram. Soc.* 2005, 88, 28-33.

[43] Y. W. Bao; X. H. Wang; H. B. Zhang; Y. C. Zhou. *J. Europ. Ceram. Soc.* 2005, 25, 3367–3374.
[44] F. Muller; T. Schneider; M. Feldmann; P. Greil. In *Werkstoffwoche 98*; G. Ziegler, et al., Ed.; Wiley-VCH, Weinheim, 1999; pp 63–68.
[45] J. Lapin; D. Tiberghien; F. Delannay. *Interm.* 2000, 8, 1429-1438.
[46] B. J. Keene. *Int. Mater. Rev.* 1993, 38, 157-192.
[47] N. Sobczak; L. Stobiersci; W. Radziwill; M. Ksiazek; M. Warmuzek. *Interf. Anal.* 2004, 36, 1067-1070.
[48] D. Munz; T. Fett. Ceramics. Springer Verlag, Heidelberg, 1999.
[49] S. Myhra; J. A. A. Crossley; M. W. Barsoum. *J. Phys. Chem. Solids* 2001, 62, 811–817.
[50] Y. Zhou; Z. Sun; X. Wang; S. Chen. *J. Phys. Condens. Matter.* 2001, 13, 10001–10010.
[51] W. Krenkel; F. Berndt. *Mater. Sci. Eng. A* 2005, 412, 177-181.
[52] A. G. Odeshi; H. Mucha; B. Wielage. *Carbon* 2006, 44, 1994-2001.
[53] F. Lamouroux; X. Bourrat; J. Sevely; R. Naslain. *Carbon* 1993, 31, 1273-1288.
[54] Laifei Cheng; Yongdong Xu; Litong Zhang; Xiaowei Yin. *Carbon* 2000, 38, 2103-2108.
[55] X. Yin; L. Cheng; L. Zhang; Y. Xu. *Carbon* 2002, 40, 905-910.
[56] W.J. Clegg; K. Kandall; N.M. Alford; D. Birchall; T.W. Button. *Nature* 1990, 347, 455-457.
[57] Z. Krstic; V.D. Krstic. *J. Europ. Ceram. Soc.* 2008; 28; 1723-1730.
[58] C. J. Gilbert; D. R. Bloyer; M. W. Barsoum; T. El-Raghy; A. P. Tomsia; R. O. Ritchie. *Scrip. Mater.* 2000, 42, 761-767.
[59] R. Radhakrishnan; C. H. Henager; J. L. Brimhall; S. B. Bhaduri. *Scrip. Mater.* 1996, 34, 1809-1814.
[60] J. Zhang; L. Wang; L. Shi; W. Jiang; L. Chen. *Scrip. Mater.* 2007, 56, 241-244.
[61] S. Li; J. Xie; L. Zhang; L. Cheng. *Mater. Sci. Eng. A* 2004, 381, 51-56.
[62] Y. Du; J. Schuster; H. Seifert; F. Aldinger. *J. Am. Ceram. Soc.* 2000, 83, 197-203.
[63] S. Fan; L. Zhang; Y. Xu. *Comp. Sci. Tech.* 2007, 67, 2390-2398.
[64] G.M. Samsonov; I.M. Vinitskii; In: Handbook of Refractory Compounds, IFI Plenum, New York, 1980, pp 217.
[65] O. Dezellus ; S. Jacques ; F. Hodaj ; N. Eustathopoulos. *J. Mater. Sci.* 2005, 40, 2307–2311.
[66] T. El-Raghy; P. Blau; M. W. Barsoum. *Wear* 2000, 238, 125-130.
[67] Y. Zhang; G.P. Ding; Y.C. Zhou; B.C. Cai. *Mater. Lett.* 2002, 55, 285-289.
[68] A. Souchet; J. Fontaine; M. Belin; T. Le Mogne; J.-L. Loubet; M.W. Barsoum. *Tribol. Lett.* 2005, 18, 341-352.
[69] J. Jiang; F.H. Stott; M.M. Stack. *Wear* 2004, 256, 973-985.
[70] T. Aizawa; A. Mitsuo; S. Yamamoto; T. Sumitomo; S. Muraishi. *Wear* 2005, 259, 708–718.
[71] J.H. Meng; J.J. Lu; J.B. Wang; S.R. Yang. *Mater. Sci. Eng. A* 2006, 418, 68-76.
[72] S. Gupta; D. Filimonov; T. Palanisamy; M.W. Barsoum. *Wear* 2008, 265, 560-565.

In: MAX Phases: Microstructure, Properties and Applications ISBN 978-1-61324-182-0
Editors: It-Meng (Jim) Low and Yanchun Zhou

Chapter 4

$Ti_3Sn_{(1-X)}Al_XC_2$ MAX PHASE SOLID SOLUTIONS: FROM SYNTHESIS TO MECHANICAL PROPERTIES

S. Dubois**, *G. P. Bei, V. Gauthier-Brunet, C. Tromas, P. Gadaud and P. Villechaise
Institut P', UPR 3346, Université de Poitiers-ENSMA-CNRS,
Département de physique et mécanique des matériaux
Bât. SP2MI, Bd Pierre et Marie Curie, BP 30179,
86962 Futuroscope Chasseneuil Cedex, France

ABSTRACT

$Ti_3Sn_{(1-x)}Al_xC_2$ MAX phase solid solutions are successfully synthesized from different reactant mixtures by using Hot Isostatic Pressing. Rietveld refinement allows to carefully characterize their structures. The octahedrons and trigonal prims distortion parameters and the Ti-C-Ti and Ti-A distances are calculated as a function of the Al content. Mechanical properties of solid solutions are studied from nanoindentation experiments. It is shown that solid solution hardening is not operative in this system. Hardness and elastic modulus are found to vary non-monotonically from Ti_3SnC_2 to Ti_3AlC_2, such a result is discussed in terms of Ti-A and Ti-C-Ti bond stiffness.

1. INTRODUCTION

The ternary carbides and nitrides with the general formula $M_{n+1}AX_n$ - where n=1, 2 or 3, M is an early transition metal, A is an A-group element (from IIIA to VIA), and X is either C or N are of significant interest [1-6]. The hexagonal nanolaminated crystal structure of these so-called MAX phases consists of edge-sharing M_6X octahedra intercalated with layers of pure A-group elements. In the M_2AX or 211's phases, every third layer is an A layer [1-3]; in

* Corresponding author: sylvain.dubois@univ-poitiers.fr.

the M_3AX_2 (or 312's)[3] and M_4AX_3 (or 413's) [7, 8], every fourth and fifth layer, respectively, is an A-group element layer. There are about sixty MAX phases, but only seven are M_3AX_2.

MAX phases possess a remarkable set of properties which derive from their nanolayered structure through the strong covalent M-X bonds which are interleaved with A layers through weak M-A bonds. Indeed, they combine some characteristics of metals and ceramics: for example, they demonstrate high strength and stiffness at high temperature, but at the same time, they exhibit high thermal and electrical conductivities and are machinable [9]. They are also thermal shock resistant and shows high resistance to oxidation.

Among the 312 phases, Ti_3AlC_2 and its solid solutions have attractive properties in terms of low density and excellent high-temperature oxidation resistance. Indeed, the oxidation kinetics of Ti_3AlC_2 and Ti_2AlC have been determined and excellent high-temperature oxidation resistance [10-12] has been demonstrated. It is known that protective Al_2O_3 scale does not form during high temperature oxidation process of TiAl alloys [13-16]. Thus, oxidation resistance of such alloys is not satisfying. However and besides the much lower Al content, a continuous and protective Al_2O_3 scale forms onto Ti_3AlC_2 and Ti_2AlC surfaces during high temperature oxidation in air [10-12]. Such an excellent oxidation resistance has been discussed in terms of oxide scale adherence onto the substrate [17] and in terms of microstructure of the Ti_3AlC_2/Al_2O_3 interface [18, 19].

The MAX phases can form a large number of isostructural solid solutions by substitutions at the M, A and X sites. Examples of solid solutions on the M site are $(Ti,V)_2AlC$, $(Ti,Cr)_2AlC$, $(Ti,Nb)_2AlC$ and $(Cr,V)_2AlC$ [20, 21]. Sun *et al.* [22] and Wang and Zhou [23] have demonstrated solid solutions strengthening and bulk moduli enhancement for $(M_1,M_2)_2AlC$ (M_1 and M_2 = Ti, V, Cr) as compared to the end members. Many A-site MAX phase solid solutions also exist; $Ti_3Si_xAl_{1-x}C_2$ and $Ti_3Si_xGe_{1-x}C_2$ being the most studied[24-26]. Ganguly *et al.* [24] have evidenced that the hardness values of $Ti_3Si_{(1-x)}Ge_xC_2$ lies in between those of Ti_3SiC_2 and Ti_3GeC_2, thus concluding that solid solution hardening effect is not operative in this system. MAX carbonitrides, i.e. $M_{n+1}A(C,N)_n$ phases are the most important example of MAX phase solid solutions on the X site. $Ti_2AlC_{0.5}N_{0.5}$ has been reported by Barsoum *et al.* [27] to be significantly harder than either of its end members Ti_2AlC and Ti_2AlN, and there may be other ratios of C and N with even better properties.

Recently, non pure Ti_3SnC_2 MAX phase has been produced by Hot Isostatic Pressing (HIP) using Fe as additive [28]. In the present work, we synthesized a series of $Ti_3Sn_{(1-x)}Al_xC_2$ solid solutions by HIP starting with different reactant mixtures. The cell parameters, atom positions and impurity content were determined by the Rietveld method as a function of the Al content. The microstructure was observed by Scanning Electron Microscopy (SEM) and the composition was determined by Energy Dispersive X-ray Spectroscopy (EDXS). Nanoindentation was used to measure the intrinsic mechanical properties, i.e. elastic modulus and hardness, of the solid solutions.

2. Experimental Details

Powders of titanium (150-250μm, 99.5% purity), tin (2-20 μm, 99% purity), aluminum (45-150 μm, 99.5% purity), titanium carbide (<45 μm, 98% purity) and carbon (graphite, <20μm) were used as the reactant powder materials. Sub-stoichiometric TiC (i.e. : $TiC_{0.66}$)

was also used as a reactant ; it was synthesized by mechanical alloying starting with a mixture of Ti and 0.66C [29,30]. Different Ti:Sn:Al:C, Ti:Sn:Al:TiC or $TiC_{0.66}$:Sn:Al ratios were chosen in order to produce a large set of $Ti_3Sn_{(1-x)}Al_xC_2$ solid solutions (i.e.: with different Al content). Moreover, either (Ti+C) sub-stoichiometry or (Sn+Al) over-stoichiometry were used in this study. Indeed, it is well known that some *A* element is vaporized during powder metallurgy processing (HIP, Hot Pressing or reactive sintering) of the reactant mixture.

Powders were thoroughly mixed for 1 hour in a turbula in order to prepare uniform reactant powder mixtures. The different mixtures were cold-compacted into cylindrical steel dies using a uniaxial pressure of 800 MPa. The green density, evaluated from weight and geometric measurements, ranged from 77 to 81 % of the theoretical density. The green body was then sealed under vacuum in a pyrex container. The resulting compact was then placed in a HIP chamber and subjected to the following temperature and pressure cycle:

- The sample was heated to 1450°C under Ar in 1h.
- The HIP was pressurized with Ar to 50 MPa in 2h.
- Once the processing temperature was reached, the sample was held at 1450°C for 2 hours and 50 MPa for 1h before cooling to room temperature and atmospheric pressure.

After HIP-ing, samples were machined to remove the encapsulating glass container and sliced using a diamond wheel. Samples were thus grinded using silicon carbide paper and then polished with a diamond suspension. Finally, a chemo-mechanical polishing (CMP) has been performed using a neutral suspension of alumina particles. Such a chemo-mechanical polishing allows to produce a very flat surface and to avoid any work hardening due to conventional grinding.

Phase identification was performed by X-Ray Diffraction (XRD) using a Bruker D501 diffractometer with Cu-K_α radiation. XRD data were refined using the MAUD software [31] in order to extract the lattice parameters of the different $Ti_3Sn_{(1-x)}Al_xC_2$ solid solutions and the sample composition. Microstructures were examined by SEM (SEM, JEOL 5600LV). EDXS (EDXS, Oxford Isis 300) was used to determine the Sn over Al and the Ti over (Al+Sn) ratios, it thus allows extracting the Al content in the $Ti_3Sn_{(1-x)}Al_xC_2$ solid solutions.

A Nano Hardness Tester (NHT) apparatus from Coatings and Surface Measurements (CSM) Instruments (Peseux, Switzerland) equipped with a Berkovich indenter was used to perform nanoindentation tests. The equivalent indenter method [32,33] was used to analyse the nanoindentation curves and to extract the hardness and Young's modulus values. The shape of the indenter has been carefully calibrated for true penetration depth as small as 20 nm both by indenting fused silica samples of known Young's modulus (72 GPa) and by direct Atomic Force Microscopy (AFM) observation of the indenter. In order to study hardness variation with penetration depth, 9 loads (1, 2, 3, 6, 10, 15, 30, 150 and 300 mN) have been used. Since most of the samples were not single phase samples, an hardness cartography and a statistical analysis method were used, for each load and each MAX phase solid solution sample.

The Young's modulus has also been determined by means of the dynamic resonant method [34] in bending mode on beam specimen. Experiments are performed at 1K/mn under high vacuum (10^{-4} Pa) from 150 up to 1400 K without any harmful contact, the sample being maintained horizontally between steel wires located at the vibration nodes. Furthermore, excitation and detection are insured by an electrostatic device (capacitance created between the sample and a unique electrode [35]).

The Young's modulus E of a bulk sample in bending mode is directly determined by the relation:

$$E = 0.9464 \rho F^2 \frac{L^4}{h^2} T\left(h/_L, \nu\right)$$

where F is the resonance frequency, ρ the density, ν Poisson's ratio, h and L, the beam thickness and span length, and T(h/L, ν) a correcting factor close to 1 [35,36]. The accuracy of the method is better than 1 % with regular specimen dimensions.

3. Results and Discussion

3.1. Microstructural Characterization

$Ti_3Sn_{(1-x)}Al_xC_2$ samples were synthesized from different reactant mixtures: 3Ti + Sn + 2C + 0.1Fe [28], 3Ti + *y*Sn + *z*Al + 2C, 2.8$TiC_{0.66}$ + 0.8Sn + 0.4Al and 1.8TiC + *y*Sn + *z*Al + Ti. Figure 1 shows scanning electron micrographs obtained, in back-scattered mode, on samples (b), (c) and (f) (see table 1 which gives the relationship between sample name and initial reactant mixtures). In figure 1c, MAX phase lamellas are obvious with an average dimension in the range 60-70 μm in length and 10-20 μm in width. EDXS performed on the lamella indicates the mean mole ratio of $\frac{Ti}{Al+Sn} = 3.01$ and $\frac{Al}{Sn} = 3.94$, hence the chemical composition of the solid solution is $Ti_3Al_{0.8}Sn_{0.2}C_2$. One can also notice in figures 1a and 1b that $Ti_3Sn_{(1-x)}Al_xC_2$ (grey) phases always appear in between $Ti_2Sn_{(1-x)}Al_xC$ (light grey) and TiC (dark grey) phases. Such a result strongly suggests that the 211 and TiC phases react to form the 312 solid solution [37]. SEM results demonstrate that TiC, $Ti_2Sn_{(1-x)}Al_xC$ and $Ti_3Sn_{(1-x)}Al_xC_2$ are prone to co-exist. Such a feature is not surprising since close structural relationships between Ti_3AlC_2, Ti_2AlC and TiC [38]; Ti_3SiC_2, Ti_5Si_3 and TiC [39] and Ti_2AlN and TiN [40] have been demonstrated by using High Resolution Transmission Electron Microscopy and Selected-Area Electron Diffraction. Nevertheless, it has been shown that, depending on the composition of the reactant mixture, the reaction mechanisms of Ti_3AlC_2 formation are different [39, 41-43]. In such a context, our experiments, whose goal consists in the synthesis of a large set of different $Ti_3Sn_{(1-x)}Al_xC_2$ solid solutions, do not allow to get the reaction mechanisms of $Ti_3Sn_{(1-x)}Al_xC_2$ formation and the influence of the reactant mixture composition. Lattice parameters, compositions and Al content (x) of the obtained $Ti_3Sn_{(1-x)}Al_xC_2$ solid solutions are given in table 1, together with the different initial reactant mixtures. As an example, the Rietveld refined pattern obtained on sample e is shown in figure 2 ; the weighted reliability factor is 11.7%. It can be noticed, in table 1, that only samples (c), (f) and (g) contain more than 95 vol. % of $Ti_3Sn_{(1-x)}Al_xC_2$ solid solutions. Samples (f) and (g), which respectively contain 80 and 100 at% of Al as A element, appear as pure sample at the XRD detection limit.

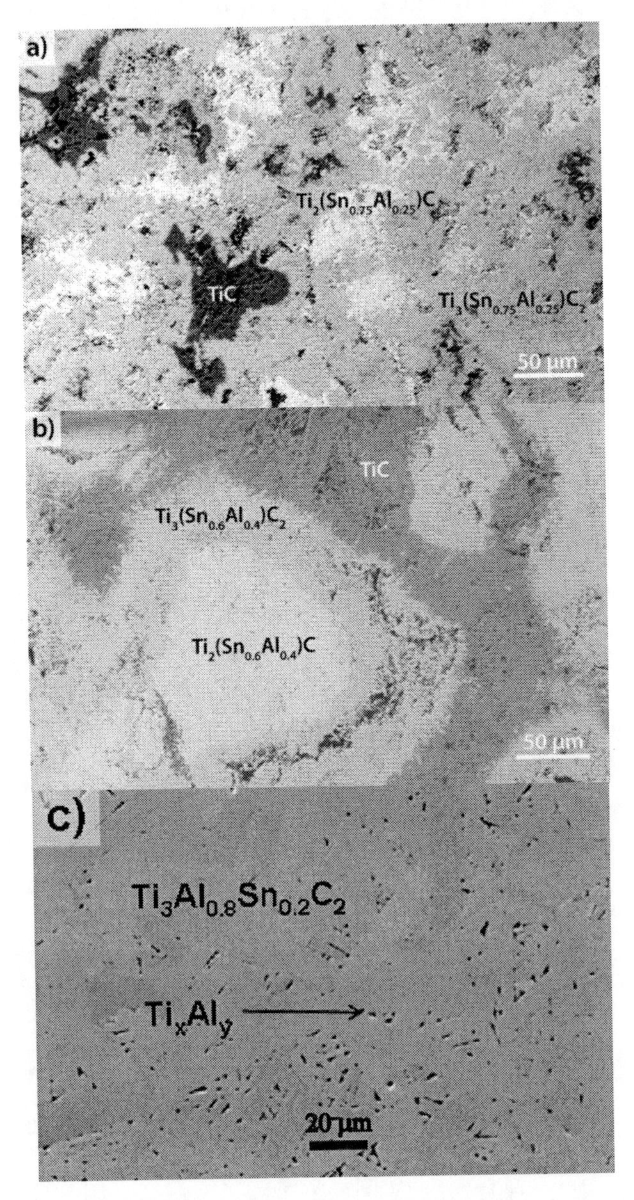

Figure 1. Scanning electron images recorded in backscattered mode on (a) sample (b), (b) sample (c), (c) sample (f).

Thus, the formation mechanism of $Ti_3Sn_{(1-x)}Al_xC_2$, with x=0.8, is very likely the same as the formation mechanism of Ti_3AlC_2 from TiC, Al and Ti [37,41]. TiC reacts with Al, Ti and Sn to form either $Ti_3Sn_{0.2}Al_{0.8}C_2$ or $Ti_2Sn_{0.2}Al_{0.8}C$ [37]. At high enough temperature, $Ti_2Sn_{0.2}Al_{0.8}C$ can react with TiC to form $Ti_3Sn_{0.2}Al_{0.8}C_2$. Nevertheless, reaction syntheses allow having a set of solid solutions where the Al content varies in the range 0-1. Figure 3 shows the variation of the c/a ratio as a function of the Al content. It shows that the lattice contraction of $Ti_3Sn_{(1-x)}Al_xC_2$ is strongly anisotropic, namely, much more reduction along the a axis than along the c axis. c/a variation is consistent with the formation of an ideal solid solution and follows the Vegard's law. Indeed, according to this law, unit cell parameters vary linearly with composition for a continuous substitutional solid solution in which atoms that substitute for each other are randomly distributed. This law is valid for ideal solid solutions when the lattice parameters of the pure components differ by less than 5% [44] which is the case for Ti_3AlC_2 and Ti_3SnC_2 MAX phases. Moreover, as in $V_2Ga_{(1-x)}Al_xC$ [45], the z-parameter of Ti_{II} (Z_{Ti-II}) follows Vegard's law (not shown).

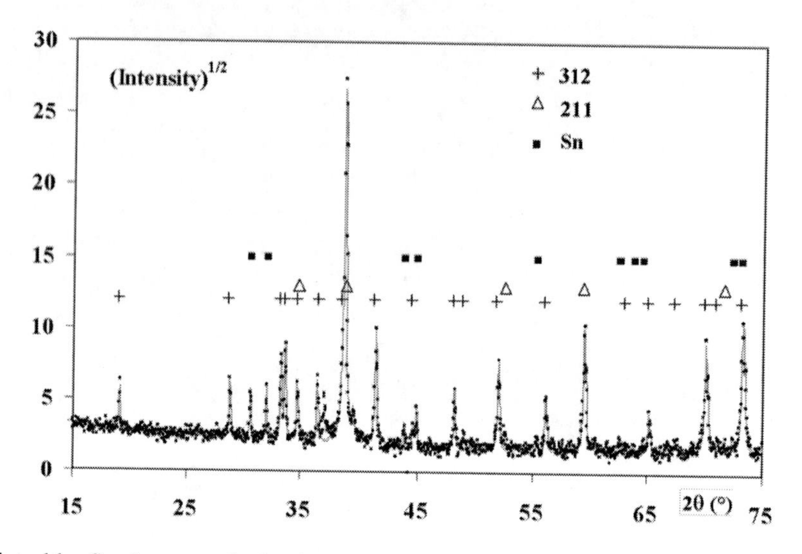

Figure 2. Rietveld refined pattern obtained on sample (e). Black squares: experimental data, full line: Rietveld refinement.

Table 1. Initial reactant mixture, lattice parameters, composition and Al content (x) in the different $Ti_3Sn_{(1-x)}Al_xC_2$ solid solutions

Initial reactant mixture	lattice parameters of $Ti_3Sn_{(1-x)}Al_xC_2$	Rietveld Refinement (vol %)	x value from EDXS experiments (at. %)
3Ti + Sn + 2C + 0.1 Fe [28]	a = 3.1366 Å c = 18.650 Å Z_{TiII} = 0.1197	80.5 % of 312 16.2 % of TiC 3.3 % of impurities	Sample a 0
3Ti + 0.8Sn + 0.3Al + 2C	a = 3.1227 Å c = 18.613 Å Z_{TiII} = 0.1297	57.6 % of 312 33.7 % of 211 8.7 % of TiC	Sample b 25
3Ti + 0.6Sn + 0.6Al + 2C	a = 3.1052 Å c = 18.604 Å Z_{TiII} = 0.1297	97 % of 312 3 % of 211	Sample c 38
$2.8TiC_{0.66}$ + 0.8Sn + 0.4Al	a = 3.1114 Å c = 18.598 Å Z_{TiII} = 0.1267	41.8 % of 312 43.5 % of 211 14.7 % of TiC	Sample d 40
1.8TiC + 0.6Sn + 0.6Al + Ti	a = 3.1053 Å c = 18.579 Å Z_{TiII} = 0.1317	96 % of 312 3 % of 211 1 % of Sn	Sample e 50
1.8TiC + 0.2Sn + Al + Ti	a = 3.0891 Å c = 18.603 Å Z_{TiII} = 0.1315	98 % of 312 2 % of impurities	Sample f 80
1.9 TiC + Al + Ti	a = 3.0779 Å c = 18.579 Å Z_{TiII} = 0.1326	100% of 312	Sample g 100

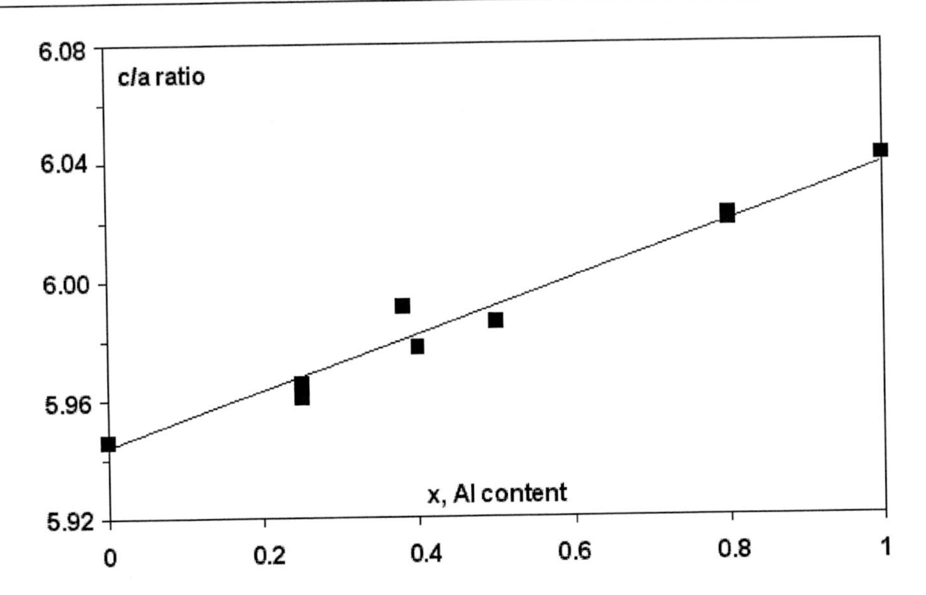

Figure 3. c/a ratio versus Al content in $Ti_3Sn_{(1-x)}Al_xC_2$ solid solutions.

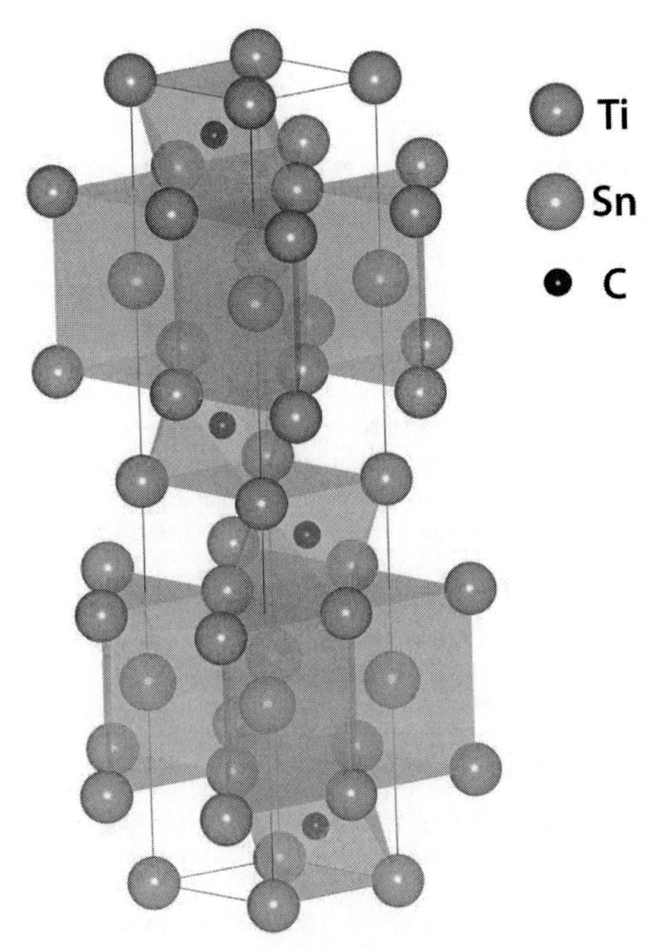

Figure 4. Unit cell of the Ti_3SnC_2 phase. [Ti_6C] octahedrons and [Ti_6Sn] trigonal prisms are represented. Drawn with VESTA[48].

Following previous work [46,47], the MAX crystal unit cell can be considered as constituted of $[M_6X]$ octahedrons and $[M_6A]$ trigonal prisms. Figure 4 shows that the stacking of two octahedrons and one trigonal prism allows describing the 312 MAX phases [48]. A cubic octahedron is the unit block of the binary MX but it loses its fourfold axis in ternary MAX and it results in a relaxation owing to this reduced symmetry. The parameter O_d, defined as the ratio of the distances between two faces not in the basal planes (d_1) and two opposite faces contained in the basal planes (d_2) [47], allows to estimate the non cubic distortion of the octahedron. It is given, in a 312 MAX phase, by:

$$O_d = \frac{d_1}{d_2} = \frac{\sqrt{3}}{2\sqrt{Z_{Ti}^2\left(\frac{c}{a}\right)^2 + \frac{1}{12}}}$$

The distortion parameter P_d, defined as the ratio of the M-M distance and the M-A distance [46], allows estimating the distortion of the trigonal prism. It is given, in either a 211 or 312 MAX phases, by:

$$P_d = \frac{1}{2\sqrt{\left(\frac{1}{4} - Z_{Ti}\right)^2\left(\frac{c}{a}\right)^2 + \frac{1}{3}}}$$

For 312 MAX phases, an ideal packing of hard spheres of equal diameter leads to a ratio $c/a = 8\sqrt{\frac{2}{3}} \approx 6.532$ and $O_d = P_d = 1$ for ideal octahedron (cubic) and trigonal prism. From the position of Ti atoms (Z_{Ti}), and a and c lattice parameters given in table 1, one can easily determine the O_d and P_d distortion parameters. Figure 5 shows the O_d and P_d variations as a function of the Al content in the $Ti_3Sn_{(1-x)}Al_xC_2$ solid solution. As expected, it can be concluded that both polyhedron are distorted whatever the Al content in the $Ti_3Sn_{(1-x)}Al_xC_2$ MAX phase. Ti_3SnC_2 shows strong distortion of the octahedron; however, Ti_3AlC_2 shows small distortion of the octahedron. The trigonal prism distortion does not seem to change a lot with the Al content of the solid solution. Finally, the octahedron and trigonal prism distortion are quite similar for Al content in the range 0.4-0.5 (the ratio O_d/P_d departs from unity by 4.10^{-4}). As previously shown [3], anisotropic deformation, i.e. much more reduction along the *a*-axis than along the *c*-axis of the $[Ti_6C]$ octahedrons occurs in ternary compounds. O_d is larger in Ti_3SnC_2 than in Ti_3SiC_2 [3] and Ti_3AlC_2. Figure 6 shows the variation of the Ti-C-Ti (Figure 6a) and Ti-A (Figure 6b) distances as a function of the Al content. The Ti-C-Ti distance monotonically increases with the Al content. These distances are moreover strongly modified compared with those in TiC (4.3247 angströms). However, the Ti-A distance decreases with the Al content increase. These results indicate that the distortion is accommodated by shrinking of the Ti-C bonds in the $[Ti_6C]$ octahedrons and it differs from the one obtained by Gamarnik *et al.* who mentioned that accommodation occurs by Ti-C bond rotation [49]. Such a shrinking of the Ti-C-Ti distances will affect the Ti-C bond stiffness and may thus modify the elastic properties of the ternary compounds. The small Ti-Ti spacing, observed in Ti_3SnC_2, would increase direct bonding between Ti sheets whereas higher values, observed in Ti_3AlC_2, would favour the formation of C 2p-Ti ds bonds.

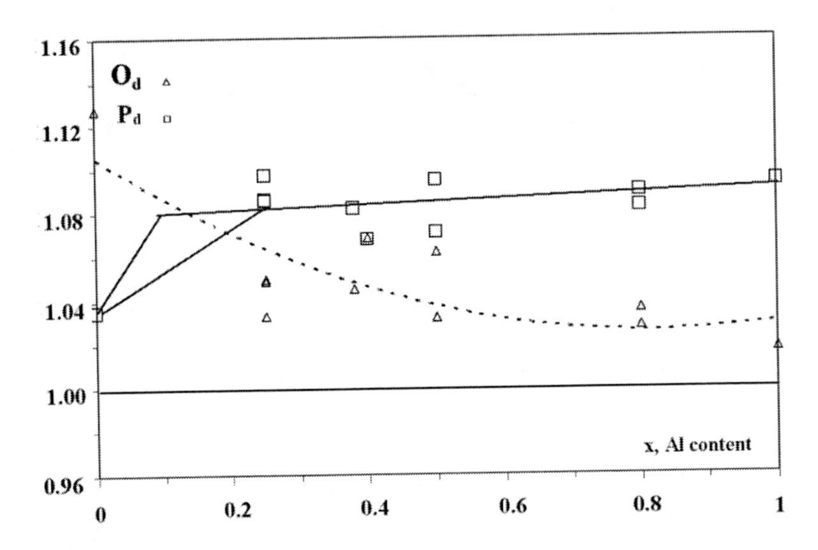

Figure 5. Octahedron (O_d, open triangles) and trigonal prism (P_d, open squares) distortion parameters versus Al content in $Ti_3Sn_{(1-x)}Al_xC_2$ solid solutions.

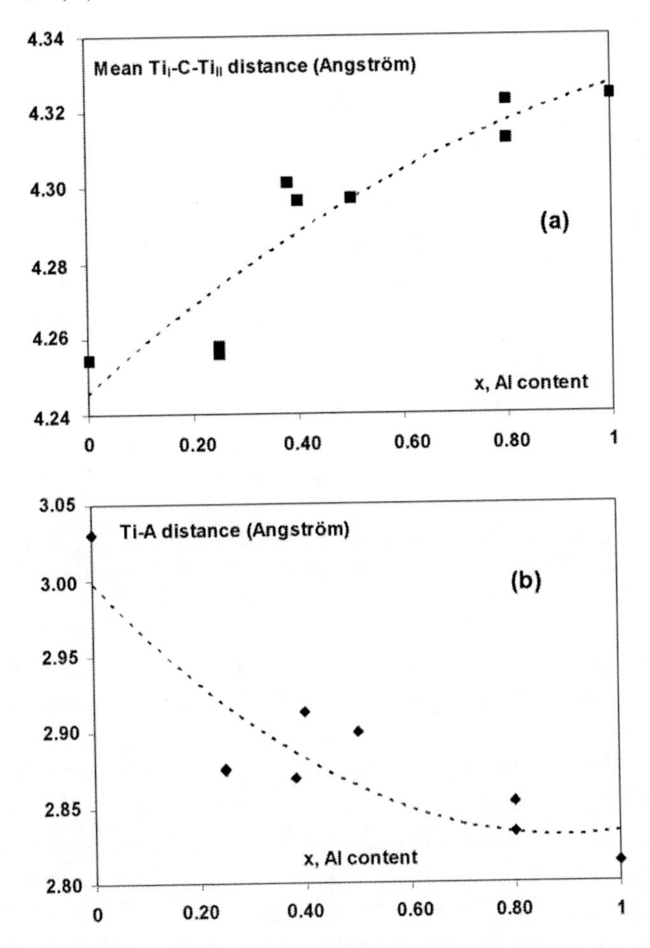

Figure 6. (a) Mean Ti_I-C-Ti_{II} and (b) Ti-A distances versus Al content in $Ti_3Sn_{(1-x)}Al_xC_2$ solid solutions.

3.2. Mechanical Properties

Nanoindentation tests were performed on different $Ti_3Sn_{(1-x)}Al_xC_2$ samples in order to determine elastic modulus and hardness values as a function of the Al content. Two different methods were used to statistically determine Young's modulus and hardness values of MAX phase solid solutions.

The first method consists in a statistical analysis of the hardness histogram and in an identification of the phases using optical microscopy. The second method consists in a statistical analysis of the hardness values chosen after phase identification by using SEM.

(A) Hardness and Young's Modulus of Ti_3SnC_2: Hardness Cartography Associated with Indent and Grain Identifications Using Optical Microscopy [50]

In order to study hardness variation with penetration depth, 9 loads (1, 2, 3, 6, 10, 15, 30, 150 and 300 mN) have been used. As the sample is composed of five different phases, a statistical analysis has been performed, for each load, over 100 nanoindentations.

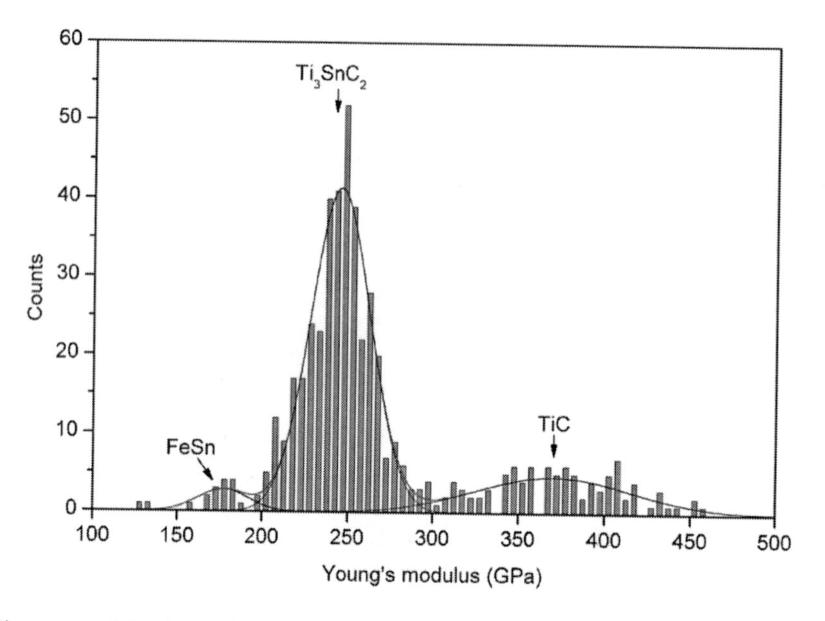

Figure 7. Histogram of elastic modulus determined from 500 indents realized with a 3 mN load on the Ti_3SnC_2 sample. Gaussian deconvolution of the histogram gives the 3 peaks shown on the figure.

In the case of the 3 mN load, a more detailed analysis, performed over 500 indents, is provided here. The histogram of elastic modulus, shown in figure 7, and obtained for a 3mN load, presents three peaks centred at 370, 245 and 177 GPa. The main peak at 245 GPa is unambiguously attributed to the Ti_3SnC_2 MAX phase. This value is moreover in good agreement with results obtained by nanoindentation in MAX phase thin films [51] or by ultrasound methods [52]. The peak at 370 GPa is ascribed to the TiC phase. This peak is centred at a lower value than those often reported for TiC [53] and presents a large full width at half maximum (90 GPa). Such an effect is likely related both to the presence, in the sample, of other phases, softer than TiC and thus elastically deformed below the TiC tested grains and to the small size of TiC grains that can furthermore be substoichiometric. Finally, the peak at

177 GPa very likely results from the minority phases, and from artefacts induced when the indent is localized on a grain boundary. In figure 7, gaussian peaks deconvolution leads to area ratios of 74% for Ti_3SnC_2, 22% for TiC and 4% for the minority phase. These values are in reasonable agreement with the volume fractions determined from Rietveld refinement [50].

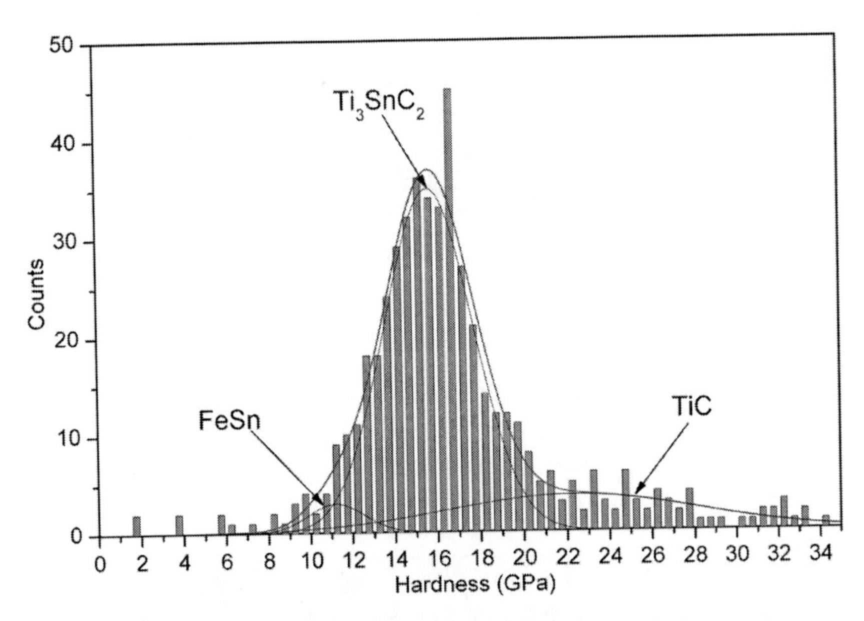

Figure 8. Histogram, with a bin size of 500MPa, of hardness values determined from 500 indents realized with a 3 mN load on the Ti_3SnC_2 sample. Gaussian deconvolution of the histogram gives the 3 peaks shown on the figure.

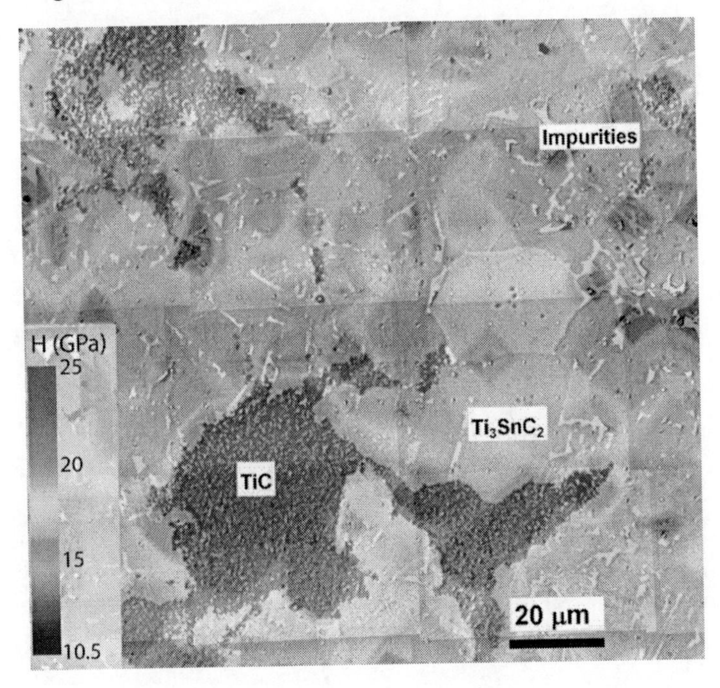

Figure 9. Hardness cartography, deduced from a bicubic interpolation between measurements points, superimposed to optical micrography of the Ti_3SnC_2 sample surface.

The histogram of hardness values, shown in figure 8, presents three peaks centred at 8, 15.7 and 25 GPa. For a better understanding of the hardness histogram, hardness cartography has been established. 400 of the studied 3 mN indents were organized in a regular square shaped pattern, with a 7 µm step size. The grains of the sample, as well as the indents, have thus been identified by optical microscopy with polarized light. Hardness cartography, obtained from a bicubic interpolation in between measurement points, is plotted in figure 9 and superimposed to an optical microscopy image of the indented area. This cartography clearly demonstrates the correlation between hardness value and the nature of the indented phase. It thus explains the 3 peaks observed in figure 8. The highest peak, centred at 15.7 GPa, is thus unambiguously ascribed to the Ti_3SnC_2 hardness.

For each indentation load, the same statistical analysis has been performed over 100 indents to identify the Ti_3SnC_2 peak on hardness and elastic modulus histograms and to extract a mean value.

(B) Hardness and Young's Modulus of $Ti_3Sn_{(1-X)}Al_xc_2(X<>0)$: Hardness Values Selected after Phase Identification of the Different Grains Using SEM

In the following section, experimental details of the different analyses are given in the case of the $Ti_3Sn_{0.2}Al_{0.8}C_2$ sample which contains (see table 1): $Ti_3Sn_{0.2}Al_{0.8}C_2$ (98 vol. %), and impurities (Ti_xAl_y intermetallics, about 2 vol%).

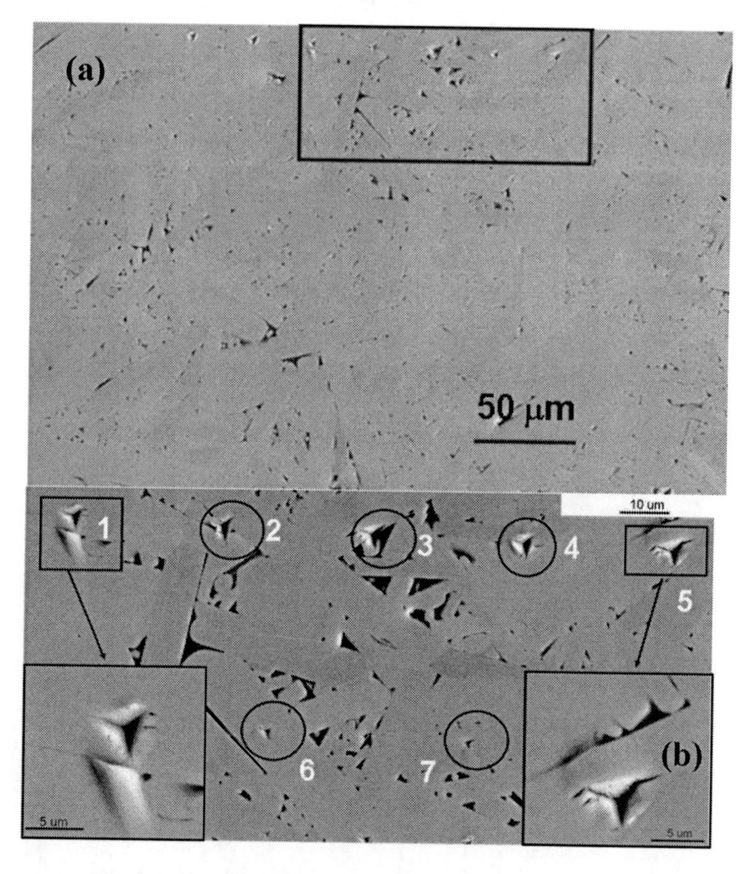

Figure 10. (a) SEM image of a part of the indent network performed on $Ti_3Al_{0.8}Sn_{0.2}C_2$ (sample f). (b) enlarge view of the indents shown in the rectangle region of (a).

A network of indents was performed on the surface of the solid solution sample; such an indent network is shown, as an example, in figure 10a. SEM observation of the indent network allows checking that indents are performed inside a $Ti_3Sn_{0.2}Al_{0.8}C_2$ single grain, inside grain boundaries or inside an impurity (Ti_xAl_y). Figure 10b shows, as an example, that indents 1, 2, 4 and 5 can be considered as not being performed in single grain. Indent 3 is performed in a Ti_xAl_y region. However, indents 6 and 7 can be considered as being performed in a single grain of the 312 MAX phase solid solution. In this case, only hardness and Young's modulus values extracted from indent 6 and 7 are considered in the statistics. From hardness and Young's modulus values measured on $Ti_3Sn_{0.2}Al_{0.8}C_2$ single grain, one can get a mean value for intrinsic hardness and Young's modulus of the MAX phase solid solution.

For each indentation load and each $Ti_3Sn_{(1-x)}Al_xC_2$ solid solution, the same analysis has been performed over 20 indents to measure the $Ti_3Sn_{(1-x)}Al_xC_2$ intrinsic hardness and elastic modulus values and to extract, for each property, a mean value.

(C) $Ti_3Sn_{(1-X)}Al_xc_2$ Hardness and Young's Modulus Variations with Tip Penetration Depth

Hardness and Young's modulus values are plotted in figure 11 as a function of the tip penetration depth and for different $Ti_3Sn_{(1-x)}Al_xC_2$ MAX phase solid solution. The Indentation Size Effect (ISE) observed on the hardness vs penetration depth curves is well fitted, for each solid solution, by the Nix and Gao model [54]. One can notice that a deviation from the Nix and Gao model is observed for the largest penetration depth.

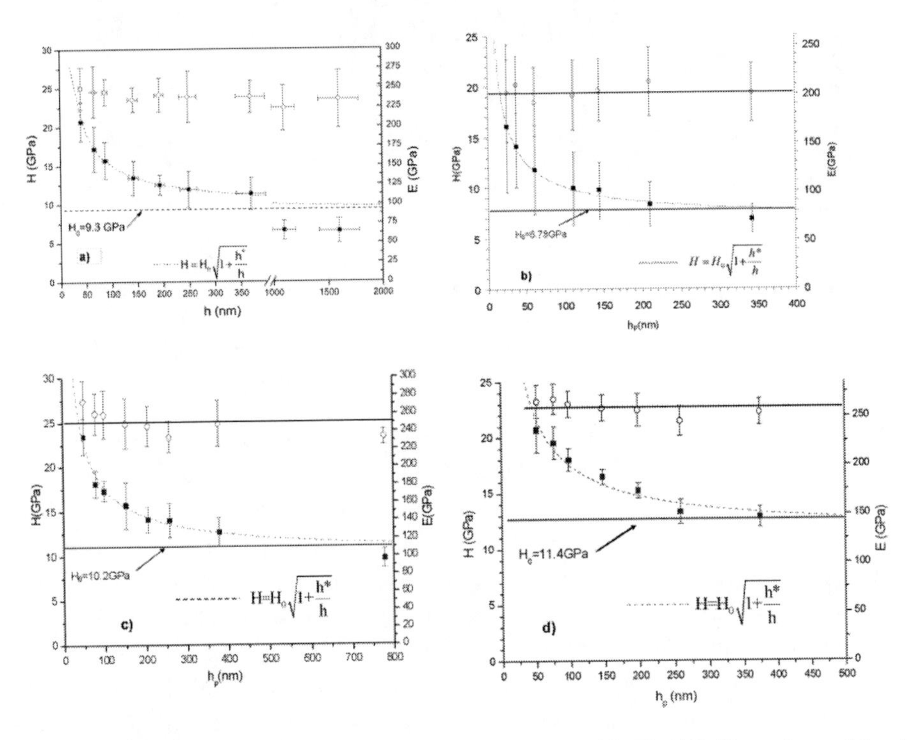

Figure 11. (a) Ti_3SnC_2, (b) $Ti_3Sn_{0.75}Al_{0.25}C_2$, (c) $Ti_3Sn_{0.8}Al_{0.2}C_2$, (d) Ti_3AlC_2 Young's modulus (open circle) and hardness (closed squares) as a function of the indenter penetration depth. Dashed lines: fit of the Indentation Size Effect (see text).

This model leads to the intrinsic hardness values (H_o) of the $Ti_3Sn_{(1-x)}Al_xC_2$ MAX phase solid solution. For each $Ti_3Sn_{(1-x)}Al_xC_2$ sample, the Young's modulus remains constant whatever the penetration depth. Such a result attests that the ISE is not an artefact due to the indenter shape calibration. For the largest penetration depth where a deviation of the Nix and Gao model is observed (see Figure 11a), AFM observations, shown as an example in figure 12, demonstrate that, for such high loads, several grains are involved in the deformation process during the indent formation. It implies that grain boundaries play an important role in macroscopic deformation. In such a situation, which is close to microindentation test conditions, MAX phases hardness is underestimated as it is the case for hardness measurements performed using micro indentation testing.

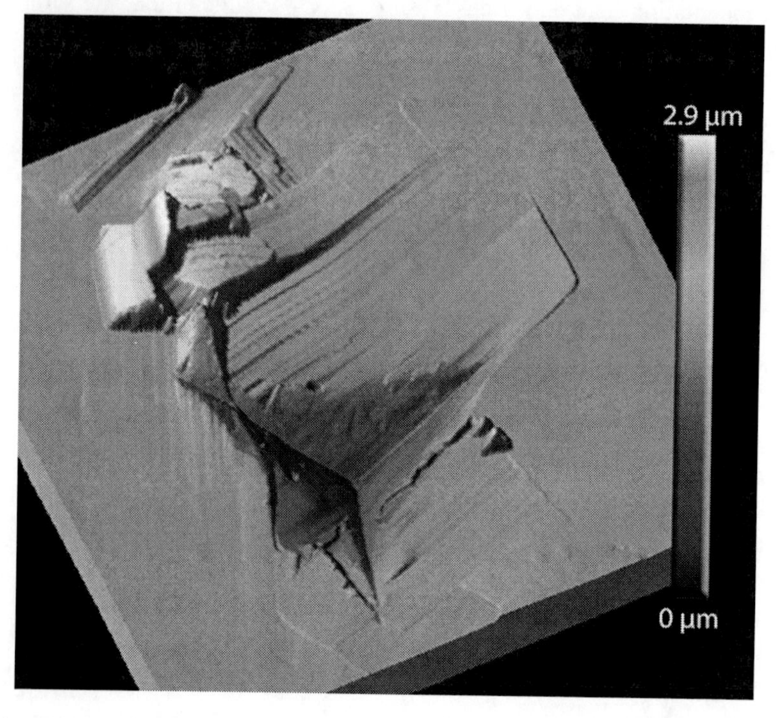

Figure 12. 300 mN indent in Ti_3SnC_2 observed by AFM (23x23 μm^2 topography image in 3D). Slip lines are observed in several grains, as well as grain pull-outs near the indent.

(D) $Ti_3Sn_{(1-x)}Al_xc_2$ Hardness and Young's Modulus Variations with Al Content

Figure 13 shows the hardness and elastic modulus variations as a function of the Al content in the $Ti_3Sn_{(1-x)}Al_xC_2$ solid solutions. First of all, one can notice that intrinsic hardness values (in the range 7-11 GPa) deduced from nanoindentation experiments performed as a function of load are larger than the currently determined microhardness values [55-57] (in the range 3-7 GPa for Ti_3AlC_2). Nevertheless, our value is in compatible agreement with the ones determined by nanoindentation performed on Ti_3AlC_2 thin films[51] or on Ti_3SiC_2 bulk material [58]. It is also shown, in figure 13, that hardness and elastic modulus does not vary monotonically with Al content; they first slightly decrease until x=0.25 and then increase up to x=1. Thus, solid solution hardening effect is not operative in this system. The same kind of results has been obtained by Ganguly *et al.* [24] in $Ti_3Si_{(1-x)}Ge_xC_2$ and by Zhou *et al.* [25] in $Ti_3Al_{1-x}Si_xC_2$ solid solutions. Based on a quite limited set of data, it seems that only substitutions on the X sites lead to solid solution hardening [27]; those on the M sites [23,59] and *A* sites, as shown here and in references 24 and 25, do not.

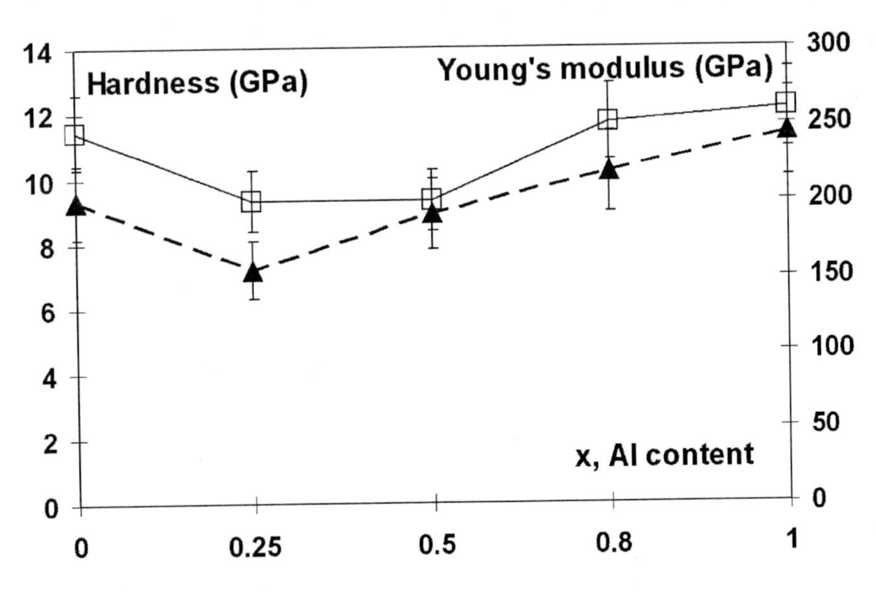

Figure 13. Hardness (closed triangles) and Young's modulus (open squares) versus Al content in $Ti_3Sn_{(1-x)}Al_xC_2$ solid solutions.

As MAX phases consist of "soft" *M-A* and "hard" *M-X* bonds, it is reasonable that the strengthening of the softer bond leads to an increase of the elastic modulus. The effect of intercalation of a weak *M-A* bond are twofold: first, enhancement of room temperature ductility, toughness and damage tolerance, and second, degradation of strength and hardness. Therefore, optimizations of elastic stiffness and strength of MAX phases have to focus on properly strengthening of the M-A bonds by tuning the chemical composition and the valence electron concentration [23,46,60-62]. It has been shown theoretically that the increase of *p* electrons from *A* atoms when moving rightward along the periodic table results in an increase of the bulk moduli up to s^2p^3 column [59]. The shift of the Ti *d-A p* hybrid to lower energy and its higher filling leads to stronger bonds. Nevertheless, Al ($3s^23p$) and Sn ($5s^25p^2$) do not lie in the same period of the Mendeleiëv's periodic table. Hug shows that Ti-Sn (2.975 Å) and Ti-Al (2.898 Å) bonds have about the same bond stiffness in Ti_2AlC and Ti_2SnC MAX phases [26]. In $Ti_3Sn_{(1-x)}Al_xC_2$, Ti-A bond lengths differ more significantly and so Ti-A bond stiffness may also differs. As a consequence, the slight difference observed in the elastic modulus of the $Ti_3Sn_{(1-x)}Al_xC_2$ solid solutions may likely be explained by the strength of the Ti-A bonds but more work is needed to confirm such hypothesis.

(E) $Ti_3Sn_{0.8}Al_{0.2}C_2$ Young's Modulus Variations with Temperature

Figure 14 shows the temperature variation of the $Ti_3Sn_{0.2}Al_{0.8}C_2$ Young's modulus. In the temperature range investigated, the Young's modulus decreases slowly and almost linearly with increasing temperature. The loss rate of Young's modulus, in this temperature range, is about 30.8 MPa.°C^{-1} which is much lower than that of $Ti_3Si_{(1-x)}Al_xC_2$ (49 MPa.°C^{-1}) [63]. The room temperature Young's modulus is 268 GPa which corresponds to a 6% difference with the elastic modulus that we get from nanoindentation experiments. Such a slight difference is within the standard deviation of the nanoindentation experiments. Indeed, a single oriented grain is under study during nanoindentation experiments and even if about 20 values have been extracted to get a mean value among the different orientations, it is possible that the statistics is not high enough to recover the mean Young's modulus along all crystallographic orientations. As shown previously for Ti_3AlC_2, Ti_3SiC_2 [64], Nb_2AlC and Nb_4AlC_3 [65], the $Ti_3Sn_{0.2}Al_{0.8}C_2$ Young's modulus does not vary significantly with temperature in the temperature range under study. Radovic *et al.* [52] suggested that the critical temperature for the accelerated decrement of the elastic modulus is close to the Brittle-to-Ductile Transition Temperature (BDTT) for Ti_3SiC_2 and Ti_2AlC. For most MAX phases, such a BDTT is about 1000°C.

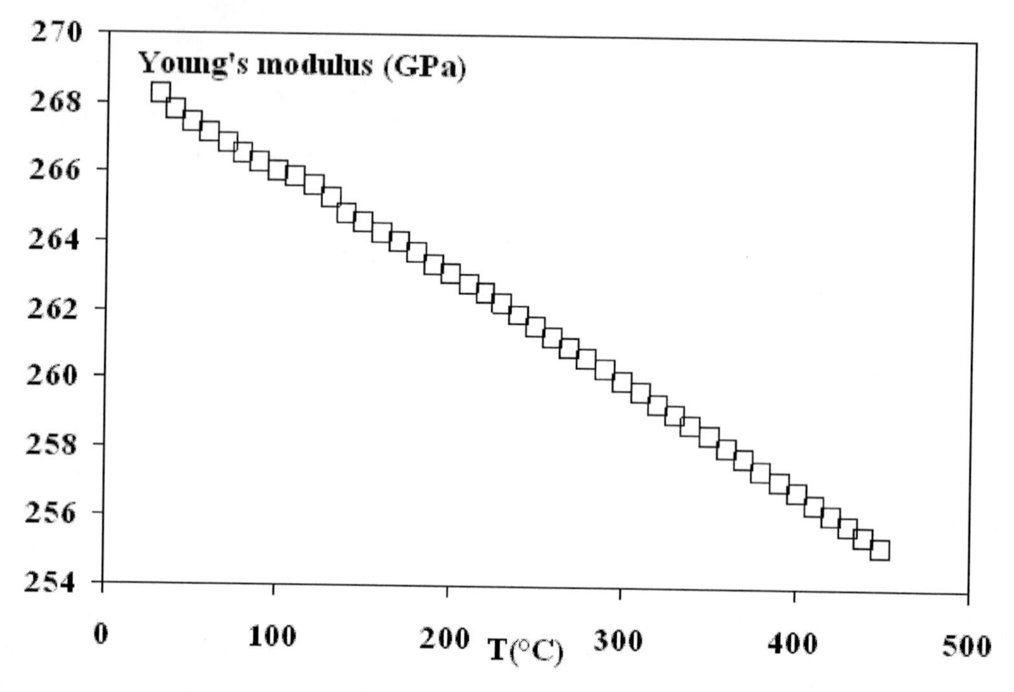

Figure 14. $Ti_3Sn_{0.2}Al_{0.8}C_2$ Young's modulus, measured by means of dynamic resonant method in bending mode, as a function of temperature.

4. SUMMARY

$Ti_3Sn_{(1-x)}Al_xC_2$ MAX phase solid solutions have been successfully synthesized from different reactant mixtures by HIP. It is shown that 312 solid solutions are likely formed from reaction between TiC and 211 solid solution. Rietveld refinement allows to carefully characterize the structure of the different solid solutions. It is shown that, both for Ti_3SnC_2 and Ti_3AlC_2, hexagonal unit cell is strongly distorted. Nevertheless, $[Ti_6C]$ octahedra are mainly distorted in Ti_3SnC_2 whereas $[Ti_6Al]$ trigonal prisms are mainly distorted in Ti_3AlC_2; the trigonal prism distortion does not vary significantly for $0.25<x<1$ in $Ti_3Sn_{(1-x)}Al_xC_2$. Nanoindentation experiments allow to determine hardness and elastic modulus of $Ti_3Sn_{(1-x)}Al_xC_2$ solid solutions. It is demonstrated that solid solution hardening is not operative in this system. It is demonstrated that elastic modulus and hardness values do not vary monotonically with the Al content. Nevertheless, the elastic modulus of Ti_3AlC_2 (260 GPa) is higher than the Ti_3SnC_2 one (245 GPa). Assuming that the strengthening of the weaker Ti-A bonds would increase the elastic modulus of the MAX phases, such an increase may be attributed to the Ti-Al stiffer bonds present in Ti_3AlC_2 than the Ti-Sn bonds in Ti_3SnC_2. More work is needed to confirm such an hypothesis. Finally, it is also shown that Young's modulus of the $Ti_3Sn_{0.2}Al_{0.8}C_2$ solid solutions slightly vary with temperature in the range 20-450°C.

ACKNOWLEDGMENTS

The authors thank A. Baudet for technical assistance and T. Cabioc'h for fruitful discussions. This work was financially supported by the "Agence Nationale de la Recherche" in the PLASMAX project (ANR-07-MAPR0015).

REFERENCES

[1] W. Jeitschko, H. Nowotny and F. Benesovsky, "Kohlen-stoffhaltige ternare Verbindungen (H-Phase)", *Monatsh. Chem.* 94, 672 (1963).

[2] W. Jeitschko, H. Nowotny and F. Benesovsky, "Die H-phasen Ti_2TlC, Ti_2PbC, Nb_2InC, Nb_2SnC und Ta_2GaC", *Monatsh. Chem.* 95, 431 (1964).

[3] W. Jeitschko, and H. Nowotny, "Die Kristallstructur von Ti_3SiC_2-Ein Neuer Komplxcarbid-Typ", *Monatsh. Chem.* 98, 329 (1967).

[4] H. Wolfsgruber, H. Nowotny, and F. Benesovsky, "Die Kristallstuktur von Ti_3GeC", *Monatsh. Chem.* 98, 2403 (1967).

[5] V. H. Nowotny, "Strukturchemie einiger Verbindungen der Übergangsmetalle mit den elementen C, Si, Ge, Sn", *Prog. Solid State Chem.* 5, 27 (1971).

[6] M. W. Barsoum, and T. El-Raghy, "Synthesis and characterization of a remarkable ceramic: Ti_3SiC_2", *J. Am. Chem. Soc.* 79, 1953-1956 (1996).

[7] M. W. Barsoum, L. Farber, I. Levin, A. Procopio, T. El-Raghy, and A. Berner, "HRTEM of Ti_4AlN_3; or $Ti_3Al_2N_2$ Revisited", *J. Am. Ceram. Soc.* 82, 2545-2547 (1999).

[8] C. J. Rawn, M. W. Barsoum, T. El-Raghy, A. Procopio, C. M. Hoffmann, and C. Hubbard, "Structure of Ti_4AlN_{3-x} - a layered $M_{n+1}AX_n$ nitride", *Mater. Res. Bull.* 35, 1785 (2000).

[9] M. W. Barsoum, "The $M_{N+1}AX_N$ phases: A new class of solids thermodynamically stable nanolaminates", *Prog. Solid State Chem.* 28, 201 (2000).

[10] X. H. Wang and Y. C. Zhou, "Oxidation behaviour of Ti_3AlC_2 at 1000-1400°C in air." *Corros. Sci.* 45, 891-907 (2002).

[11] X. H. Wang and Y. C. Zhou, "High temperature oxidation of Ti_2AlC in air." *Oxid. Met.* 59, 303-320 (2003).

[12] M. Sundberg, G. Malmqvist, A. Magnusson and T. El Raghy, "Alumina forming high temperature silicides and carbides." *Ceram. Int.* 30, 1899-1905 (2004).

[13] S. Becker, A. Rahmel, M. Schorr and M. Schütze, "Mechanism of isothermal oxidation of intermetallic TiAl and of TiAl alloys." *Oxid. Met.* 38, 427-433 (1992).

[14] R. G. Reddy, X. Wen and M. Divakar, "Isothermal oxidation of TiAl alloy." *Metall. Mater. Trans. A*, 32, 2357-2361 (2001).

[15] M. S. Chu and S. K. Wu, "The improvement of high temperature oxidation of Ti-50Al by sputtering Al film and subsequent inter-diffusion treatment." *Acta Mater.* 51, 3109-3120 (2003).

[16] M. P. Brady and P. F. Tortorelli, "Alloy design of intermetallics for protective scale formation and for use as precursors for complex ceramic phase surfaces." *Intermetallics* 12, 779-789 (2004).

[17] Z. J. Lin, M. Zhuo, Y. C. Zhou, M. Li and J. Y. Wang, "Microstructures and adhesion of the oxide scale formed on titanium aluminium carbide substrates." *J. Am. Ceram. Soc.* 89, 2964-2966 (2006).

[18] Z. J. Lin, M. J. Zhuo, Y. C. Zhou, M. S. Li, J. Y. Wang, "Interfacial microstructure of Ti_3AlC_2 and Al_2O_3 oxide scale." *Script. Mater.* 54, 1815-1820 (2006).

[19] Z. J. Lin, M. Li and Y. C. Zhou, "TEM investigations on layered ternary carbides." *J. Mater. Sci. Technol.* 23, 145-165 (2007).

[20] J.C. Schuster, H. Nowotny and C. Vaccaro, "The ternary systems: Cr-Al-C, V-Al-C, and Ti-Al-C and the behavior of H-phases (M_2AlC)." *J. Solid State Chem.* 32, 213 (1980)

[21] S. Gupta and M.W. Barsoum, "Synthesis and oxidation of V_2AlC and ($Ti_{0.5}$, $V_{0.5}$)AlC in air." *J. Electrochem. Soc.* 151, 1-6 (2004).

[22] Z. Sun, R. Ahuja, J. M. Schneider, "Theoretical investigation of the solubility in ($M_xM'_{2-x}$)AlC (M and M'=Ti, V, Cr)." *Phys. Rev. B*, 68, 224112 (2003).

[23] J. Y. Wang, Y. C. Zhou, "Ab initio elastic stiffness of nano-laminate (M_xM_{2-x})AlC (M=Ti, V, Nb and Cr) solid solution." *J. Phys. Condens. Matter*, 16, 2819-2827 (2004).

[24] A. Ganguly, T. Zhen, M. W. Barsoum, "Synthesis and mechanical properties of Ti_3GeC_2 and $Ti_3(Si_xGe_{1-x})C_2$ (x=0.5, 0.75) solid solutions." *J. Alloys and Compds.* 376, 287-295 (2004).

[25] Y.C. Zhou, J.X. Chen, J.Y. Wang, "Strengthening of Ti_3AlC_2 by incorporation of Si to form $Ti_3Al_{1-x}Si_xC_2$ solid solutions." *Acta Materialia*, 54, 1317-1322 (2006).

[26] P. Finkel, B. Seaman, K. Harrell, J. Palma, J. P. Hettinger, S. E. Lofland, A. Ganguly, M. W. Barsoum, Z. Sun, S. Li and R. Ahuja, "Electronic, thermal, and elastic properties of $Ti_3Si_{1-x}Ge_xC_2$ solid solutions." *Phys. Rev. B*, 70, 085104 (2004).

[27] M. W. Barsoum, M. Ali, T. El-Raghy, "Processing and characterization of Ti_2AlC, Ti_2AlN, and $Ti_2AlC_{0.5}N_{0.5}$." *Metall. Mater. Trans. 31A,* 1857-1865 (2000).

[28] S. Dubois, T. Cabioc'h, P. Chartier, V. Gauthier and M. Jaouen, "A new ternary nanolaminate carbide : Ti_3SnC_2." *J. Am. Ceram. Soc.* 90, 2642-2644 (2007).

[29] Z. G. Liu, J. T. Guo, L. L. Ye, G. S. Li and Z. Q. Hu, "Formation mechanism of TiC by mechanical alloying." *Appl. Phys. Lett.* 65, 2666-2668 (1994).

[30] S. Dubois, E. Heian, N. Karnatak, M. F. Beaufort, D. Vrel, "Synthesis of TiC/Ni cermets via mechanically activated self propagating high temperature synthesis." *Materials Science Forum*, 426-432, 2033-2038 (2003).

[31] L. Lutterotti, S. Matthies, and H.-R. Wenk, "MAUD: a friendly Java program for material analysis using diffraction." *IUCr: Newsletter of the CPD* 21, 14 (1999).

[32] J. Woirgard and J.C. Dargenton, "An alternative method for penetration depth determination in nanoindentation measurements." *J. Mater. Res.* 12, 2455-2458 (1997).

[33] G.M. Pharr and A. Bolshakov, "Understanding nanoindentation unloading curves." *J. Mater. Res.* 17, 2660 (2002).

[34] ASTM (2001) 'Standard Test Method for Young's Modulus, Shear Modulus and Poisson's Ratio by Impulse Excitation of Vibration', *Annual Book of ASTM Standards*, Vol. 03.01, pp. 1099-1112

[35] P. Mazot, J. de Fouquet, J. Woirgard, and J. P. Pautrot, "Mesure des constantes d'élasticité et du frottement intérieur des solides par méthode de résonance entre 20 et 1200°C", *Journal de Physique III-* May, 751-763 (1992).

[36] P. Gadaud, P. and S. Pautrot, "Characterization of the elasticity and anelasticity of bulk glasses by dynamical subresonant and resonant techniques", *Journal of Non-Crystalline Solids*, 316, 145-152 (2003).

[37] C. Peng, C. A. Wang, Y. Song, Y. Huang, "A novel simple method to stably synthesize Ti_3AlC_2 powder with high purity", *Mater. Sc. And Eng. A* 428, 54-58 (2006).

[38] Z. J. Lin, M. J. Zhuo, Y. C. Zhou, M. S. Li, J. Y. Wang, "Microstructural characterization of layered ternary Ti_2AlC", *Acta Mater.* 54, 1009-1015 (2006).

[39] Z. J. Lin, M. J. Zhuo, Y. C. Zhou, M. S. Li, J. Y. Wang, "Microstructural relationship between compounds in the Ti-Si-C system", *Scripta Mater.* 55, 445-448 (2006).

[40] Z. J. Lin, M. J. Zhuo , M. S. Li, J. Y. Wang and Y. C. Zhou, "Synthesis and microstructure of layered-ternary Ti_2AlN ceramic", *Scripta Mater.* 56, 1115-1118 (2007).

[41] A. Zhou, C. A. Wang, Y. Huang, "A possible reaction mechanism on synthesis of Ti_3AlC_2", *Mater. Sc. and Eng. A* 352, 333-339 (2003).

[42] Z. Ge, K. Chen, J. Guo, H Zhou, J. M. F. Ferreira, "Combustion synthesis of ternary carbide Ti_3AlC_2 in Ti–Al–C system", *J. Europ. Ceram. Soc.*, 23, 567-574 (2003).

[43] J. H. Han, S. S. Hwang, D. Lee, S. W. Park, "Synthesis and mechanical properties of Ti_3AlC_2 by hot pressing TiC_x/Al powder mixture", *J. Europ. Ceram. Soc.* 28, 979–988 (2008).

[44] K. T. Jacob, S. Raj and L. Rannesh, "Vegard's law: a fundamental relation or an approximation?", *International Journal of Materials Research*, 9, 776-779 (2007).

[45] J. Etzkorn, M. Ade, H. Hillebrecht, "V_2AlC, V_4AlC_{3-x} ($x \approx 0.31$), and $V_{12}Al_3C_8$: Synthesis, Crystal Growth, Structure, and Superstructure", *Inorg. Chem.* 46, 7646-7653 (2007).

[46] G. Hug, "Electronic structures of and composition gaps among the ternary carbides Ti_2MC" *Phys. Rev. B*, 74, 184113 (2006).

[47] G. Hug, M. Jaouen and M. W. Barsoum, "X-ray absorption spectroscopy, EELS, and full-potential augmented plane wave study of the electronic structure of Ti_2AlC, Ti_2AlN, Nb_2AlC, and $(Ti_{0.5}Nb_{0.5})_2AlC$", *Phys. Rev. B*, 71, 024105 (2005).

[48] K. Momma and F. Izumi, "VESTA: a three-dimensional visualization system for electronic and structural analysis." *J. Appl. Crystallogr.* 41, 653-658, 2008.

[49] M. Gamarnik, M. W. Barsoum, "Bond lengths in the ternary compounds Ti_3SiC_2, Ti_3GeC_2 and Ti_2GeC", *J. Mater. Sci.* 34, 169-174 (1999).

[50] C. Tromas, N. Ouabadi, V. Gauthier-Brunet, M. Jaouen and S. Dubois, "Mechanical properties of nanolaminate Ti_3SnC_2 carbide determined by nanohardness cartography", *J. Am. Ceram. Soc.* 93, 330-333 (2010).

[51] O.Wilhelmsson, J. P. Palmquist, E. Lewin, J. Emmerlich, P. Eklund, P. O. A. Persson, H. Hogberg, S. Li, R. Ahuja, O. Eriksson, L. Hultman, and U. Jansson,, "Deposition

and Characterization of Ternary Thin Films Within the Ti–Al–C System by DC Magnetron Sputtering", *J. Cryst. Growth*, 291, 290–300 (2006).

[52] M. Radovic, M. W. Barsoum, A. Ganguly, T. Zhen, P. Finkel, S. R.Kalidindi, and E. Lara-Curzio, "On the Elastic Properties and Mechanical Damping of Ti_3SiC_2, Ti_3GeC_2, $Ti_3Si_{0.5}Al_{0.5}C_2$ and Ti_2AlC in the 300–1573 K Temperature Range" *Acta Mater.* 54, 2757–67 (2006).

[53] C. Kral, W. Lengauer, D. Rafaja, P. Ettmayer, *J. Alloys and Compds.* 265, 215-233 (1998).

[54] W. D. Nix and H. Gao, "Indentation size effects in crystalline materials: a law for strain gradient plasticity", *J. Mech. Phys. Solids*, 46, 411-425 (1998).

[55] N. V. Tzenov, M. W. Barsoum, "Synthesis and Characterization of Ti_3AlC_2", *J. Am. Ceram. Soc.* 83, 825-832 (2000).

[56] Y. C. Zhou, J. X. Chen, J. Y. Wang, "Strengthening of Ti_3AlC_2 by incorporation of Si to form $Ti_3Al_{1-x}Si_xC_2$ solid solutions", *Acta Mater.* 54, 1317-1322 (2006).

[57] J. H. Han, S. S. Hwang, D. Lee, S. W. Park, "Synthesis and mechanical properties of Ti_3AlC_2 by hot pressing TiC_x/Al powder mixture", *J. Europ. Ceram. Soc.* 28, 979-988 (2008).

[58] J. M. Molina-Aldareguia, J. Emmerlich, J.-P. Palmquist, U. Jansson, L. Hultman, "Kink formation around indents in laminated Ti_3SiC_2 thin films studied in the nanoscale", *Scripta Mater.* 49, 155-160 (2003).

[59] I. Salama, T. El Raghy, M. W. Barsoum, "Synthesis and mechanical properties of Nb_2AlC and $(Ti,Nb)_2AlC$", *J. Alloys and Compds.* 347, 271-278 (2002).

[60] Z. M. Sun, R. Ahuja, J. M. Schneider, "Theoretical investigation of solubility in $(M_xM'_{2-x})AlC$ (M and M' = Ti, V, Cr)", *Phys. Rev. B* 68, 224112 (2003).

[61] R. S. Kumar, S. Rekhi, A. L. Cornelius, M. W. Barsoum, "Compressibility of Nb_2AsC to 41 GPa", *Appl. Phys. Lett.* 86, 111904 (2005).

[62] T. Liao, J. Y. Wang, Y. C. Zhou, "Superior mechanical properties of Nb_2AsC to those of other layered ternary carbides: a first-principles study", *J. Phys. Condens. Matter*, 18, L527-L533 (2006).

[63] Y. C. Zhou, D. T. Wan, Y. W. Bao, J. Y. Wang, "In situ processing and high-temperature properties of $Ti_3Si(Al)C_2$/SiC composites", *Int. J. Appl. Ceram. Technol.* 3, 47-54 (2006).

[64] Y. W. Bao, Y.C. Zhou, "Evaluating high-temperature modulus and elastic recovery of Ti_3SiC_2 and Ti_3AlC_2 ceramics", *Mater. Lett.* 57, 4018-4022 (2003).

[65] J. Wang, Y.C. Zhou, "Recent progress in theoretical prediction, preparation and characterization of layered ternary transition-metal carbides", *Annu. Rev. of Mater. Res.* 39, 415-443 (2009).

In: MAX Phases: Microstructure, Properties and Applications ISBN 978-1-61324-182-0
Editors: It-Meng (Jim) Low and Yanchun Zhou

Chapter 5

FABRICATION OF Ti_3SiC_2 BY SPARK PLASMA SINTERING

Weibing Zhou[1,2], Bingchu Mei[2] and Jiaoqun Zhu[1]

[1]School of materials and Science Engineering, Wuhan University of Technology
[2]State Key Laboratory of Advanced Technology for Materials Synthesis and Processing, Wuhan University of Technology, P.R. China

ABSTRACT

In this chapter, a novel sintering technique called spark plasma sintering (SPS) is introduced. It is a versatile technique to rapidly sinter a number of materials including metals, ceramics, polymers, and composites in a period of minutes. In this method, as-received powders are placed in a graphite die, pressed uniaxially, and then heated by passing a high pulsed current through the powders and/or the die. Sintering occurs within minutes. It is suggested that electro-discharging among particles can activate the particle surface and assist sintering. Using this technique, polycrystalline bulk Ti_3SiC_2 samples can be obtained by one of three reactions; (1) $3Ti + SiC + C \rightarrow Ti_3SiC_2$; (2) $3Ti + Si + 2C \rightarrow Ti_3SiC_2$ and (3) $Ti + Si + 2TiC \rightarrow Ti_3SiC_2$. To improve the purity of samples, we used aluminium as sintering aid. The results showed that addition of aluminum can considerably accelerate the synthesis reaction of Ti_3SiC_2 and fully dense, essentially single-phase polycrystalline Ti_3SiC_2 could be successfully obtained by sintering 2TiC/1Ti/1Si/0.2Al powders at 1250~1300°C. The process parameters in the sintering process revealed that addition of aluminium decreased the temperature for the synthesis reaction of Ti_3SiC_2. The density of the obtained Ti_3SiC_2 was 97.8% of the theoretical value. It was shown that spark plasma sintering is an effective method to synthesis Ti_3SiC_2.

1. INTRODUCTION

Spark plasma sintering (SPS), also known as field assisted sintering technique (FAST) or pulsed electric current sintering (PECS) or pulse discharge sintering (PDS), is a newly developed process which makes possible sintering high quality materials in short periods by charging the intervals between powder particles with electrical energy and high sintering pressure [1]. SPS systems offer many advantages (*e.g.* rapid sintering, less sintering additives, uniform sintering, low running cost, easily operation, *etc.*) over conventional systems using hot press (HP) sintering, hot isostatic pressing (HIP) or atmospheric furnaces. In addition, the SPS process can be used for sintering and densification of many advanced materials such as FGM (functionally graded materials), fine ceramics, amorphous materials, target materials, thermoelectric generator, nano-composites [2-8].

The main characteristic of the SPS is that the pulsed DC current directly passes through the graphite die, as well as the powder compact, in case of conductive samples. Therefore, the heat is generated internally, in contrast to the conventional hot pressing, where the heat is provided by external heating elements. This facilitates a very high heating or cooling rate (up to 1000 k/min), hence the sintering process generally is very fast (within a few minutes). The general speed of the process ensures it has the potential of densifying powders with nanosize or nanostructure while avoiding coarsening which accompanies standard densification routes. Whether plasma is generated has not been confirmed yet, especially when non-conductive ceramic powders are compacted. It has, however, been experimentally verified that densification is enhanced by the use of a current or field. Figure 1 shows the schematic of spark plasma sintering process.

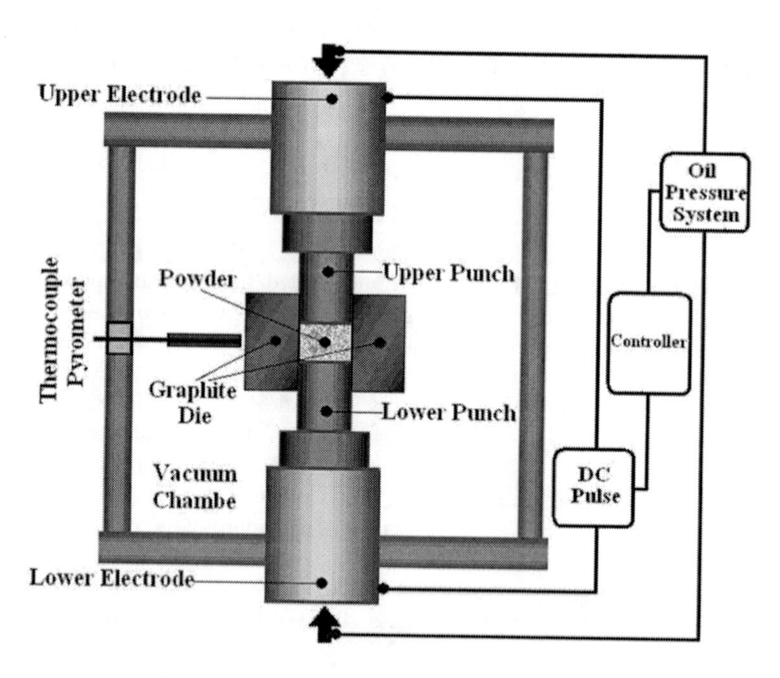

Figure 1. Schematic of spark plasma sintering process.

2. FABRICATION OF Ti_3SiC_2 BY SPARK PLASMA SINTERING

Zhang *et al.* [9] synthesized bulk Ti_3SiC_2 from Ti/SiC/TiC powders at 1250 °C to 1450°C using the pulse discharge sintering (PDS) technique. The purity of the resulting Ti_3SiC_2 was as high as 92 vol. %, and the relative density was higher than 99% when the sintering temperatures was above 1350°C. They observed three microstructures of Ti_3SiC_2, i.e. fine, coarse and duplex grains, depending on the sintering temperatures and time. The work indicated that Ti/SiC/TiC powders could be employed as the initial materials to fabricate Ti_3SiC_2. In the following year, they also obtained high-purity Ti_3SiC_2 material by sintering Ti/Si/C powders through pulse discharge sintering technique [10]. The results showed that high purity Ti_3SiC_2 could not be obtained from the Ti/Si/C powder with molar ratio of 3:1:2 and Ti_3SiC_2 preferred to form at relatively low sintering temperature for a short time. When 5Ti/2Si/3C and 3Ti/1.5Si/2C powders were sintered for 15 min, the TiC content was respectively decreased to 6.4 and 10 wt% at 1250–1300 °C. The corresponding relative density of the samples sintered from 5Ti/2Si/3C powder was calculated to be as high as 99% at the temperature above 1300 °C. It is suggested that low-temperature rapid synthesis of Ti_3SiC_2 would be possible through the PDS technique, provided that the composition of the starting powders should be adjusted to be off-stoichiometric ratio from 3:1:2.

Gao et al. [11] demonstrated that Ti_3SiC_2 could be rapidly synthesized and simultaneously consolidated from mixtures of $Ti/Si/_2TiC$ by spark plasma sintering (SPS) technique. Ti_3SiC_2 with 2 wt pct TiCx was produced at the sintering temperatures between 1250 oC to 1300oC. They also detected the preferential grain growth of Ti_3SiC_2 along the crystallographic basal plane and an anisotropic hardness, which were due to the fact that these platelet grains tended to align perpendicular to the loading surface.

3. Fabrication of High-Purity Ti_3SiC_2 by SPS with Al as Additives

In our research work, high purity Ti_3SiC_2 material was obtained by spark plasma sintering technique. As we know, Ti_3SiC_2 has the same structure and similar properties as those of Ti_3AlC_2. Moreover, a liquid as an Al–Si alloy containing 0–44%Si forms at 1000°C in the Al–C–Si–Ti quaternary system [12]. Thus, addition of aluminum in the starting material may improve the synthesis of Ti_3SiC_2 via a solid–liquid reaction. The objective of this work was to prepare high-purity Ti_3SiC_2 by spark plasma sintering Ti/TiC/Si/Al.

All the work was conducted using powder mixtures of titanium (99.0% pure, 10.6 μm), Si (99.5% pure, 9.5 μm), Al (99.8% pure, 12.8 μm) and TiC (99.2% pure, 8.4 μm). In brief, the admixture with a designed composition was firstly mixed in ethanol for 24 h, then was filled into a graphite die of 20 mm in diameter and finally sintered in spark plasma sintering system (Model SPS-1050, Japan). The samples were heated at a rate of 80 °C /min until the requisite temperature was reached, the soaking time was 8 minutes.

The sintered sample was polished and the density was measured by Archimedes' method. The sintered products were characterized by XRD using a rotating anode X-ray diffractometer (Model D/MAX-RB, RIGAKU Corporation, Japan). Scans were made with Cu Kα radiation (40 kV and 50 mA) at a rate of 1°/min, using a step of 0.02°. The lattice parameters were measured using silicon as the internal standard. The microstructures of the samples were investigated via scanning electron microscope (SEM) (Model JSM-5610LV, JEOL Ltd., Japan) equipped with energy-dispersive spectroscopy (EDS). The micro-hardness was measured using a Leitz Microhardness Tester (Leitz Wetzlar, Germany) at 1 N with a loading time of 30 s. The hardness was calculated by averaging at least 10 measurements.

3.1. Effect of Al Content on Synthesis of Ti_3SiC_2

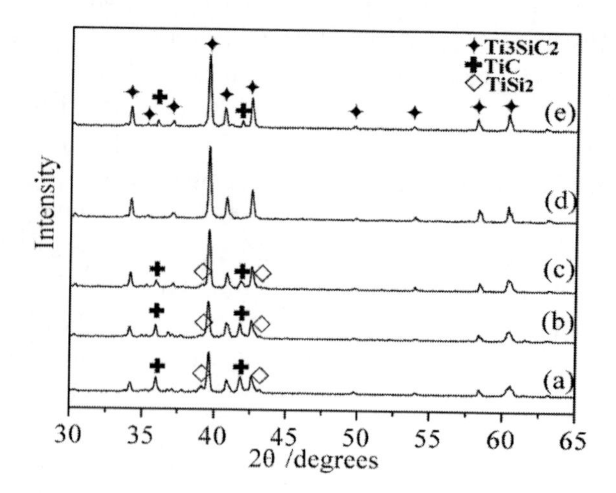

Figure 2. X-ray diffraction patterns of samples fabricated by SPS at 1250°C from different starting compositions in molar ratios of (a) TiC:Ti:Si = 2:1:1.2; (b) TiC:Ti:Si:Al = 2:1:1.15:0.05; (c) TiC:Ti:Si:Al = 2:1:1.1:0.1; (d) TiC:Ti:Si:Al = 2:1:1:0.2; (e) TiC:Ti:Si:Al = 2:1:0.9:0.3.

The loss of Si might occur by evaporation when the sample was heated in Ar or vacuum, resulting in formation of quite a large amount of TiC in the product. High purity Ti_3SiC_2 could be synthesized by using a starting composition with 20 mol% more Si than suggested by stoichiometry of Ti_3SiC_2 [13]. In sample (a), with the composition of TiC:Ti:Si = 2:1:1.2, the main phases were Ti_3SiC_2, TiC and $TiSi_2$.

In contrast to sample (a), samples (b), (c), (d), (e), had the starting composition of Ti_3Si_1.2-xAlxC_2, where x= 0.05, 0.1, 0.2, or 0.3, respectively. Their X-ray diffraction patterns revealed that the amount of TiC decreased and the content of Ti_3SiC_2 increased with the addition of aluminum. For sample (b) and (c) with the addition of 0.05 mol and 0.1 mol of Al, the main phase was Ti_3SiC_2, and the peaks of TiC ($2\theta = 36°$) and $TiSi_2$ were also been identified by X-ray diffraction (Figure 2). For sample (d), which had the addition of 0.20 mol of aluminum, their products were of pure Ti_3SiC_2, and no phase but Ti_3SiC_2 was identified by X-ray diffraction. But in sample (e), with 0.30 mol of aluminum, the X-ray diffraction peaks of Ti_3SiC_2 became weakened, and phase TiC reappeared at the same time.

From above results, TiSi2 was found to be intermediate phase during the synthesis process. A solid-liquid reaction mechanism between TiSi2 liquid phase and TiC particles was proposed to explain the rapid formation of Ti_3SiC_2. The result is consistent with other researchers. Gao et al. [11] obtained high purity Ti_3SiC_2 by spark plasma sintering Ti/2TiC/Si mixture powders at 1250°C for 10mins under the pressure of 60 MPa. But in our experiment, high purity (> 97%) Ti_3SiC_2 was prepared in relative lower pressure (30 MPa) and shorter time (8 min). The interesting results might be explained as follows: firstly, Al has a relatively low melting point; the sample formed a fraction of liquid that favored the diffusion of both Ti and Si atoms, resulting in the rapid formation of Ti_3SiC_2. Secondly, when the SPS technique was applied on the Ti/TiC/Si/Al powder, pulse electric field may activate the surface of the powder particles resulting in an easy sintering process. Meanwhile, possible plasma occurred between the particles may locally increase the temperature too much higher a level than the controlled temperature, which could effectively promote the synthesis reaction.

3.2. The Influence of Sintering Temperature

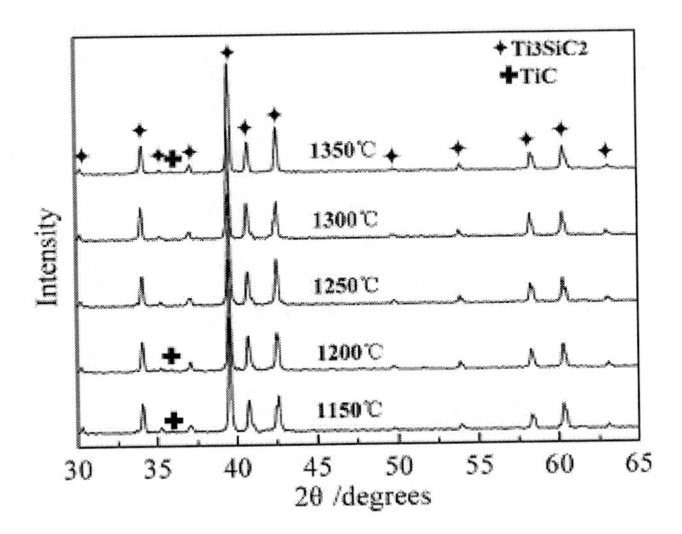

Figure 3. X-ray diffraction patterns of samples sintered at different temperatures of the starting powders by SPS with the molar ratios TiC:Ti:Si: Al = 2:1:1:0.2.

Shown in Figure 3 are the X-ray diffraction patterns of samples obtained from the mixture of raw materials of 2TiC + 1Ti + 1Si + 0.2Al (moles) at different temperatures. When sintered at 1150~1200 ˚C, the main phase was Ti_3SiC_2, however, the peak of TiC (2θ=36º), which was very weak in contrast with those of Ti_3SiC_2, was present. When the sintering temperature reached 1250 ˚C and 1300 ˚C, the product was pure Ti_3SiC_2; no phase but Ti3SiC2 was identified by X-ray diffraction. But when sintered at 1350 ˚C, the peak of TiC reappeared again, which indicates Ti_3SiC_2 has decomposed according to the reaction: $Ti_3SiC_2 \rightarrow TiCx + Si\uparrow$. The results indicate that the optimal temperature for the formation of Ti_3SiC_2 is in the range 1250~1300 ˚C. The results were consistent with previous work [10, 11].

3.3. Microstructure of Ti_3SiC_2 Samples

Figure 4 shows the microstructures of Ti_3SiC_2 samples sintered at 1250˚C from different starting compositions. The secondary electron image coupled with energy-dispersive spectroscopy analysis reveals that the main phase Ti_3SiC_2 is laminate, while the impurity phase TiC and TiSi2 is in quadratic form or aspheric-shape with small size. The two samples have evidently different morphology characteristics. For sample (a) with a starting composition of 2TiC/Ti/1.2Si, the Ti_3SiC_2 grains have the sizes of less than 1 μm and 10 μm, in thickness and elongated dimension, respectively. Additionally, large quantities of poorly developed crystals of impurity phase TiC and TiSi2 existed in sample (a). However, for sample (b) with a starting composition of 2TiC/Ti/Si/0.2Al, those small TiC and TiSi2 grains disappear completely, while Ti_3SiC_2 grains are well developed, and reach the sizes of about 5 μm and 25 μm in thickness and elongated dimension, respectively. Its density was measured to be 4.43 g/cm^3, being 97.8% of theoretic density of Ti_3SiC_2. Micro-hardness of the sample was 4.0 GPa (at 1.0 N and 30 s). SEM micrographs confirm the XRD results that appropriate addition of aluminum in the starting material considerably improves Ti_3SiC_2 content of SPS samples. Moreover, addition of aluminum can also accelerate the crystal growth of Ti_3SiC_2.

Figure 4. SEM photographs of the fracture surfaces of samples, (a) 2TiC/Ti/1.2Si and (b) 2TiC/Ti/Si/0.2Al.

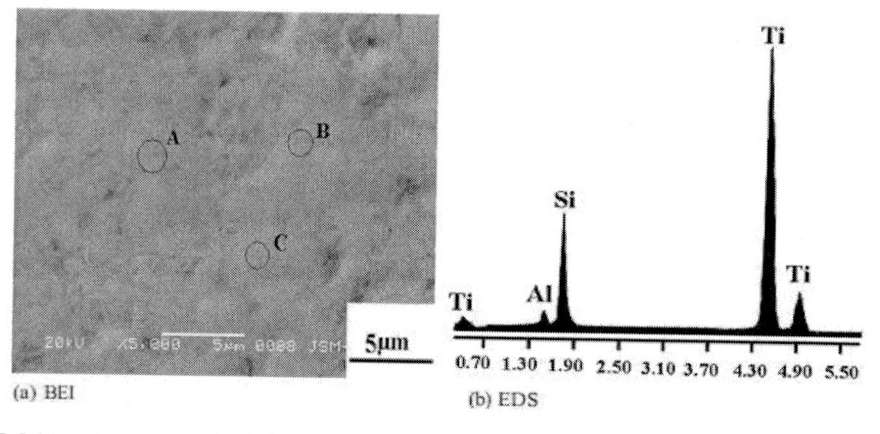

Figure 5. Microstructures of polished surfaces of sample by sintering 2TiC/1Ti/1Si /0.2 Al at 1250 °C.

Table 1. Atomic ratios of Ti, Si and Al in the microscope of $Ti_3(Si/Al)C_2$

Atomic Wt%	Ti	Si	Al	Si / Al
A	72.63	23.98	3.39	7.07
B	71.68	24.85	3.47	7.16
C	72.06	24.33	3.61	6.74
The whole	72.34	24.10	3.56	6.77

Figure 5(a) shows the backscattered electron image of the polished surface of sample sintered at 1250°C, the corresponding energy-dispersive spectrum (EDS) is shown in Figure

5(b). The chemical analysis results as listed in Table 1 revealed that the atomic ratio of Si to Al in the sintered product (=6.77) is larger than that in the starting mixture (=5). As we know, the saturated vapor tension of Al at the high temperature is larger than that of Si at the same temperature. When a sample is heated at a high temperature, the loss rate of Al by evaporation must be larger than that of Si. Hence, it is reasonable to consider that some Si atom sites in the Ti3SiC2 have been substituted by Al to form a solid solution of $Ti_3(Si/Al)C_2$ for the above samples. The measured crystal lattice parameters are a=0.30741 nm and c=1.77441 nm, which was slightly less than the results given by Barsoum (Ref. [14] a = 0.30665 nm and c = 1.76710 nm) and Sun (Ref. [15] a = 0.3068 nm and c =1.7645 nm), since the atomic semi diameter of silicon slightly less than that of aluminum.

3.4. The Process Parameters in Spark Plasma Sintering Process

In the spark plasma sintering process, the gas in the chamber is discharged out at a constant rate. As a result, the vacuum pressure does not change during the sintering process. However, a peak in the pressure may appear if a large quantity of gas is given out from a reaction in the sample. Further insight into the effect of aluminium can be gained by analyzing the temperature dependence of the vacuum pressure in the chamber, as shown in Figure 6. For sample (a), the pressure reached a peak at a temperature of 1160 ˚C. But for sample (b), the pressure peak appeared at a lower temperature of 1100 ˚C, and it was more pronounced than that of sample (a). These results support the notion that an appropriate addition of aluminum both decreased the reaction temperature as well as accelerated the reaction speed of $2TiC + Ti + Si \rightarrow Ti_3SiC_2$.

Figure 7 shows the temperature dependence of Z-axis displacement and Figure 8 shows the time dependence of Z-axis displacement. It can apparently be seen that addition of aluminum can improve the densification of sample and accelerated the reaction speed.

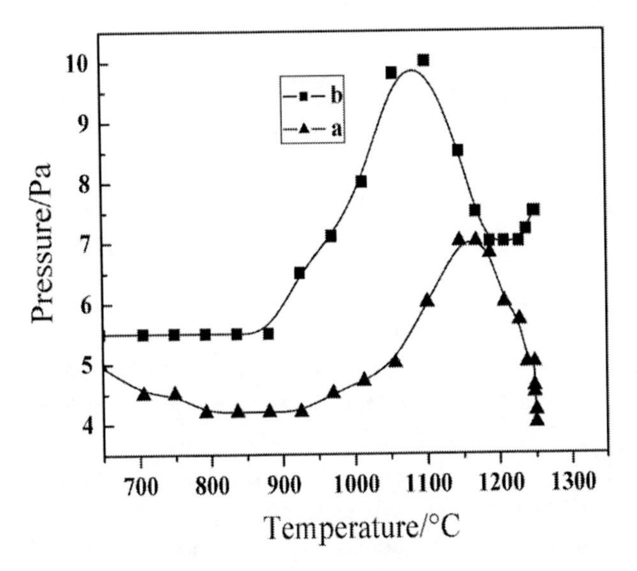

Figure 6. The temperature dependence of vacuum, (a) 2TiC/Ti/1.2Si and (b) 2TiC/Ti/Si/0.2Al.

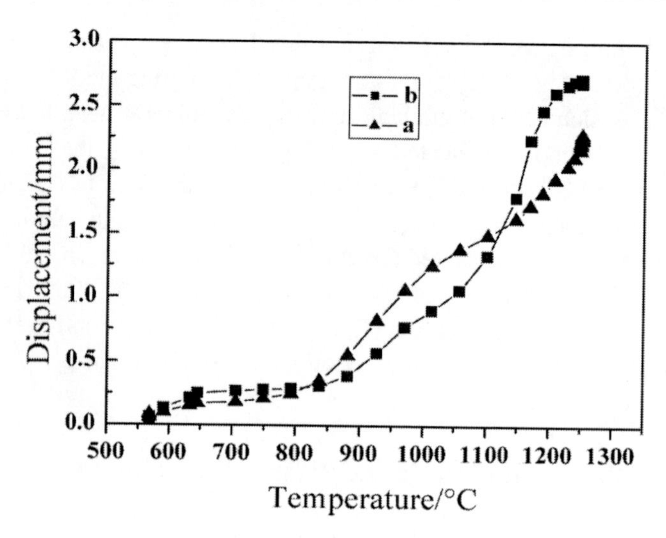

Figure 7. The temperature dependence of Z-axis displacement, (a) 2TiC/Ti/1.2Si and (b) 2TiC/Ti/Si/0.2Al.

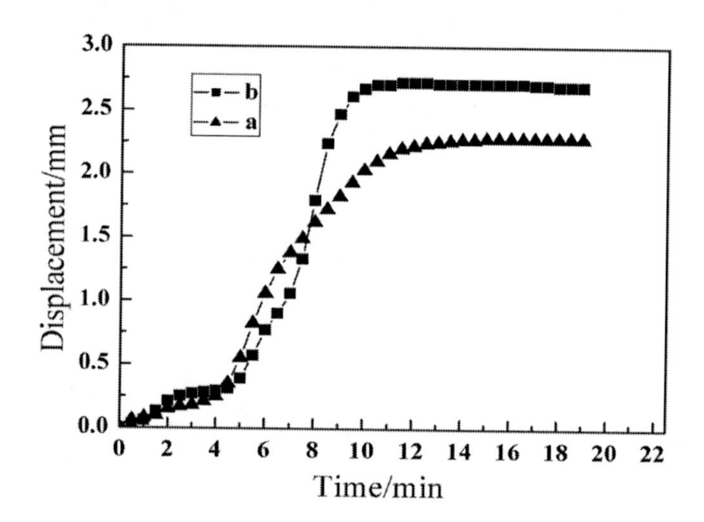

Figure 8. The time dependence of Z-axis displacement (sintered at 1250°C), (a) 2TiC/Ti/1.2Si and (b) 2TiC/Ti/Si/0.2Al.

CONCLUSIONS

The SPS method is an efficient way for the synthesis and simultaneous densification of Ti3SiC2 at a relatively low temperature in a soaking duration of minutes. Firstly, the formation of Ti_3SiC_2 is highly enhanced by the above-mentioned effects of plasma generation, surface purification, and resistance heating in the SPS process. Secondly, the synthesis of Ti_3SiC_2 is probably quickened by using the Ti/Si/2TiC powders due to the inclusion of Si to form the Ti-Si liquid phase, which may accelerate the diffusion-controlled reaction process as compared with the 3Ti/SiC/C powders. Appropriate addition of aluminum accelerates the reaction synthesis, and favors the crystal growth of Ti_3SiC_2. Fully dense, essentially single phase Ti_3SiC_2 can be fabricated by spark plasma sintering of 2TiC/1Ti/1Si/0.2Al powder mixture at the temperature of 1250~1300°C under the pressure of 30 MPa for 8min.

REFERENCES

[1] K. Matsugi, T. Hatayama, O. Yanagisawa, *J. Jpn. Inst. Metals,* 59 (1995) 740-745.

[2] Y.C. Wang, Z.Y. Fu, Study of temperature field in spark plasma sintering. *Materials Science and Engineering B,* 90 (2002) 34–37.

[3] K. Wangun, T.Y.Tan, Fu Zhengyi, Study on atom diffusion under the treatment by pulse current heating. *Materials Science and Engineering B,* 135 (2006)154–161.

[4] J.J. Sun, J.B. Li, G.L. Sun, W.G. Qu, Synthesis of dense NiZn ferrites by spark plasma sintering. *Ceram. Int.*, 28 (2002) 855–858.

[5] L. Gao, H.Z. Wang, et al., Fabrication of YAG–SiC nanocomposites by spark plasma sintering, *J. Eur. Ceram. Soc.* 22 (2002) 785–789.

[6] B.C. Mei, Y. Miyamoto, Investigation of TiAl/Ti_2AlC composites prepared by spark plasma sintering. *Mater. Chem. Phys.*, 75 (2002) 291–295.

[7] W. Pan, L.D. Chen, A. Okubo, T. Hirai, Tough multilayered α-β Si_3N_4 ceramics prepared by spark plasma sintering. *Mater. Lett.*, 49 (2001) 239–243.

[8] D.S. Perera, M. Tokita, S. Moricca, Comparative study of fabrication of Si_3N_4/SiC composites by spark plasma sintering and hot isostatic pressing. *J. Eur. Ceram. Soc.*, 18 (1998) 401–404.

[9] Z.F. Zhang, Z.M. Sun, H. Hashimoto, T. Abe, A new synthesis reaction of Ti_3SiC_2 through pulse discharge sintering Ti/SiC/TiC powder. *Scripta. Mater.*, 45 (2001) 1461-1467.

[10] Z.F. Zhang, Z.M. Sun, H. Hashimoto, T. Abe, Application of pulse discharge sintering (PDS) technique to rapid synthesis of Ti_3SiC_2 from Ti/Si/C powders. *J. Eur. Ceram. Soc.*, 22 (2002) 2957-2961.

[11] N.F. Gao, J.T. Li, D. Zhang, Y. Miyamoto, Rapid synthesis of dense Ti_3SiC_2 by spark plasma sintering. *J. Eur. Ceram. Soc.*, 22 (2002) 2365-2370.

[12] J.C. Viala, N. Peillon, F. Bosselet, J. Bouix, Phase equilibria at 1000°C in the Al-C-Si-Ti quaternary system: An experimental approach. *Mater. Sci. Eng.* A, 229 (1997) 95-113.

[13] J.F. Li, F. Sato, R.Watanabe, Synthesis of Ti_3SiC_2 polycrystals by hot-isostatic pressing of the elemental powders. *J. Mater. Sci. Lett.*, 18 (1999) 1595-1597.

[14] M.W. Barsoum, The $M_{n+1}AX_n$ phases: a new class of solids; thermodynamically stable nanolaminates. *Prog. Solid State Chem.*, 28 (2000) 201–281.

[15] Z.M. Sun, Y. Zhang, Y.C. Zhou, Synthesis of Ti_3SiC_2 powders by a solid-liquid reaction process. *Scr. Mater.*, 41 (1998) 61-66.

In: MAX Phases: Microstructure, Properties and Applications ISBN 978-1-61324-182-0
Editors: It-Meng (Jim) Low and Yanchun Zhou

Chapter 6

SYNTHESIS OF THERMAL-SPRAYED MAX PHASE COATINGS

Yao Chen[a] and Arvind Agarwal[b]

[a]School of Mechanical and Electric Engineering, Soochow University, Suzhou, P.R. China

[b]Plasma Forming Laboratory, Mechanical and Materials Engineering, Florida International University, Miami, US

ABSTRACT

The applications, potential advantages and challenges in synthesizing MAX phase coatings by thermal spray techniques are reviewed in this chapter. MAX phases are known for their unique combination of metallic and ceramic properties. MAX phases are expected to be promising coating materials for many applications such as jet engines, aircrafts, and petrochemical installations with their ability to withstand harsh service conditions including high temperature and repeated thermal shock. However, the development of thermal sprayed MAX coatings on large engineering components has been very slow when compared to bulk MAX phases. Although thermal spray techniques are scaled-up processes for coating deposition, the non-availability of commercial pure MAX phase powder feedstock appears to be the main reason for lack of such coatings. This chapter summarizes examples of thermal sprayed MAX phase coatings in the literature and also provides future research direction in the field.

1. INTRODUCTION

The $M_{n+1}AX_n$ phases or MAX phases (n = 1, 2, 3), nanolayered ternary metal carbides or nitrides, where M is an early transition metal, A is an A-group element, and X is C and/or N [1,2], were first discovered in the late 1960s [3]. However, it was only in 1990s when Barsoum and EI-Raghy [4] made the impressive achievements by synthesizing pure Ti_3SiC_2 phase with unique mechanical and physical properties. Since then much efforts have been devoted to the deeper understanding of fundamentals and discovery of new MAX phases, crystal structure, nucleation, growth mechanisms, mechanical and physical properties. Over 50 MAX phases have so far been discovered, such as M_2AX phase (211), M_3AX_2 phase (312), M_4AX_3 phase (413) and their corresponding solid solutions [5-9]. MAX phases are characterized by the unique combination of metallic and ceramic properties. They usually exhibit high specific stiffness, superb machinability, good damage tolerance, good corrosion resistance, excellent thermal shock resistance, and excellent thermal and electrical conductor. It is also observed that MAX phases undergo fully reversible dislocation-based deformation due to kink and shear band deformation as well as delamination within grains [10]. Therefore, MAX phases are candidate materials for high-temperature structural applications, protective coatings to resist corrosion, oxidation, thermal shock and abrasion, electrical contacts, tunable damping films for microelectromechanial systems [10]. This chapter is a critical review of the thermal sprayed MAX phase coatings from the material science perspective. The details on bulk synthesis techniques, mechanical and physical properties of MAX phases can be found in other chapters of this book.

2. Significance of Development of MAX-Phase Coatings

There are several components in jet engines, automobiles, aircrafts, and petrochemical installations that experience extreme service conditions including high temperature and repeated thermal shock. MAX phases are, therefore, considered the promising coating materials for these applications [10-11]. Much work on MAX phase synthesis have been done through different processing techniques such as self-propagation high-temperature synthesis (SHS) [12-15], arc-melting and post annealing [16], solid-state reactive sintering [17-18], in-situ hot pressing/solid-liquid reaction synthesis [19-22], hot pressing [23-29], hot isostatic pressing (HIP) [4], spark plasma sintering (SPS) [30-34], pulse discharge sintering (PDS) [35-38], pressureless sintering [5, 39-41] and mechanical alloying (MA) [42]. Most of these techniques fabricate MAX phase into a bulk structure but not as coatings. There are very few studies on synthesis of MAX phase as coatings when compared to bulk structures.

Most efforts on MAX phase coatings are restricted to the synthesis using thin-film processing methods such as Physical Vapor Deposition (PVD) [43-52], Chemical Vapor Deposition (CVD) [53-58], and Solid-Sate Reactions [59-61]. Recently, Eklund *et al.* [10] has published a comprehensive review article on the synthesis of MAX phase films. For most PVD and CVD methods to synthesize MAX phase films, the substrate temperature is usually required to be in the range of 800-1000℃, making the temperature-sensitive substrate materials excluded in these processing techniques [10]. In addition to relatively slow deposition rate, most PVD and CVD procedures also require high vacuum level (10^{-7} Torr), and the vacuum-chamber dimensions restrict the development of MAX phase deposition on the large-sized engineering components for aerospace and automobile applications. Therefore, for PVD and CVD methods, homogeneous deposition of MAX phase on the surface of large-sized components at lower deposition temperature are two major challenges [62]. Hence, there is a need to develop other coating techniques which can deposit or synthesize MAX phase coatings on large engineering components in the non-vacuum environment at a much faster deposition rate. Thermal spray methods are ideally suited for such purpose.

3. Brief Review of Thermal Spray Techniques

Thermal spray techniques are well-established line-of-sight surface coating processing techniques, which have been widely employed on the deposition of a wide rage of thick material coatings (thickness ~ 50 μm – few inches) in the fields of automobile, aerospace, gas turbine technology, electronics, telecommunications, air and space navigation, and medical equipment for thermal, wear, corrosion, and oxidation protection [63, 64]. The fundamental of the thermal spray techniques is that the metallic or non-metallic powders or their mixture can be fully melted and/or partially melted in a heat source, and then are accelerated to high velocity to impinge upon the surface-prepared substrate, following the experience of flattening, deformation and rapid solidification to form the splats [64-66]. As a consequence, the deposit is built-up by successive impingement and interbonding among the splats. The final microstructure and properties of the thermal sprayed coatings are strongly controlled by the individual splat morphology, the adhesion between the splats and substrate as well as between these individual splats. A variety of processing parameters such as average surface temperature and velocity of in-fight powder particles, plasma power, primary and secondary gas flow rates, powder feed rate, stand-off distance, substrate material and their surface characteristics should be optimized to obtain high performance coatings [63].

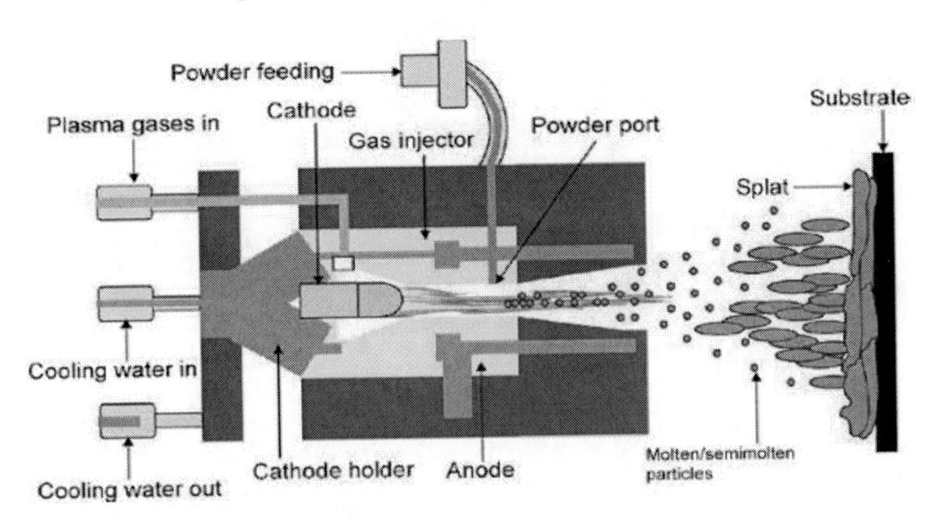

Figure 1. Schematic of plasma spray operation [63].

Depending upon the variety of the feedstock materials and heat sources, thermal spray techniques are mainly categorized into [64]: (1) Plasma spray (PS), where the heat source is a plasma jet produced by high-voltage electrodes (as shown in Figure 1), (2) High velocity oxyfuel spray (HVOF), where combustion of fuel gas produces the required heat energy (as indicated in Figure 2), (3) Wire arc spray, as depicted in Figure 3, where the arc is produced between two consumable electrodes, and (4) Cold spray, as indicated in Figure 4, where the coating deposition occurs primarily due to large plastic deformation in the splats at temperature much below the melting point. It is necessary to note that rapidly solidified microstructures obtained in PS and HVOF process make them most promising for tailoring the mechanical properties, especially in achieving excellent grain refinement up to nanosize grains. Moreover, recent development of PS and HVOF in the near-net-shape forming has been triggered by increasing interest in fabrication of components with more complicated shapes [69].

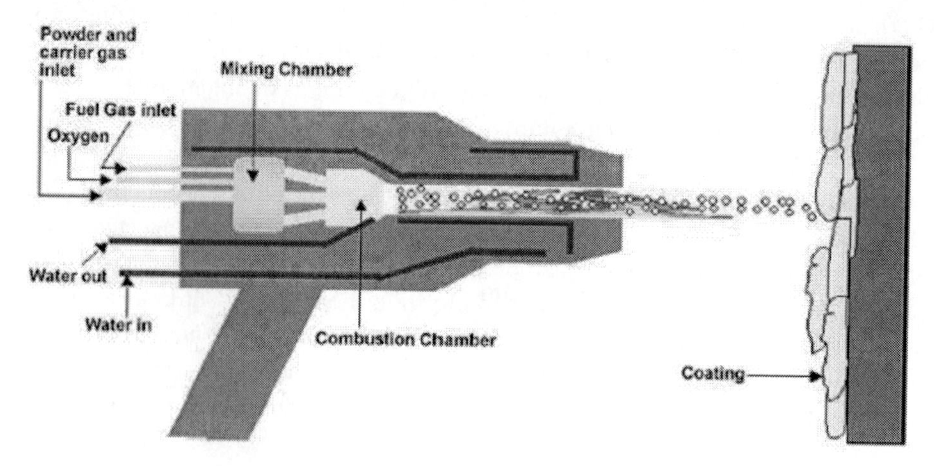

Figure 2. Schematic of high velocity oxyfuel spraying [63].

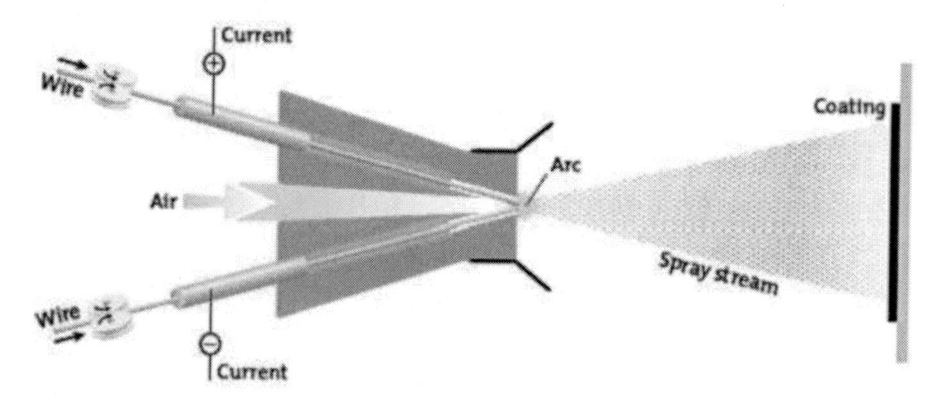

Figure 3. Schematic of wire arc spray [67].

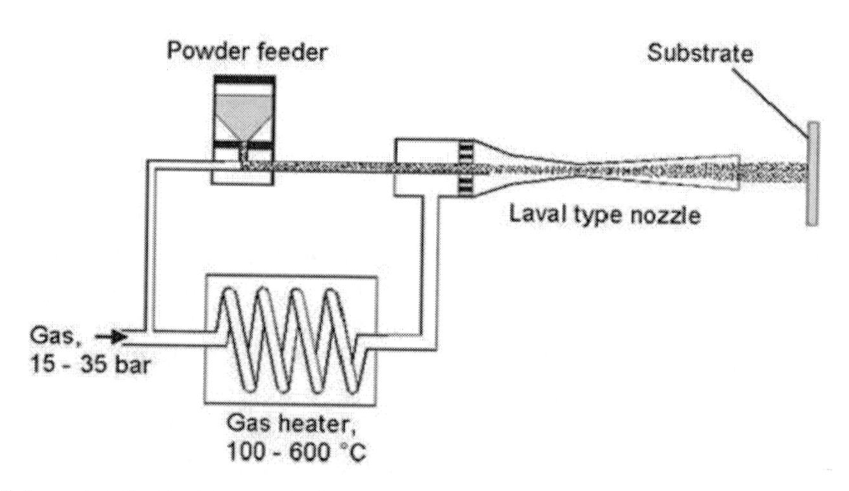

Figure 4. Schematic of typical cold spray process [68].

4. Challenges in Fabrication of MAX Coatings Using Thermal Spray Techniques

The pioneering work by Barsoum and Knight on thermal sprayed MAX phase coatings comprising of ceramic compounds based on the general formula of M_2AX and M_3AX_2 for improved corrosion, oxidation and wear resistance was a patent issued in 2001 [70]. It was stated in the patent that the feedstock ceramic powders based on M_2AX and M_3AX_2 for thermal spray were produced using conventional means, i.e., mechanical crushing of bulk MAX phase samples [70]. It is well known the temperature within the thermal spray gun is usually above 2000 - 2500°C, which is higher than the decomposition temperature of the MAX phases (850 - 2300°C). Therefore, it is of significance to control the residence time of in-flight MAX-phase feedstock particles in the plasma plume. Unfortunately, little information on the thermal sprayed MAX phase coatings is available in the open literature since then. *The sluggish development in the thermal sprayed MAX phase coatings may be attributed to the lag in the MAX phase powder feedstock manufacturing which is critical and mandatory for thermal spray processes.* It is well understood that the preferred particle shape for thermal spray should be spherical in order to have good flowability and to reach an optimal condition of melting and continuous spraying. The fused and crushed particles produced using mechanical crushing of bulk materials generally exhibit angular and irregular morphology, may lead to limited flowability and subsequently, degradation of properties of thermal sprayed coatings. Is there an alternative route to the deposition of thick MAX phase coatings without using fused MAX phase particles? It is interesting and significant to explore the possibility of in-situ reaction synthesis of MAX phase coating using thermal spraying of powder mixture, which is usually used as precursor materials in synthesizing bulk MAX phases. In this case, we would be able to clearly distinguish the fundamental difference between majority of the synthesis techniques of bulk MAX phases and the in–situ reaction synthesis of MAX phase coatings through thermal spray. *The former processing methods are relatively close to the thermodynamics equilibrium [10], while the latter processing is far from the thermodynamic equilibrium.* This difference ascribes to the intrinsic characteristics of extremely rapid heating rate of these in-flight particles within the thermal spray gun and the extremely cooling rate experienced by the fully melted and/or partially melted particles when they impact upon the substrate or previously deposited layers. Some of the examples of thermal sprayed MAX phase coatings are described in the next section.

5. Examples of Thermal Sprayed MAX Phase Coatings

5.1. MAX Phase Coatings through HVOF

As compared with plasma jet, HVOF possesses relatively lower particle temperature and higher in-flight particle velocity, which contribute to lower surface temperature and shorter residence time of the feedstock powders, and hence decrease in the degree of melting and oxidation. Therefore, HVOF is a promising technique with an advantage of reducing the oxidation and dissociation of MAX phase feedstock particles in comparison with plasma spraying [71]. Frodelius *et al.* [71] fabricated Ti_2AlC coating with a thickness of ~ 130 μm via HVOF spraying of crushed Ti_2AlC powders in the shapes of spherical agglomerates and/or flake-like from MAXTHAL 211® (Ti_2AlC) bulk material. The powder feedstock had an average particle size of ~ 38 μm (hereafter referred to S) and ~ 56 μm (hereafter referred to L), as shown in Figure 5. It is evident from the X-ray diffraction results of Ti_2AlC powders, as shown in Figure 6, that phase constituents of the powders are mainly Ti_2AlC phase with a small amount of Ti_3AlC_2. Figures 7 and 8 show X-ray diffraction results of the coatings sprayed at different total gas flows using powder S and L, respectively. The information deciphered from the XRD results are as follows: (1) in addition to the majority phase of Ti_2AlC and Ti_3AlC_2, all the coatings contain the impurity phases such as TiC and Al-Ti alloy (Al_xTi_x), (2) the relative amount of Ti_3AlC_2 in the sprayed coating is higher than in the feedstock powders, indicating that high temperature during the HVOF process induced the phase transformation of Ti_2AlC powders, (3) ratio of Ti_2AlC and TiC increases with the decreasing total gas flow, because lower gas flow leads to lower power of the flame with less heating of the powders, and (4) higher amount of MAX phases can be retained in the coatings sprayed with L powders under the given spraying condition. Figure 9 (a) shows the cross-section of coatings sprayed with powder S at the lowest gas flow whereas Figure 9 (b) shows the coating deposited with powder L at the highest gas flow. It is found that the coating sprayed with powder S has a dense microstructure, while the coating sprayed with powder L has a higher density of cracks and some delamination at the interface.

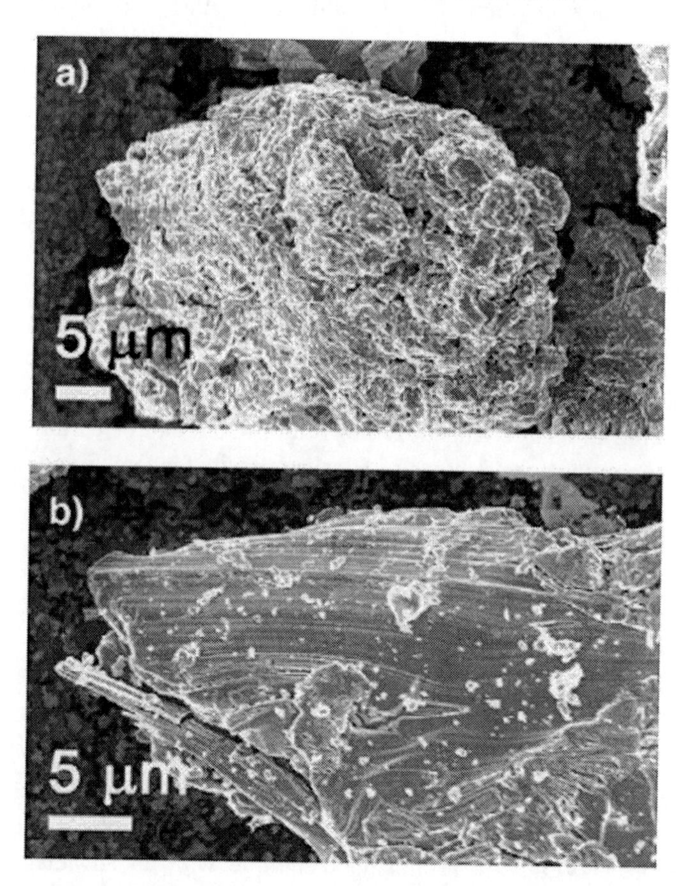

Figure 5. SEM images of the Ti_2AlC powder S, showing (a) conglomerate and (b) flake-like grain. [71].

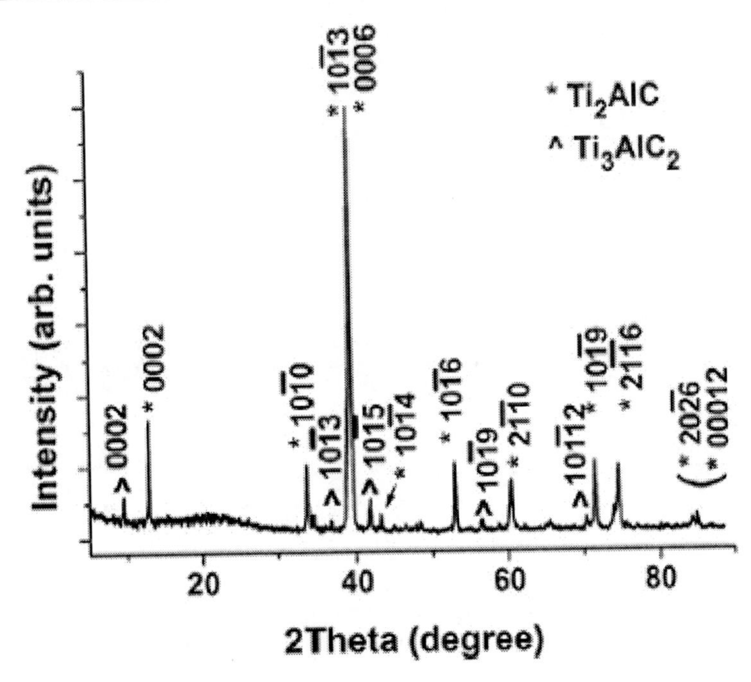

Figure 6. X-ray diffractogram of Ti_2AlC powder L. Reference positions of Ti_2AlC and Ti_3AlC_2 reflections are indicated [71].

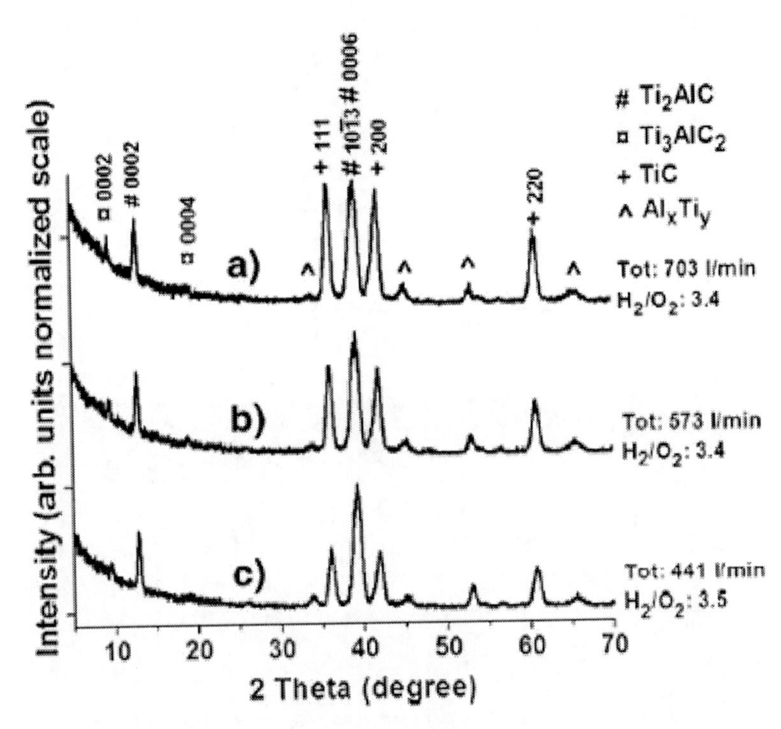

Figure 7. X-ray diffractograms from coating sprayed with powder S. The total gas flow, the ratio between hydrogen and oxygen gas flows, and the phase identification are indicated for the diffractograms (a) to (c) [71].

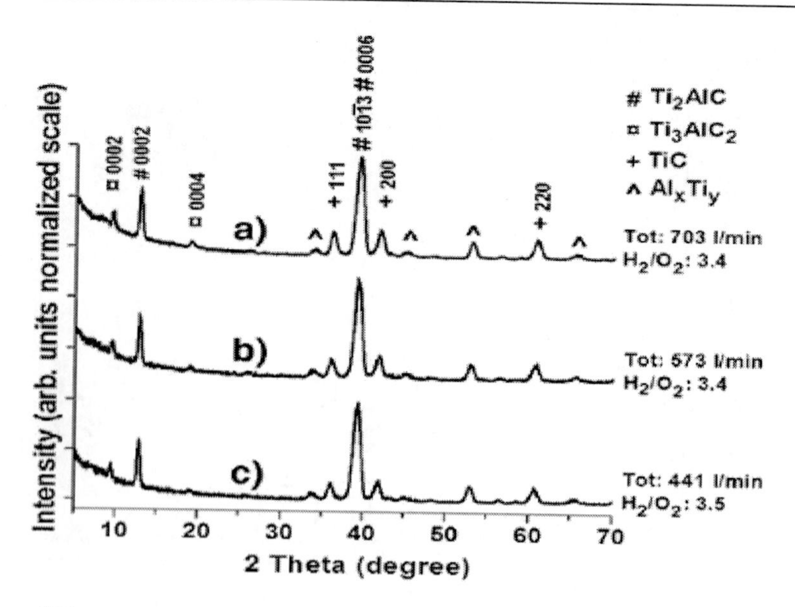

Figure 8. X-ray diffractograms from coatings sprayed with powder L. The total gas flow, the ratio between hydrogen and oxygen gas flows, and the phase identification are indicated for the diffractograms (a) to (c) [71].

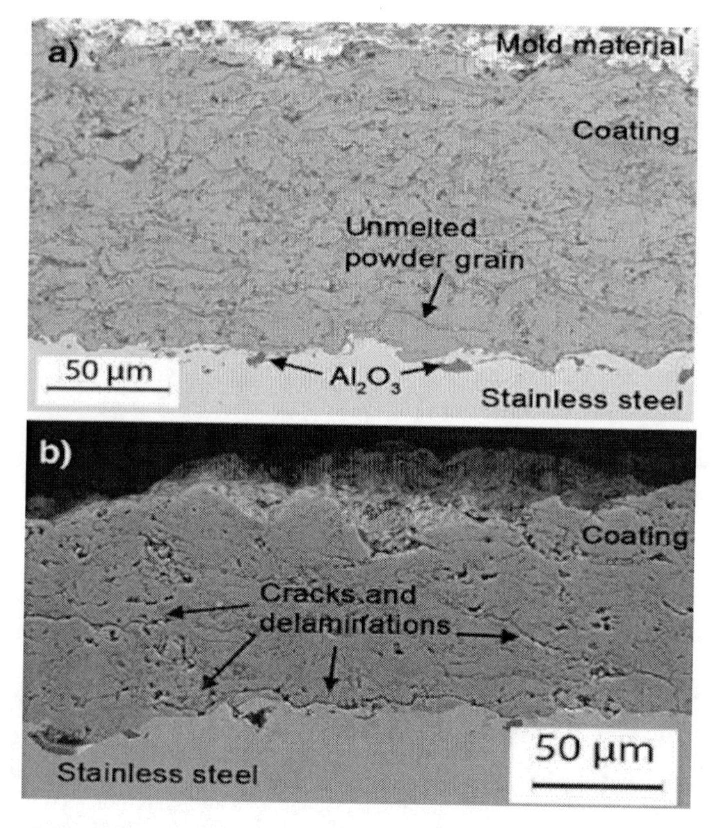

Figure 9. Cross-sectional SEM images of coatings sprayed with a) powders S at the lowest gas flow and b) powders L at the highest gas flow. [71].

Frodelius *et al.* [71] proposed a model considering the process of diffusion, melting and solidification of the feedstock powders during HVOF spraying, as shown in Figure 10. One of phase transformation mechanism during HVOF spraying of Ti_2AlC feedstock particles is correlated to the outward-diffusion of Al (i.e., Al diffuses from the core to the outer shell of the particles) taking place preferentially along the basal planes of the Ti_2AlC due to the most open and least bonded pathway for Al. The continuous outward-diffusion of Al leads to the transformation of Ti_2AlC to Ti_3AlC_2. The depletion of Al also results in the formation of TiC and Al_xTi_x, as depicted in Figure 10a. The other mechanism is related to the rapid solidification experienced by partially melted particles impacting upon the substrate, as shown in Figure 10b. However, one possibility, which can also lead to the occurrence of phase transformation during HVOF spraying of fused Ti_2AlC powder, is neglected in this research, i.e., decomposition of Ti_2AlC powders under high-temperature conditions would also contribute to phase transformation.

It is reported that the MAX phases do not melt congruently, but decompose according to the following reactions [10]:

$$M_{n+1}AX_n(S) \rightarrow M_{n+1}X_n(S) + A(g)$$
$$M_{n+1}AX_n(S) \rightarrow M_{n+1}X_n(S) + A(l)$$

Also, the decomposition of MAX phases is initiated by loss of the "A" element. The decomposition temperatures of MAX phases are strongly affected by the environment and impurities such as C vacancies and O interstitials, and therefore the decomposition temperatures vary over a wide range; from 850 -1800°C for bulk Cr_2GaN and, above 2300°C for Ti_3SiC_2 [10]. It is reported [72] that the surface temperature of the in-flight Ti_2AlC particles (~2111°C) would reach its decomposition temperature and subsequently induce the phase transformation.

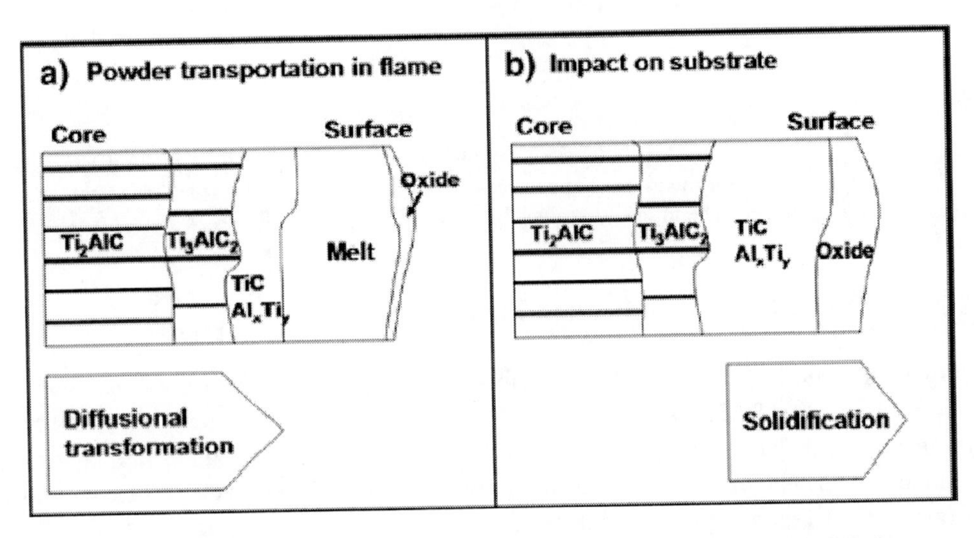

Figure 10. Illustration of the phase transformation occurring in a Ti_2AlC grain by (a) diffusion processes during powder transportation through the flame and (b) solidification upon impact on the substrate. Mechanical effects are neglected [71].

X-ray pole figure was also measured for the case of coatings sprayed with powder L, as indicated in Figure 11. It is evident that the Ti_2AlC grains have a preferred orientation with (0001) basal planes aligned with their normal perpendicular to the substrate surface due to smearing during energetic impact of the textured Ti_2AlC powder, which has a dominant ductility along the basal planes.

The hardness of the coating sprayed with powder S at the highest total gas flow was 5.3±1.1GPa, whereas the hardness of the coating sprayed with powder S at the lowest total gas flow decreased to 3.6±0.9 GPa. The lower hardness corresponds to the decrease in the amount of TiC in the sprayed coating. The measured hardness of HVOF Ti_2AlC coating is lower than that of Ti_2AlC bulk material (~5.5 GPa). Porosity and weak bonding between splats in the coating contribute largely to the decrease in hardness. Bending test on the coating sprayed with powder S indicated that coatings were well adherent to the substrate with cracks perpendicular to the bending direction and free from flaking except at the edges.

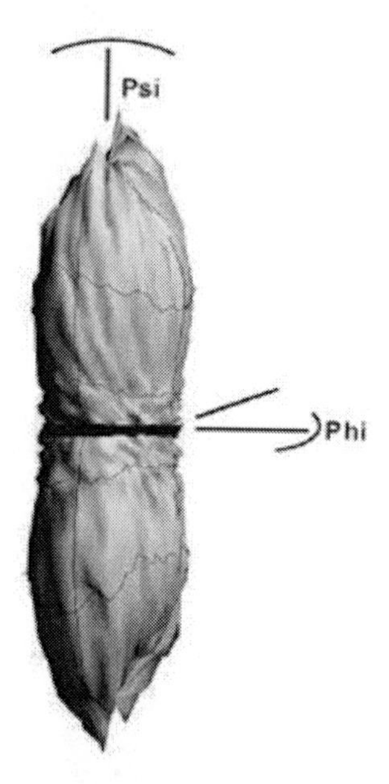

Figure 11. The pole figure of Ti_2AlC 0002 for a coating sprayed with powder L at the highest total gas flow. The scanning angles for the given 2θ position are 0-360° (phi) and 0-90° (psi). The coating is positioned in the phi-plane. The intensity scale is linear [71].

5.2. MAX Phase Coatings Through Plasma Spray

Agarwal's group [11] employed tribochemically mixed Ti-SiC-C powder with composition of 74 wt.% Ti, 20 wt% SiC and 6 wt% graphite in the particle size of 40-50 μm as the feedstock material for the synthesis of Ti_3SiC_2 protective coatings using plasma spray technique. Plasma spraying was carried out at two different power levels, using SG-100 gun (Praxair Surface Technologies, Indianapolis, IN). The processing parameters are listed in the Table 1. The temperature and velocity of the powder particles exiting from the plasma plume were measured using in-flight diagnostic Accuraspray G3 optical sensor (Tecnar Automation Ltd., QC, Canada).

Table 1. Plasma spraying parameters used for synthesizing Ti_3SiC_2 coatings [11]

Sample ID	Power (kW)	Primary gas, Ar (slm*)	Secondary gas, He (slm*)	Carrier gas, Ar (slm*)	Standoff distance (mm)
TSC24	24	32.1	59.5	19.8	101.6
TSC32	32	32.1	59.5	19.8	101.6

* Standard liters per minute.

Figure 12a shows the morphology of Ti-SiC-C feedstock powder. Corresponding elemental X-ray maps of powder particle are shown in Figure 12(b-d). It is clearly seen that tribochemically blended powder is not homogeneously mixed and may lead to chemical segregation in the plasma sprayed coatings. Meanwhile, XRD results of the powder feedstock (not included here) indicated the presence of α-Ti, SiC and graphite phases, and no impurity or oxide was observed.

Figure 13 shows the XRD results of plasma sprayed TSC24 and TSC32 coatings. Ti_3SiC_2 phase is observed in both coatings, proving that MAX phase can be synthesized during plasma spraying through the reactions between the precursor powders under high-temperature processing condition. In addition to Ti_3SiC_2 phase, Ti_5Si_3, TiC, TiO, α-Ti and β-Ti are also found in the plasma sprayed coatings. The volume fraction of different phases in as-sprayed coating is quantified in Table 2, using the concept that volume fraction of a phase corresponds to the ratio of diffracted intensity of this phase to that of all phases. The temperature and velocity of particles exiting from the plasma gun was measured using in-flight diagnostic sensor (Table 3), implies that the average surface temperature of in-flight particles increases with plasma power, whereas average velocity of these in-flight particles is almost constant.

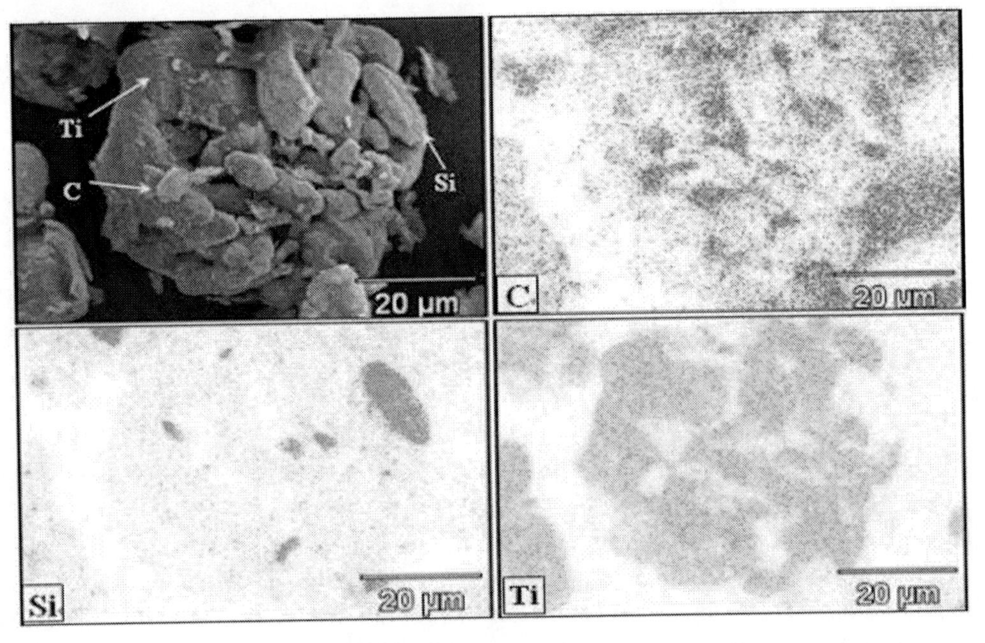

Figure 12. (a) SEM micrograph of the blended Ti-SiC-C powder and corresponding X-ray maps showing (b) Ti, (c) Si and (d) distribution [11].

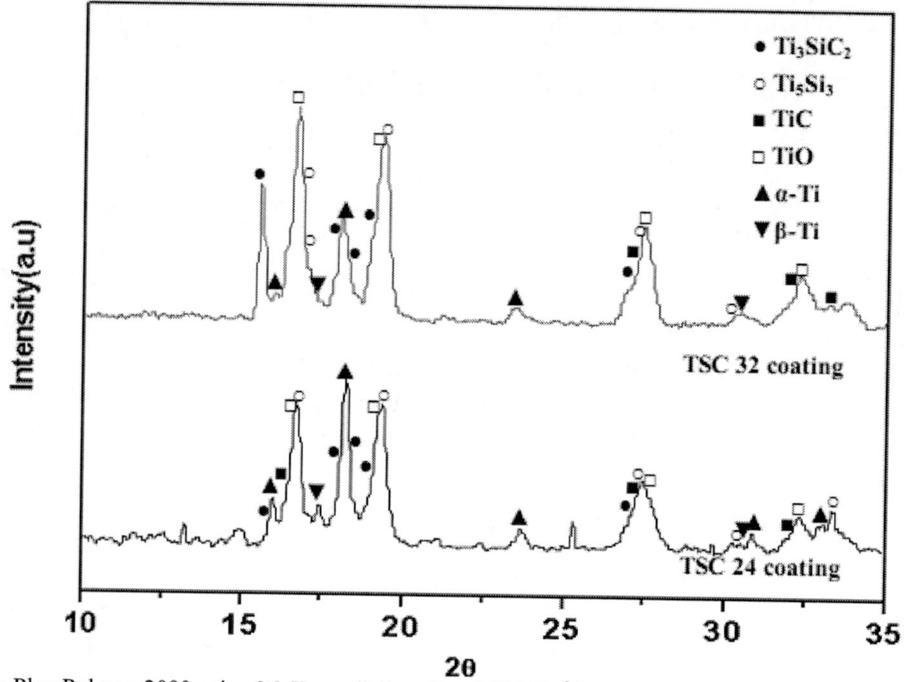

(DiffracPlus Release 2000 using MoKα radiation (λ = 0.70930 Å)).

Figure 13. XRD patterns of plasma sprayed TSC 24 and TSC 32 coatings [11].

Table 2. Relative weight percentages of the phases present in the coatings [11]

Sample ID	Ti_3SiC_2	TiC	Ti_5Si_3	TiO	α - Ti	β – Ti
TSC24	15.75	11.11	24	18.51	16	7.1
TSC32	19.06	14.28	18.25	23.01	8.9	4

Table 3. In-flight particle diagnostic data for plasma sprayed coatings [11]

Sample ID	Temperature (K)	Velocity (m/s)	Residence Time (s)	Cooling Rate (K/s)
TSC24	2898	114	8.91×10^{-4}	3.25×10^{6}
TSC32	3052	116	8.62×10^{-4}	4.39×10^{6}

Figure 14 shows the high magnification view of the coatings along with the EDS spectrum. It is clearly seen from the EDS results (Figure 14c and Figure 14d) that the locations 1 and 2 in Figure 14a correspond to TiC and Ti rich-Si solid solution, respectively. From the growth morphology, the plate-like phase is identified as Ti_3SiC_2 phase due to its large ratio of lattice parameter (c/a) [25]. The platelets are too small to be analyzed by EDS. The average platelet length of Ti_3SiC_2 phase is 800 nm with an aspect ratio of 4-5. Similar platelet morphology of Ti_3SiC_2 phase has been observed by other researchers [25]. Also, SEM micrographs exhibited microstructure with multiple characteristics, i.e. Ti_3SiC_2 platelets are surrounded by TiC, followed by titanium dendrites, and the outer layer is Ti_{rich}-Si solid solution. The absence of Ti_5Si_3 phase, as detected by XRD results, is attributed to the fact that silicides are easily dissolved by the HF etchant [25] used in this study.

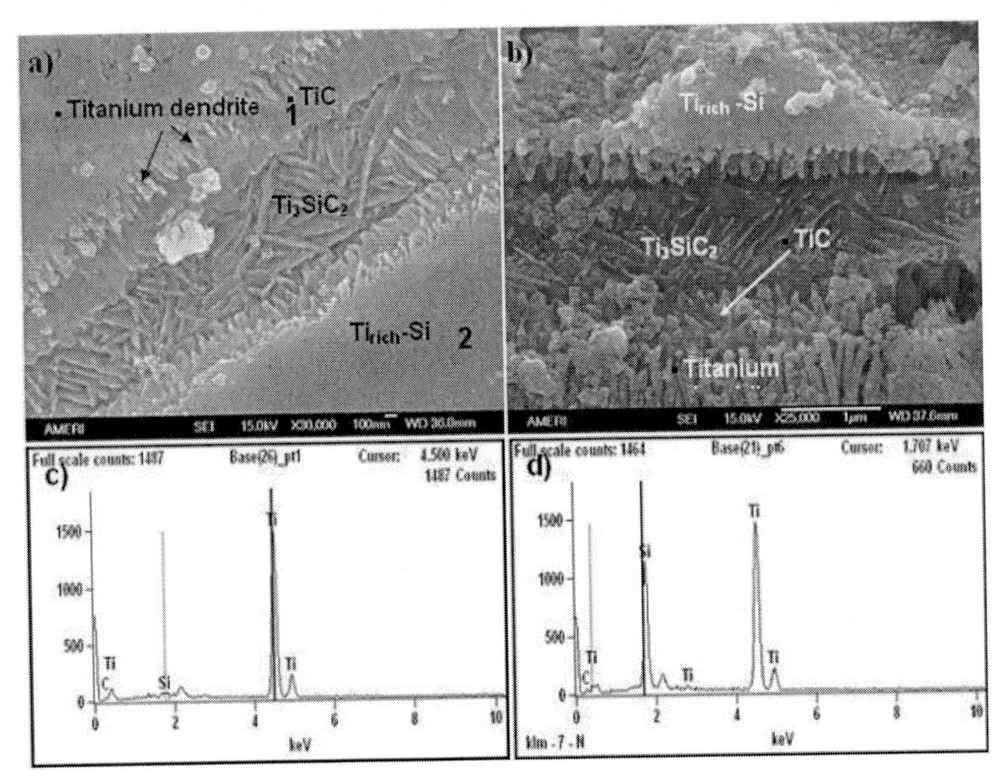

Figure 14. SEM micrographs showing the multiple characteristics of microstructure of plasma sprayed coating (a) TSC24 and (b) TSC32. EDS results for TSC24 coating from location 1 and 2 are shown in (c) and (d) respectively [11].

The most significant result of this research is that Ti_3SiC_2 phase can be reactively synthesized in a very short time (~millisecond) using plasma spray technique. Therefore it is necessary to understand the reaction mechanism of Ti_3SiC_2 formation using Ti-SiC-C powder feedstock under rapid processing conditions of plasma spraying. The reaction mechanism has been elucidated based on the diffusion calculations and microstructural and phase characterization.

The diffusivity (D) and diffusion distance (x) in Ti-Si-C system has been estimated using equations 1 and 2, respectively.

$$D = D_0 \exp(-Q / RT) \quad (1)$$

$$x \approx \sqrt{Dt} \quad (2)$$

in which D_0 is diffusion constant, Q is the activation energy, R is the universal gas constant, T is the temperature (in degrees Kelvin) and t is the diffusion time. The diffusion constants (D_0) and activation energy (Q) values at different temperature ranges are obtained from the literature [73]. Based on the experimentally obtained residence time (t) of powder particles in the plasma plume (Table 3), diffusion distance (x) for different species in Ti-Si-C system is estimated using equations (1) and (2), respectively. The results for different diffusing species in Ti-SiC-C system are listed in Table. 4. It is strongly indicated that carbon diffuses rapidly in Si than that in Ti. Moreover, the diffusivity of carbon in SiC is very low. Gibbs free energies of formation (G_f) of different phases in Ti-Si-C system were also calculated [11], as shown in Figure 15, in which Gibbs free energy of TiC formation is more negative in the temperature ranging from 500 to 3000 K. Therefore, it is safe to assume that carbon reacts favorably with Ti at the start of the reaction. The reaction mechanism of Ti_3SiC_2 synthesis by plasma spraying using Ti-SiC-C powder feedstock is proposed as following:

Firstly, two simultaneous reactions including (i) SiC sublimation: SiC → Si (l/g) + C (s) and (ii) TiC formation: Ti + C → TiC_x occur favorably in the plasma synthesis of Ti_3SiC_2. SiC sublimation occurs because the average temperature of in-flight particles varies between the range of 2898-3052K (as listed in Table 3), and SiC sublimation was reported to occur around 2540-3150K [74]. The evidence of SiC sublimation is also corroborated by TEM micrograph (Figure 16), in which particles are very fine with a diameter of 20-40 nm. Corresponding SAED pattern (Figure 16) illustrates these nanosize powders as TiC and SiC. It should be noted that the original particle size of SiC is ~ 40-50 μm. The significant decrease in particle size (from 40 μm to 40 nm) sufficiently implies that SiC sublimes and vapors of Si condense to recombine with the free carbon to form nanosize SiC powder. The existence of nanosize TiC particles indicates that Ti reacts with carbon. The Gibbs free energy of TiC formation is -172 kJ/mol, as shown in Figure 15.

Secondly, Ti-Si solid solution forms in plasma sprayed coatings, as shown from the SEM micrograph and EDS analysis in Figure 12. The formation of Ti-Si solid solution is ascribed to the combination of extremely high heating rate and rapid solidification along with the plasma spray, because Lis *et al.* [75] reported that metallic reactants (Ti, Si) can rapidly melt and form a Ti-Si liquid phase under the condition of high heating rate of the reactants (500 K/s). It is necessary to note that during first step of the reaction sublimation of SiC will lead to depletion of Si content due to escape as vapor phase at 3000K. Hence, Ti_{rich}-Si liquid phase forms during second step as the following reaction Ti (l) + Si (l, depletion) → Ti_{rich}-Si (l).

Table 4. Diffusion distance of different species in Ti-Si-C system

System	Temperature Range (K)	D_o (cm²/sec)	Q (kJ/mol)	Computed diffusivity (cm²/sec)	Computed diffusion distance (μm)
C in Si	1323- 1673	1.9	12.98	0.583 – 0.747	220-249
C in Ti	1173- 1473	1.3×10^{3}	348.6	3.88×10^{-13} - 5.6×10^{-10}	1.8×10^{-4} - 6.8×10^{-3}
	1473- 1673	77.8	340.2	$8.6 \times 10^{-13} – 1 \times 10^{-9}$	$2.6 \times 10^{-4} – 0.9 \times 10^{-2}$
	1673- 1873	3.02×10^{-3}	84	$7.2 \times 10^{-6} – 1.3 \times 10^{-5}$	0.773 – 1.039
C in SiC	2123- 2453	8.62×10^{5}	713.7	$2.3 \times 10^{-12} – 5.4 \times 10^{-10}$	$4.4 \times 10^{-4} – 6.7 \times 10^{-3}$
Ti in Si	1223- 1473	10^{-8}	1.79	8.3×10^{-9} - 8.6×10^{-9}	$2.6 \times 10^{-2} – 2.7 \times 10^{-2}$

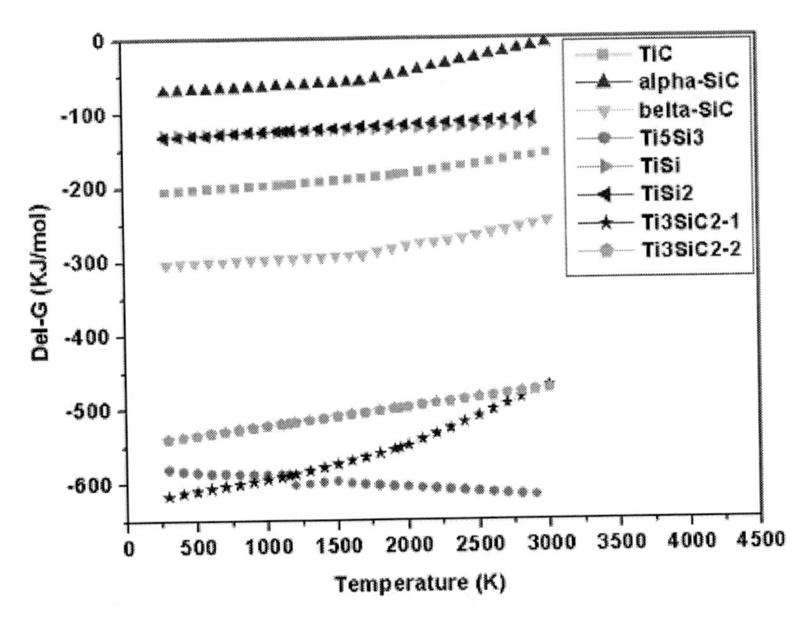

Figure 15. Gibbs free energy of formation of different phases in Ti-Si-C system as a function of temperature [11].

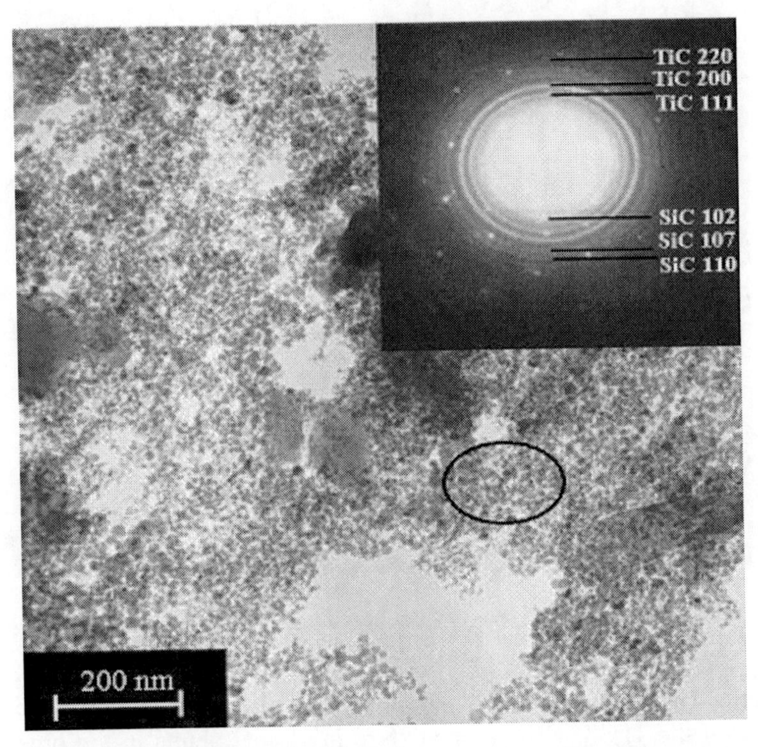

Figure 16. TEM micrograph showing nanosize particles of TSC24 coating and corresponding selected area electron diffraction (SAED) pattern [11].

With further cooling of particles exiting from the plasma plume, TiC reacts with Ti_{rich}-Si liquid to form Ti_3SiC_2, i.e., TiC_x + Ti-Si (l) → Ti_3SiC_2. The reaction temperature range for the formation of Ti_3SiC_2 is around 1500–1900K, which is supported by previous research done by Sun *et al.* [40] and Riley *et al.* [76] using pulse discharge sintering and SHS process respectively, with precursor powders of Ti-SiC-C. Meanwhile, it is evidenced both from Figure 14a, Figure 14b and Figure 17 that Ti_3SiC_2 always coexists with TiC, which sufficiently implies either incomplete reaction of TiC_x + Ti-Si (l) → Ti_3SiC_2 under the condition of rapid cooling rate or the epitaxial nucleation of Ti_3SiC_2 on the crystallographic plane of TiC, in which the latter possibility is proven by the TiC seed layer for the growth of Ti_3SiC_2 film [10]. Based on the binary Ti-Si phase diagram, eutectic reaction of L (Ti-11atom%Si) → Ti-5atom%Si (solid solution) + Ti_5Si_3 occurs at approximately 1609K, which is within the temperature range of Ti_3SiC_2 (1500–1900K). The eutectic reaction dissipates some Ti-Si solid phase, leading to depletion in Ti-Si liquid phase in the peritectic reaction for the formation of Ti_3SiC_2. Hence, Ti_3SiC_2 phase is often found to grow along with TiC phase, as indicated in Figure 17. Also, the eutectic products of Ti_5Si_3 and titanium solid solution are supported by XRD patterns (Figure 13) and EDS result (Figure 14d).

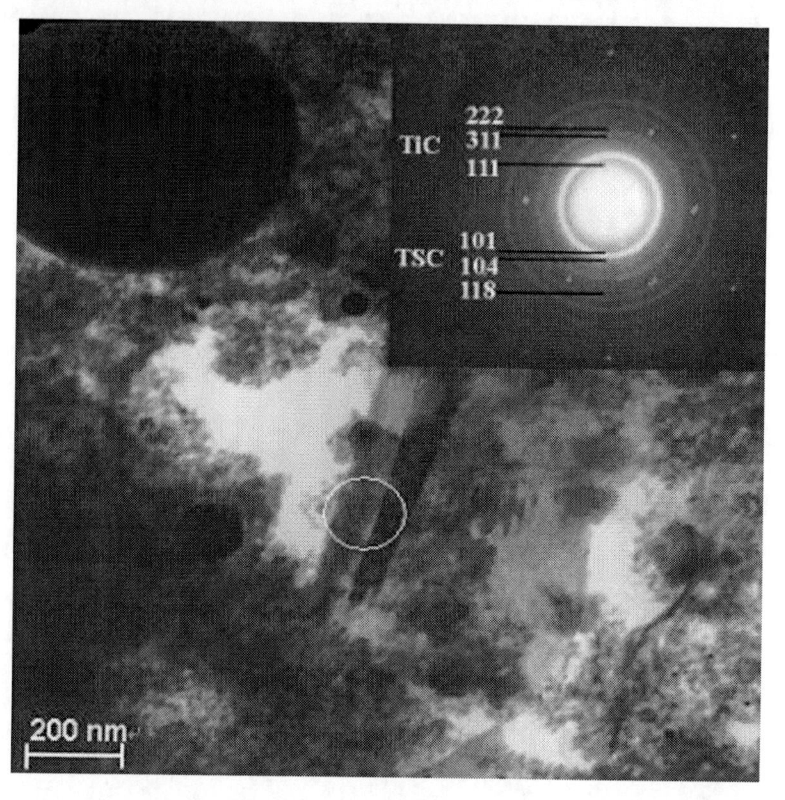

Figure 17. TEM bright field image showing plate-like Ti_3SiC_2 phase and the corresponding SAED ring patterns in sample TSC32 (TSC – Ti_3SiC_2) [11].

It is highlighted that only 15-19 vol.% Ti_3SiC_2 phase was formed due to the short residence time of the Ti-SiC-C powder mixture in plasma plume in the order of ~ 10^{-4} s, which greatly constrains the nucleation and growth of Ti_3SiC_2 phase. There is a remaining issue with the synthesis of MAX coatings using thermal spray of non-MAX phase powders to achieve sufficiently phase-pure MAX phase coatings.

6. OUTLOOK

Although remarkable achievements have been obtained focusing on the synthesis of pure MAX phases and understanding of their physical, mechanical and chemical properties in the last few years, the development of thermal sprayed MAX coatings on large engineering components for improving corrosion and oxidation properties has been very slow since the first patent related to the thermal spraying of 211 or 312 phases in 2001. From the viewpoint of engineering applications, therefore, much effort should be put into the synthesis of pure *MAX* phase coatings in the future.

Indeed, the possibility of synthesizing pure *MAX* phase coatings on large engineering components through thermal spray techniques faces two problems for the materials community; the first is development of pure *MAX* phase powder feedstock for thermal spray. Secondly, optimization of thermal spray processing window is necessary when powder mixture for synthesis of bulk *MAX* phases is thermally sprayed. From the results obtained by Agarwal's group [11], the residence time of the Ti-SiC-C powder mixture in plasma plume is a very important factor to achieve higher volume fraction of Ti_3SiC_2 in the as-sprayed coating. In other words, more sufficient reaction between these powders under high-temperature processing conditions is the key to fabricating phase-pure *MAX* phase coatings in this case. Hence, the plasma processing variables including higher plasma power (i.e. increase in the surface temperature of these in-flight particles), low carrier gas flow rate (i.e. increase in the residence time of particles in the plasma plume) and low powder feed rate (i.e. increase in the surface temperature of these particles) should be carefully optimized. Additionally, powder pretreatment such as spray drying would also assist in homogenous reaction throughout the coating, which might contribute to obtain higher volume of *MAX* phase in the thermal sprayed coatings. Processing parameters must be controlled to prevent overheating of in-flight particles to prevent the phase transformation of the *MAX* phase, whereas some degree of particle melting is necessary to achieve a sufficient level of particle adhesion and cohesion.

The application of Ti_3SiC_2 as a structural component has also been restricted due to its low hardness (4 GPa) and poor wear resistance [77]. It has been proven that the hardness of Ti_3SiC_2 can be significantly increased by incorporating TiC [78], Al_2O_3 [79] and/or SiC [80] to form composite materials. Among these reinforcing phases, TiC is one of the most promising high-temperature structural materials because of its unique combination of high modulus, high melting point (3160°C), high hardness (28 GPa), good erosion resistance, good thermodynamic stability [81] and small thermal expansion mismatch [79] with Ti_3SiC_2. Additionally, TiC phase usually coexists and shows special orientation relationship with Ti_3SiC_2 phase during the synthesis process of bulk Ti_3SiC_2 [82]. It is, therefore, expected to develop *MAX* phase composite coatings for such practical applications requiring the combination of good wear resistance and mechanical properties.

Since *MAX* phases show sufficient ductility, a cold-spray technique may also be utilized to deposit *MAX* phase coatings. However, it can be achieved only if the pre-alloyed *MAX* phase powder is available as feedstock. The source of bonding is the kinetic energy achieved by the particles under high-pressure accelerating gas at temperatures much below the melting point of the spray particles. Since cold spray involves temperatures much lower than the melting points of spray particles, the coating experiences little or no phase changes and grain growth during deposition. Moreover, undesirable effects like oxidation, nitriding, decarburizing and any decomposition are also avoided in the process.

Note: All figures have been included with permission from the publisher.

References

[1] M.W. Barsoum. *Prog. Solid State Chem.* 2000, 28, 201-281.

[2] M.W. Barsoum; T. EI-Raghy. *American Scientists.* 2001, 89, 334-343.

[3] W. Jeitschko; H. Nowotny, *Monatsh Chem.* 1967, 98, 329–337.

[4] M.W. Barsoum; T. EI-Raghy. *J. Am. Ceram. Soc.* 1996, 79, 1953-1956.

[5] S. Amini; MW. Barsoum; T. EI-Raghy. *J. Am. Ceram. Soc.* 2007, 90, 3953-3958.

[6] P. Eklund; J.P. Palmquist; J. Howing; D.H. Trinh; T. EI-Raghy; H. Högberg; L. Hultman. *Acta Mater.* 2007, 55, 4723-4729.

[7] S. Dubois; T. Cabioc'h; P. Chartier; V. Gauthier; M. Jaouen. *J. Am. Ceram. Soc.* 2007, 90, 2642-2644.

[8] C.F. Hu; F.Z. Li; J. Zhang; J.M. Wang; J.Y. Wang; Y.C. Zhou. *Script. Mater.* 2007, 57, 893-896.

[9] M.W. Barsoum; H.I. Yoo; I.K. Polushina; V.Yu. Rud; T. EI-Raghy. *Phys. Rev. B.* 2000, 62, 10194-10198.

[10] P. Eklund; M. Beckers; U. Jansson; H. Högberg; L. Hultman. *Thin Solid Films.* 2010, 518, 1851-1878.

[11] V. Pasumarthi, Y. Chen, S.R. Bakshi, A. Agarwal. *J. Alloys Compd.* 2009, 484, 113-117.

[12] D.P. Riley; E.H. Kisi; D. Phelan. *J. Eur. Ceram. Soc.* 2006, 26, 1051-1058.

[13] D.P. Riley; E.H. Kisi; T.C. Hansen. *J. Am. Ceram. Soc.* 2008, 91, 3207-3210.

[14] C.L. Yeh; Y.G. Shen. *J. Alloys Compd.* 458(2008)286-291.

[15] C.L. Yeh; Y.G. Shen. *J. Alloys Compd.* 2008, 461, 654-660.

[16] S. Arunajatesan; A.H. Carim. *J. Am. Ceram. Soc.* 1995, 78, 667-672.

[17] C. Racault; F. Langlais; R. Naslain. *J. Mater. Sci.* 1994, 29, 3384-3392.

[18] S.L. Yang; Z.M. Sun; H. Hashimoto; T. Abe. *J. Alloys Compd.* 358(2003)168-172.

[19] Z.M. Sun; Y.C. Zhou. *Script. Mater.* 1999, 41, 61-66.

[20] X.H. Wang; Y.C. Zhou. *J. Mater. Chem.* 2002, 12, 455-460.

[21] Y. Zou; Z.M. Sun; S. Tada; H. Hashimoto. *Ceram. Int.* 2008, 34, 119-123.

[22] Y.M. Luo; Z.M. Zheng; X.N. Mei; C.H. Xu. *J. Cryst. Growth.* 2008, 310, 3372-3375.

[23] R. Radhakrishnan; J.J. Williams; M. Akinc. *J. Alloys Compd.* 1999, 285, 85-88.

[24] Y.M. Luo; W. Pan; S.I. Li; J. Chen; R.G. Wang; J.Q. Li. *Mater. Lett.* 2002, 52, 245-247.

[25] T. EI-Raghy; M.W. Barsoum; *J. Am. Ceram. Soc.* 1999, 82, 2849-2854.

[26] Y.C. Zhou; Z.M. Sun; S.Q. Chen; Y. Zhang. *Mater. Res. Innov.* 2(1998)142-146.

[27] K. Tang; C.A. Wang; Y. Huang; Q.F. Zan; X.G. Xu; *Mater. Sci. Eng. A.* 2002, 328, 206-212.

[28] J.H. Han; S.S. Hwang; D. Lee; S.W. Park. *J. Eur. Ceram. Soc.* 2008, 28, 979-988.

[29] S. Amini; Z. Zhou; S. Gupta; A. DeVillier; P. Finkel; M.W. Barsoum. *J. Mater. Res.* 2008, 23, 2157-2165.
[30] J.F. Zhang; L.J. Wang; W. Jiang; L.D. Chen. *J. Alloys Compd.* 2007, 437, 203-207.
[31] Z.F. Zhang; Z.M. Sun; H. Hashimoto; T. Abe. *Script. Mater.* 2001, 45, 1461-1467.
[32] N.F. Gao; J.T. Li; D. Zhang; Y. Miyamoto. *J. Eur. Ceram. Soc.* 2002, 22, 2365-2370.
[33] W. Pan; S.L. Shi. *J. Eur. Ceram. Soc.* 2007, 27, 413-417.
[34] J.F. Zhang; L.J. Wang; W. Jiang; L.D. Chen. *Mater. Sci. Eng. A.* 2008, 487, 137-143.
[35] Z.F. Zhang; Z.M. Sun; H. Hashimoto. *Metall. Mater. Trans. A.* 2002, 33, 3321-3328.
[36] S. Konoplyuk; T. Abe; T. Uchimoto; T. Takagi. *J. Mater. Sci.* 2005, 40, 34093413.
[37] S.L. Yang; Z.M. Sun; Q.Q. Yang; H. Hashimoto. *J. Eur. Ceram. Soc.* 2007, 27, 4807-4812.
[38] Y. Zou; Z.M. Sun; S. Tada; H. Hashimoto. *Mater. Res. Bull.* 2008, 43, 968-975.
[39] Murugaiah; A. Souchet; T. EI-Raghy; M. Radovic; M. Sundberg; M.W. Barsoum. *J. Am. Ceram. Soc.* 2004, 87, 550-556.
[40] Z.M. Sun; S.L. Yang; H. Hashimoto; S. Tada; T. Abe. *Mater. Trans.* 2004, 45, 373-375.
[41] A.G. Zhou; M.W. Barsoum; S. Basu; S.R. Kalindini; T. EI-Raghy. *Acta Mater.* 2006, 54, 1631-1639.
[42] J.F. Li; T. Matsuki; R. Watanabe. *J. Mater. Sci.* 2003, 38, 2661-2666.
[43] J. Emmerlich; H. Högberg; S. Sasvari; P.O.Å. Persson; L. Hultman; J.P. Palmquist; U. Jansson; J.M. Molina-Aldareguia; Z. Czigany. *J. Appl. Phys.* 2004, 96, 4817-4826.
[44] P. Eklund; A. Murugaiah; J. Emmerlich; Z. Czigany; J. Frodelius; M.W. Barsoum; H. Högberg; L. Hultman. *J. Cryst. Growth.* 2007, 304, 264-269.
[45] C. Walter; C. Martinez; T. EI-Raghy; J.M. Schneider. *Steel Res. Int.* 2005, 76, 225-228.
[46] J.M. Schneider; D.P. Sigumonrong; D. Music; C. Walter; J. Emmerlich; R. Iskandar; *J. Mater. Script. Mater.* 2007, 57, 1137-1140.
[47] T. Joelsson; A. Hörling; J. Birch; L. Hultman. *Appl. Phys. Lett.* 2006, 86, 111913.
[48] T. Joelsson; A. Flink; J. Birch; L. Hultman. *J. Appl. Phys.* 2007, 102, 074918.
[49] U. Helmersson; M. Lattemann; J. Bohlmark; A.P.Ehiasarian; J.T. Gudmundsson. *Thin Solid Film.* 2006, 513, 1-24.
[50] V. Kouznetsov; K. Macak; J.M. Schneider; U. Helmersson; I. Petrov. *Surf. Coat. Technol.* 1999, 122, 290-293.
[51] J. Rosén; L. Ryves; P.O.Å. Persson; M.M.M. Bilek. *J. Appl. Phys.* 2007, 101, 056101.
[52] T. Zehnder; J. Matthey; P. Schwaller; A. Klein; P.A. Steinmann; J. Patscheider. *Surf Coat Technol.* 2003, 163, 238-244.
[53] J. Nickl; K.K. Schweitzer; P. Luxenburg. *J. Less-Common Met.* 1971, 26, 335-and.
[54] T. Goto; T. Hirai. *Mater. Res. Bull.* 1987, 22, 1195-1201.
[55] C. Racault; F. Langlais; R. Naslain; Y. Kihn. *J. Mater. Sci.* 1994, 29, 3941-3948.
[56] J. Lu; U. Jansson. *Thin Solid Film.* 2001, 396, 53-61.
[57] S. Jacques; H. Di-Murro; M.P. Berthet; H. Vincent; *Thin Solid Film.* 2005, 478, 13-20.
[58] H. Fakih; S. Jacques; M.P. Berthet; F. Bosselet; O. Dezellus; J.C. Viala. *Surf. Coat. Technol.* 2006, 201, 3748-3755.
[59] B. Veisz; B. Pécz. *Appl. Surf. Sci.* 2004, 233, 360-365.
[60] S. Tsukimoto; T. Sakai; M. Murakami. *J. Appl. Phys.* 2004, 96, 4976-4981.
[61] Z.C. Wang; S. Tsukimoto; M. Saito; Y. Ikuhara. *Phys. Rev. B.* 2009, 79, 045318.
[62] C. Walter; D.P. Sigumonrong; T. EI-Taghy; J.M. Schneider. *Thin Solid Film.* 2006, 515, 389-393.

[63] V. Viswanathan; T. Laha; K. Balani; A. Agarwal; S. Seal. *Mater. Sci. Eng. R.* 2006, 54, 121-285.

[64] P. Fauchais; A. Vardelle; B. Dussoubs; *J. Thermal Spray Technol.* 2001, 10, 44-66.

[65] R.S. Lima; B.R. Marple. *J. Thermal Spray Technol.* 2007, 16, 40-63.

[66] M. Vardelle; A. Vardelle; A.C. Leger; P. Fauchais; D. Gobin, *J. Thermal Spray Technol.* 1995, 4, 50-58.

[67] Twin Wire Arc Spray Technology. EuTronic Arc Spray 4. From http://www.castolin.com/wCastolin_com/products/coating/Eutronic_arc_spray_4.php?navid=129.

[68] F. Gärtner; T. Stoltenhoff; T. Schmidt; H. Kreye. *J. Thermal Spray Technol.* 2006, 15, 223-232.

[69] Agarwal; T. McKechnie; S. Seal. *J. Thermal Spray Technol.* 2003, 12, 350-359.

[70] M.W. Barsoum, R. Knight. *U.S. Patent.* No. 6,231,969 B1.

[71] J. Frodelius; M. Sonestedt; S. Björklund; J-P. Palmquist; K. Stiller; H. Högberg; L. Hulman. *Surface Coatings and Technol.* 2008, 202, 5976-5981.

[72] J. Jiang; A. Fasth; P. Nylén; W.B. Choi. *J. Thermal Spray Technol.* 2009, 18: 194-200.

[73] T. EI-Raghy; M. W. Barsoum. *J. Appl. Phys.* 1998, 83, 112-119.

[74] F. Yudin; V. G. Borisov. *Refrac. Ind. Ceram.* 1987, 8, 499-504.

[75] J. Lis; R. Pampuch; L. Stobierski. *Int. J. SHS.* 1992, 1, 401-408.

[76] D.P. Riley; E. H. Kisi; E. Wu; A. McCallum. *J. Mater. Sci. Lett.* 2003, 22, 1101-1104.

[77] T. El-Raghy; A. Zavaliangos; M.W. Barsoum; S.R. Kalidindi. *J. Am. Ceram. Soc.* 1997, 80, 513–516.

[78] W.B. Tian; Z.M. Sun; H. Hashimoto; Y.L Du. *Mater. Sci. Eng. A.* 2009, 526, 16-21.

[79] C.L. Yeh; R.F Li; Y.G. Shen. *J. Alloys Compd.* 2009, 478, 699-704.

[80] X.H. Tong; T. Okano; T. Ikeki; T. Yano, *J. Mater. Sci.* 1995, 30, 3087–3090.

[81] W.J.J. Wakelkamp; F.J. van Loo; R. Metselaar. *J. Eur. Ceram. Soc.* 1991, 8, 135-139.

[82] Y.C. Zhou; Z.M. Sun; B.H. Yu. *Z Metallkd.* 2000, 91, 937–941.

In: MAX Phases: Microstructure, Properties and Applications ISBN 978-1-61324-182-0
Editors: It-Meng (Jim) Low and Yanchun Zhou

Chapter 7

STRUCTURAL AND ELECTRONIC PROPERTIES OF MAX PHASES

Y. Medkour[1], A. Roumili[1] and D. Maouche[2]

[1]Laboratoire d'Etudes des Surfaces et Interfaces des Matériaux Solides (LESIMS), Université de Sétif, Algérie

[2]Laboratory for Developing *New Materials* and their Characterizations, University of Setif, Algeria

ABSTRACT

The so-called MAX phases have shown unusual set of properties, ranging from the best of metallic to ceramic properties. In this chapter, we summarize the main results of structural and electronic properties of these fascinating materials. A description of the primitive unit cell of M_2AX, M_3AX_2 and M_4AX_3, and the pressure effect on the lattice parameters are done. The structural stability against increasing pressure and temperature is presented. Experimental and theoretical studies show that MAX phases are good electrical conductor where the conductivity is assured by the *d* electrons of the transition metal atoms, the stiffness of the MAX is attributed to the strong ionic, covalent bonds, and the remarkable machinability is related to its richness of metallic bonds and the nanolaminate nature of these materials.

Keywords: MAX phases; Ternary carbides; Electronic structure and bonding characteristics

1. INTRODUCTION

Ternary carbides and nitrides with the chemical formula $M_{n+1}AX_n$ show a new class of nanolaminate solids [1- 5], where M is an early transition metal, A is an A group element of the periodic table (mostly IIIA and IVA), and X is either C or N, the subscript *n* change from 1 to 3 [1]. By the end of the last century, these compounds have been a subject of several works in different laboratories throughout the world. The increase in their study is related to their numerous properties due to their rich composition (M, A and X) and to the diversity in their crystalline stoichiometries (M_2AX, M_3AX_2 and M_4AX_3). Recent works show that these compounds are light weight, oxidation resistant, damage tolerant, good electrical and thermal conductors and they have easy machinability [3, 5].

This chapter is divided into two sections; the first section, entitled *Structural properties,* is concentrated on the lattice parameters, bulk moduli, and the structural stability under pressure and high temperature. While the second section, *Electronic properties*, is focused on the electronic structure and the bonding in the MAX phases.

2. Structural Properties

2. 1. Lattice Parameters

M_2AX compounds constitute the majority of all known MAX phases, most of these phases were reported in the sixties by Nowotny [2]. They adapted an hexagonal symmetry (space group $P6_3/mmc$) prototype Cr_2AlC with atomic positions: *X* at *2a (0, 0, 0)*, *A* at *2d (1/3, 2/3, 3/4)*, and *M* at *4f (1/3, 2/3, Z), Z* is the only free internal parameter. The primitive unit cell of M_2AX contains eight atoms, their rearrangement can be described as the sequence [2]: $A_XC_MB_AC_MA_XB_MC_AB_M$ along the (0001) direction, *A, B*, and *C* denote the atomic positions and the subscript stands for the atom type in the site , as shown in Figure 1a. These materials are classified as nanolaminate compounds, owing to the M_6X layers interleaved with pure weaker layers of A element [1]. Table 1 illustrates the lattice parameters of all known MAX phases, the main results are [1, 3, 6]:

- The excellent correlation between the *a* lattice parameter of the MAX phases and the M-M distance in the corresponding binary transition metal.
- Based on the A group element, a linear relationship exists between the *c* parameters and the M-M distance.
- A decrease in both *a* and *c* while going from carbides to nitrides compounds, which can be related to the larger atomic radius of C atoms compared to N atoms.
- The good correlation between the c/a ratio and the number of M layers in the primitive unit cell.
- The occurrence of M_2AX phases decreases as the *sp* states of A atoms are more and more filled.

Up to date, five M_3AX_2 phases were synthesized; all of them are based carbide, as shown in Table 1. The primitive unit cell is presented in Figure 1b, there are two inequivalent M atoms; M_1 at *2a* (0, 0, 0) and M_2 at *4f* (1/3, 2/3, Z_{M2}), A atoms are located at *2b* (0, 0, 1/4) and X atoms at *4f* (2/3, 1/3, Z_X) [7]. Metallic elements of the unit cell are built in the sequence AB**A**BAC**A**C, where M atoms are in the cubic sequence, A (bold letters) atoms in the middle hexagonal and X occupy the interstice between M elements [8]. This arrangement lead to M_6X octahedral separated by the pure A layers, the number of M layers is equal to 6 which is very close to *c/a* ratio [1], the *a* parameter is slightly affected when comparing to the corresponding of M_2AX, experimental investigations [8] shows that the M_2-C bonds are longer than M_1-C which can be related to the environment difference between M_1 and M_2. The absence of nitrides sample in these phases could be related to the shortness of *a* and *c* that prevents N elements to occupy the interstitial site in the octahedron M_6X blocks.

For the M_4AX_3 we count six samples based carbide, and one sample based nitride Ti_4AlN_3, as shown in Table 1. As presented in Figure 1c, these phases present three internal parameters Z_{M1}, Z_{M2} and Z_{X1} where the atomic positions are M_1 at *4f* (1/3, 2/3, Z_{M1}), M_2 at *4e* (0, 0, Z_{M2}), A at *2c* (1/3, 2/3, 1/4), X_1 at *4f* (2/3, 1/3, Z_{X1}) and X_2 at *2a* (0, 0, 0) [8].The metal atoms were arranged in the sequence AB**A**BACB**C**BC for the α-type structure, and AB**A**BABA**B**AB for β-type structure [9], while X atoms occupy the octahedral voids of M atoms [8], the values of c/a ratio is ≈ 8 that is in good correlation with the number of M layers [1].

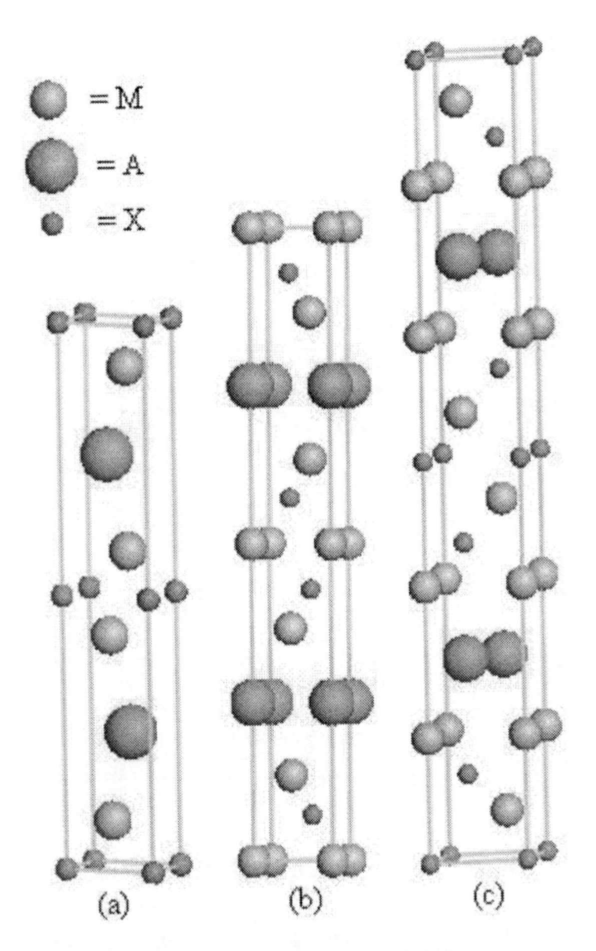

Figure 1. Primitive unit cell of: a- M_2AX, b- M_3AX_2 and c- M_4AX_3.

The ability of regular $M_{n+1}AX_n$ to form solid solution on all sites is one of the most important properties of these compounds [1], that increase the number of combination, which lead to a new and sometimes unique properties. As an example for synthesized solid solution on M sites is $(Ti_{0.5}, Zr_{0.5})InC$ [10], on A sites $Ti_3(Si_{0.5}, Ge_{0.5})C_2$ [11] and a continuous series of $Ti_2AlC_{(0.8-x)}N_x$, x=0- 0.8 occurs at 1490 ^{0}C for the X sites [12].

Table 1. Lattice parameters (in Å) and Bulk modulus (in GPa) of MAX phases, bold numbers referred to theoretical results
MAX phases that aren't referred, are taken from Ref. [1]

IIB	IIA	IVA	VA	VIA
	Al Ti_2AlC; 3.04, 13.60, 186 [13] V_2AlC; 2.91, 13.19, 201[13] Cr_2AlC; 2.86, 12.80, 165 [13] Nb_2AlC; 3.10, 13.80, 209 [13] Ta_2AlC; 3.07, 13.80, 251 [13] Ti_2AlN; 2.98, 13.61, 169 [14] Ti_3AlC_2 ; 3.07, 18.57, 156 [15] Ta_3AlC_2 [8]; 3.09, 19.15, **238** [16] Ta_4AlC_3 ; 3.08, 23.70, 261 [17] Ti_4AlN_3; 2.98, 23.37, 216 [18] Nb_4AlC_3 [19] ; 3.12, 24.12, **247** [20] $V_4AlC_{2.69}$ [21]; 2.93, 22.71, **255** [20]	**Si** Ti_3SiC_2 ; 3.06, 17.67, 206 [22] Ti_4SiC_3 ; 3.05, 22.67 [23]	**P** V_2PC; 3.07, 10.91, **226** [24] Nb_2PC; 3.28, 11.5, **227** [24]	**S** Ti_2SC; 3.21, 11.22, 191 [25] Zr_2SC; 3.40, 12.13, 186 [26] $Nb_2SC_{0.4}$; 3.27, 11.4 Hf_2SC; 3.36, 11.99, **206** [27]
	Ga Ti_2GaC; 3.07, 13.52, **146** [28] V_2GaC; 2.93, 12.84, **192** [28] Cr_2GaC; 2.88, 12.61, 188 [29] Nb_2GaC; 3.13, 13.56, **205** [28] Mo_2GaC; 3.01, 13.18, **242** [30] Ta_2GaC; 3.10, 13.57, **219** [28] Ti_2GaN; 3.00, 13.3, **157** [31] Cr_2GaN; 2.87, 12.77, **160** [31] V_2GaN; 3.00, 13.3, **177** [31] Ti_4GaC_3 [32]; 3.06, 3.44, **176** [33]	**Ge** Ti_2GeC; 3.07, 12.93, 211 [34] V_2GeC; 3.00, 12.25, 165 [35] Cr_2GeC; 2.95, 12.08, 182 [35] Ti_3GeC_2 ; 3.08, 17.76, 197 [36] Ti_4GeC_3[37];	**As** V_2AsC; 3.11, 11.3, **193** [24] Nb_2AsC; 3.31, 11.9, 224 [38]	
Cd Ti_2CdC ; 3.1, 14.41, **116** [39]	**In** Sc_2InC; Ti_2InC; 3.13, 14.06, 148 [10] Zr_2InC; 3.34, 14.91, 127 [10] Nb_2InC; 3.17, 14.73, **195** [40] Hf_2InC; 3.30, 14.73, **136** [41] Ti_2InN; 3.27, 14.83 Zr_2InN; 3.27, 14.83	**Sn** Ti_2SnC; 3.16, 13.67, 152 [10] Zr_2SnC; 3.35, 14.57, **149**[42] Nb_2SnC; 3.24, 13.80, 180 [10] Hf_2SnC; 3.32, 14.38, 169 [10] Hf_2SnN; 3.31, 14.30, **171** [43] Ti_3SnC_2 [44]; 3.13, 18.65 ,**192** [45]		
	Tl Ti_2TlC; 3.15, 13.98, **125** [46] Zr_2TlC; 3.36, 14.78, **120** [46] Hf_2TlC; 3.32, 14.62, **131** [46] Zr_2TlN; 3.30, 14.71	**Pb** Ti_2PbC; 3.20, 13.81 Zr_2PbC; 3.38, 14.66 Hf_2PbC; 3.55, 14.46		

2. 2. Structural Stability under Pressure and Bulk Modulus

Structural stability under pressure of these materials has been intensively experienced by B. Manoun and co-workers using a synchrotron X-ray radiation source and a diamond anvil cell [10, 11, 13, 14, 17, 18, 35], the main result of their works is the good stability of all studied M_2AX phases against increasing hydrostatic pressure, till 50 GPa. Another quantity that has an important role in various technological applications is the bulk modulus, which reflects the resistance of a material to volume change [47, 48], in Table 1 we have reported the available experimental and theoretical values (theoretical values are in bold numbers). The first remark is the softness of the M_2AX phases comparing to the corresponding binary transition metal carbides and nitrides [1]. The relative volume change against increasing pressure is presented in Figure 2a for TaC, Ta_2AlC and Ta_4AlC_3, it is clearly shown that the binary carbides MC are more resistant to volume change than the corresponding MAX. Inversely to the binary, replacing C by N [14] leads to decrease in *B* (see Tab. 1). The effect of substitution in M sites on the bulk modulus for M_2AlC; M=Ti, V, Cr, Nb, and Ta is shown in Figure 2b. The highest measured bulk modulus is that of Ta_2AlC, 251 GPa, and the lowest is for Zr_2InC, 127 GPa. The introduction of second type of metal elements (i.e. A element in the MAX phases) reduces the strength of the structure compared to the corresponding transition metal carbides [49- 51], Hettinger *et al.* [52] have measured the bulk modulus using velocity of sound technic, their results were ≈ 40% lower than those reported by Manoun *et al.* [13] for M_2AlC; M= Ti, V, Cr and Nb. Moreover, the bulk moduli of the solid solution were lower than their end members for $(TiV)_2AlC$ and $(TiNb)_2AlC$ [53]. Based on M elements, Sun *et al.* [30] using ab initio calculations suggest that M_2AC phases can be classified into two groups; group with VB and VIB elements roughly maintain the bulk modulus of their binary, the second group with IVB elements where the bulk modulus of the M_2AC phases significantly decreases.

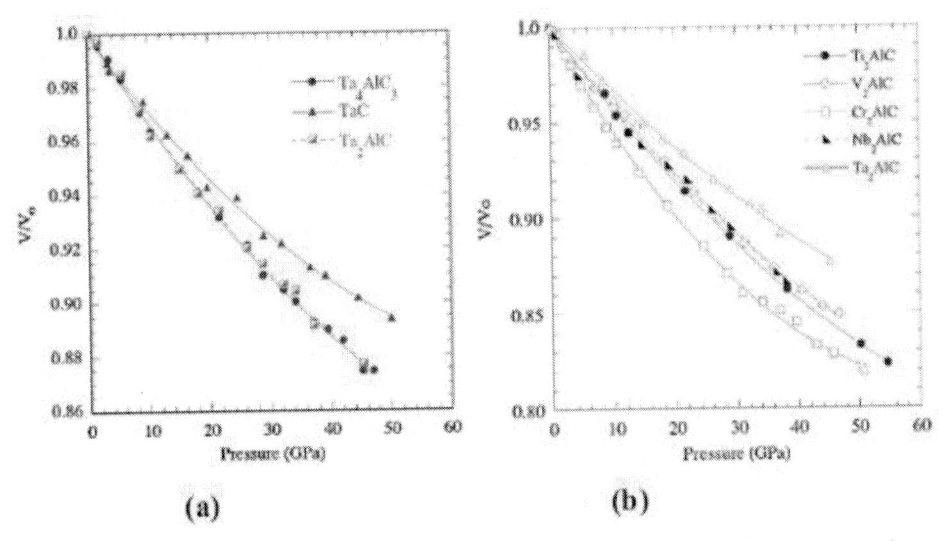

Figure 2. Relative unit cell volume change as a function of pressure for: a-TaC [1] [56], Ta_2AlC[2] [13], Ta_4AlC3[3] [17], and b-M_2AlC; M= Ti, V, Cr, Nb, and Ta[4] [13]

Since these materials have an hexagonal symmetry, the unidirectional compressibility along *a* and *c* axis must be useful in the discussion of the pressure effect, for the most part of M_2AX phases, the contraction along *a* direction was more important than *c* direction [10, 13, 14, 25], except for Nb_2(Al or As)C and Cr_2AlC [13, 38] where the opposite was observed. While the contraction is almost identical in both direction for Ta_2AlC [13] and Zr_2SC [26], as displayed in Figure 3, many theoretical works have been carried out on this exceptional

1 Reprinted from figure 3 in Ref. [56] with permission of ELSEVIER (2010).
2 Reprinted from figure 3 in Ref. [13] with permission of both; AMERICAN PHYSICAL SOCIETY and the author Bouchaib MANOUN (2010).
3 Reprinted from figure 3 in Ref. [17] with permission of AMERICAN INSTITUT OF PHYSICS (2010) .
4 Reprinted from figure 5 in Ref. [13] with permission of both; AMERICAN PHYSICAL SOCIETY and the author Bouchaib MANOUN (2010).

behaviour and show that M-X bonds are stiffer than M-A bonds [24, 41, 54, 55], the reasonable and simplest explanations of these different behaviour is based on the contribution of M-X bonds to the *a* and *c* direction [24, 54]. In the case of Nb_2(Al or As)C and Cr_2AlC the contribution of M-X bonds to the strength of *c* direction is more important than in the *a* direction, while the opposite is true for Ti_2(Al, Ge, In, Sn,) C [10, 13, 34] and others M_2AX phases.

The presence of extra-layers M_6X in M_3AX_2 comparing to M_2AX, enhances the stiffness in these compounds [16], whereas experimental results show that M_3AX_2 are less stiff than the corresponding M_2AX phases, see Table 1. The bulk modulus of Ti_3AlC_2 is 19% lower than that of Ti_2AlC [15], while that of Ti_3GeC_2 is about 7% lower than Ti_2GeC [36]. Like M_2AX phases, all examined M_3AX_2 are stable against increasing hydrostatic pressure, and no phase transformation was observed up to 50 GPa [15, 22, 36]. The linear compressibility along *a* and *c* directions, shows that the contraction of *c* axis is more important than along *a* axis. Furthermore, under non-hydrostatic pressure a phase transformation was observed from α-Ti_3(Si or Ge)C_2 to β-Ti_3(Si or Ge)C_2 [57, 58], in which the A atoms change their position to *2d* (2/3, 1/3, 1/4).

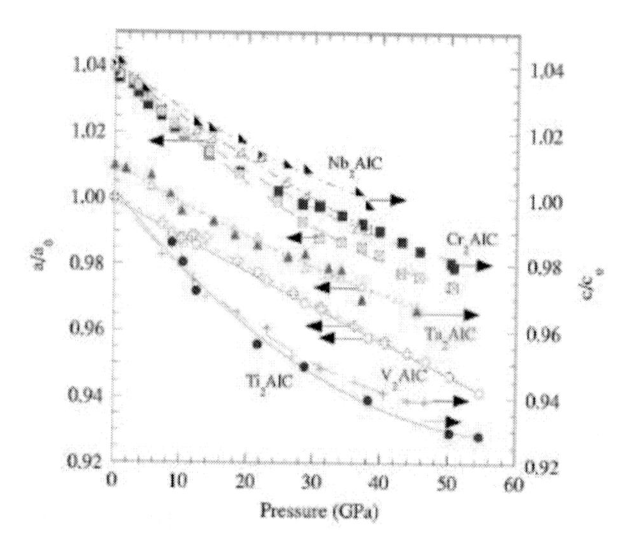

Figure 3. Relative lattice parameters changes, for M_2AlC; M= Ti, V, Cr, Nb, and Ta[5] [13].

The derivate bulk modulus of Ta_4AlC_3 from the response to hydrostatic pressure is 261 GPa [17] which is about 4% higher than the corresponding of Ta_2AlC, despite the fact that it is 24% lower than the binary TaC [56]. For the Ti_4AlN_3 [18] and Ti_2AlN [14] the increase in the bulk modulus is quite important, the former has B= 216 GPa which is about 21% higher than 169 GPa of Ti_2AlN. Du *et al.* show that the bulk modulus of M_4AlC_3 is about 84% of the corresponding binary MX [20], the increase in the measured bulk modulus from M_2AX to M_4AX_3 could be related to the increase in the M-(C-M)$_m$-C-M chains, m= 0, 2 for the M_2AX and M_4AX_3 phases respectively [16].

5 Reprinted from figure 5 in Ref. [13] with permission of both; AMERICAN PHYSICAL SOCIETY and the author Bouchaib MANOUN (2010).

2. 3. Stability at High Temperature

Most of the available works on the temperature behaviour are focused on Ti_3SiC_2 [59-62]. As the binary transition metal carbide and nitrides [63], MAX phases have a higher decomposition temperature [1], Nb_2SnC decomposes peritectically at 1390 °C into Sn and NbC_x, while NbC melts at 3600 °C [64], Barsoum and El-Raghy have studied intensively Ti_3SiC_2 [59], above 1200 °C this compound remains stable and preserves their properties. Although Ti_3SiC_2 decomposes at 2300 °C to Ti_3C_2 and Si [1]. Generally, $M_{(n+1)}AX_n$ phases do not melt, but decompose peritectically to $M_{(n+1)}X_n$ blocks and A element [1], one notes that the decomposition temperature is strongly affected by the presence of impurities and the environment conditions [1, 65].

3. Electronic Properties

Advances in ab initio electronic structure computation technique, have facilitated the understanding of many physical and chemical properties that are not yet easy to explore experimentally. Hence, their many results and illustrations of this section are based on theoretical works, along with the available experimental results.

3. 1. Band Structure

Experimental investigations show that MAX phases are characterised by a metallic-like resistivity [1, 66, 67]. Indeed, calculated band structure along the high symmetry direction in the Brillouin zone shows continuous energy dispersion, and no band gap was observed for all MAX phases [24, 68]. For the M_2AX phases, around Fermi level, X atoms do not contribute to the total density of states (TDOS) and therefore they are not involved in the conduction properties. M *d* electrons are mainly contributing to the TDOS, and should be involved in the conduction properties. While A *p* electrons do not contribute significantly. Moreover, from the curve dispersions of the band structure a weak conductivity along the *c* direction can be expected. As an example Ti_2GeC; theoretical study predicted that the conductivity along *c* axis is less than that along *a* axis [69], which was further disproved by Scabarozi *et al* [70] where the measured conductivity in *c* direction is about ≈ 1.6 that of in *a* direction. Nevertheless, the difference in the conductivity is more severe for Ti_2AlC and Ti_2AlN with a factor 2 - 2.5 [71, 72].

The valence band can be divided into two blocks:

(i) The lowest energy region composed of X *s*, A *p* states.
(ii) The bottom of the valence band is derived from X *p*, M *p*, and A *p* states.

The conduction band beyond the Fermi level is dominated by the antibonding M *d* states with little contribution from A *p* and X *p* states. For the most M_2AX phases, Fermi level locates at a minimum of the TDOS which is a criterion for the good stability of these compounds [6]. On the other hand, the connection between the value of the TDOS at the

Fermi level (N_{EF}) and the conductivity appears to be reasonable [6], with exception for (V or Cr)$_2$AlC [67] where the difference between measured and calculated N_{EF} is very important.

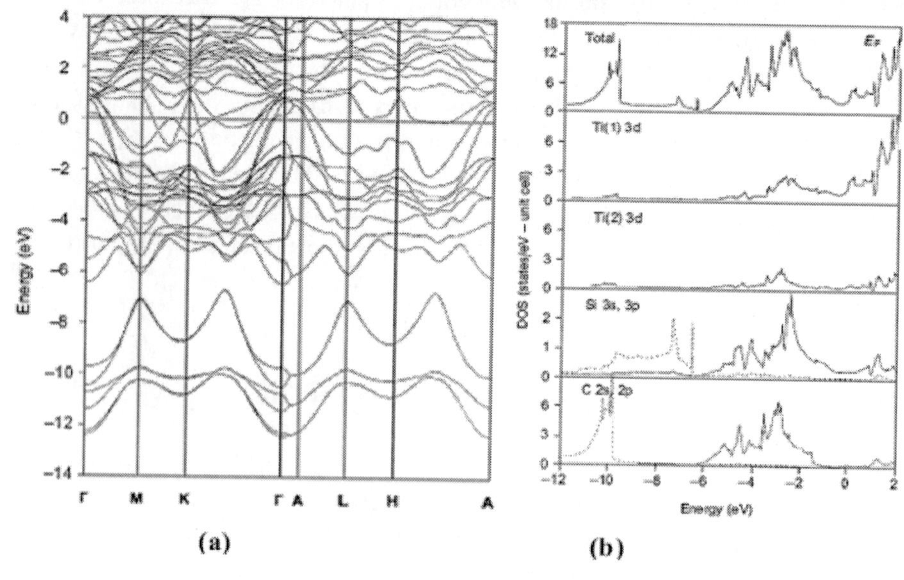

Figure 4. a- Band structure and b-TDOS and PDOS of Ti_3SiC_2 [6] [73].

The effect of additional atoms in the unit cell of M_3AX_2 on the band structure and the densities of states is shown in Figure 4a for Ti_3SiC_2 as an example [73], at lower energy the valence band is due to (C, Si) *s* states, the higher valence band is related to (C, Si) *p* states and the predominantly Ti_1, and Ti_2 *d* states, around the Fermi level the band structure is governed by the *d* states of Ti_1 as shown in Figure 4b. These later states, Ti_1 *d,* contribute to the conduction band more than Si *p* and Ti_2 *d* states. Moreover, the overlaps indicate anisotropy in the electrical conduction properties [45].

No band gap were observed for M_4AX_3 phases, the lowest valence band is essentially coming from the C *s* states, the dispersion curves below the Fermi level are predominantly of M *d* states and little contribution from C *p* and A *s, p* states [20, 23]. Around the Fermi level, numerous M *d* states lead to bonding and antibonding, the electrical conductivity is assured by the *d* states [74].

3. 2. Bonding Properties

On the other hand, different hybridised states are observed from the partial density of states [54, 55], M_2AX phases show that X *s* states are hybridised with M *d* states at lower energies level. States lying below and above Fermi level show a strong hybridisation between M *d* and both X *p* and A *p* states. A layers are bonded through *sp* hybridisation of A atoms. These hybridisation states are clearly observed in Figure 6a by the covalent- ionic bonding between the M and X atoms, and the ionic bonding between the A and M atoms [16, 24]. Based on the valence electron concentration (VEC), Emmerlich *et al.* [54] studied the

6 Reprinted from figure 2 and 3 in Ref.[73]: with permission of ELSEVIER (2010).

response of M_2AlC phases with M= Ti, V, Cr, Zr, Nb, Mo, Hf, Ta, and W) to increasing pressure up to 60 GPa. They conclude that for VEC=4, these phases have a large compressibility along *c* axis as compared to *a* axis, for VEC=5, compressibility of *a* and *c* axis are comparable and for VEC=6, *a* axis is more compressible than *c* axis. More extending series of M_2AC phases with M= Ti, V, Cr, A= Al, Si, P, and P were studied by Liao *et al.* [55], as fixed M element, the strength of these compounds increase monotonically as A element running from Al to P , then decrease for S. In these view point, Hug [6] suggests tailoring strength of Ti_2AC (with A= Al, Si, P, S, Sn, Ga, In, Ge, As, Pb, and Tl) could be based on the energy level of hybridised Ti *d*–A *p* and Ti *d*–X *p* states.

For the M_3AX_2 further hybridisation were observed, in the case of Ti_3SiC_2 [73, 74], $Ti_{1,2}$–C are bonded by a covalent *p*–*d* interaction, Si *s*–*p* states are hybridised and form a strong Si–Si bonds parallel to the basal plan, while *d*–*p* interaction of the Ti_1 and Si atoms are very weaker. In these phases, as shown in Figure 6b the M_2 atoms are bonded to both C_1 and A by *d*–*p* hybridisation. While M_1 is bonded to $C_{1,2}$ from the *d*–*p* hybridisation. Moreover, metal to metal bonds are formed from the *d*–*d* interaction. The location of hybridized M_2 *d*–C_2 *p* states at lower energy level shows that these bonds are stiffer than M_2 *p*, *d*–A *p* bonds [20, 33],

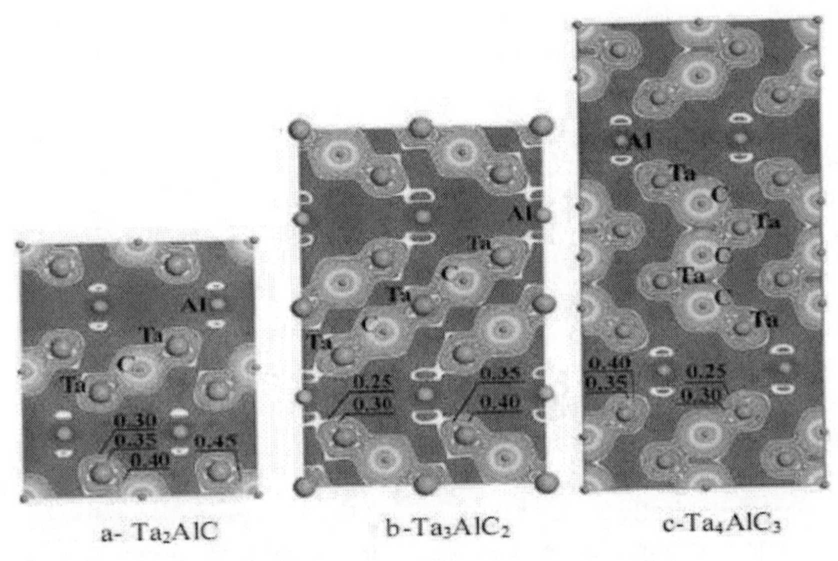

Figure 6. Valence electron density of slices on the (11-20) plane in 2x2x1 supercells of a-Ta_2AlC, b-Ta_3AlC_2, and c-Ta_4AlC_3[7], respectively [16].

Different hybridised states highlighted for M_4AX_3; at lower energy level a strong hybridisation between M *d* and X *s* states is observed, at relatively higher energy M *d* and A *p* states are hybridised, A slices are formed by their hybridised *s*, *p* states [20, 75, 76], The exceptional stiffness of these compounds is related to the extra M– *X* bonds (see Figure 6c) [16].

Further information about the stiffness of bonds in MAX phases could be achieved from the pressure dependence of the normalized bonds length M–X and M–A [24, 54]. It is clearly shown that M–C bonds are stiffer than M-A bonds as presented in Figure 7a, this can

7 Reprinted from figure 9 in Ref. [16] with permission of John Wiley and Sons (2010).

explained by the stronger covalent bonds related to M *d*–C *s* interaction which play an important role in the stability of the MAX phases [1]. In the M_3AX_2 phases more M–C bonds are present. The basal M_1–M_2 bonds evolved from the *d*–*d* interaction, they show an ionic character and they are the strongest followed by the M–C bonds, while M–A are the weakest [74]. As shown in Figure 7b, Nb–C are stronger than Nb-Al bonds, M_4AX_3 phases possess high fraction of M–C bonds which increase the stiffness of these compounds [20].

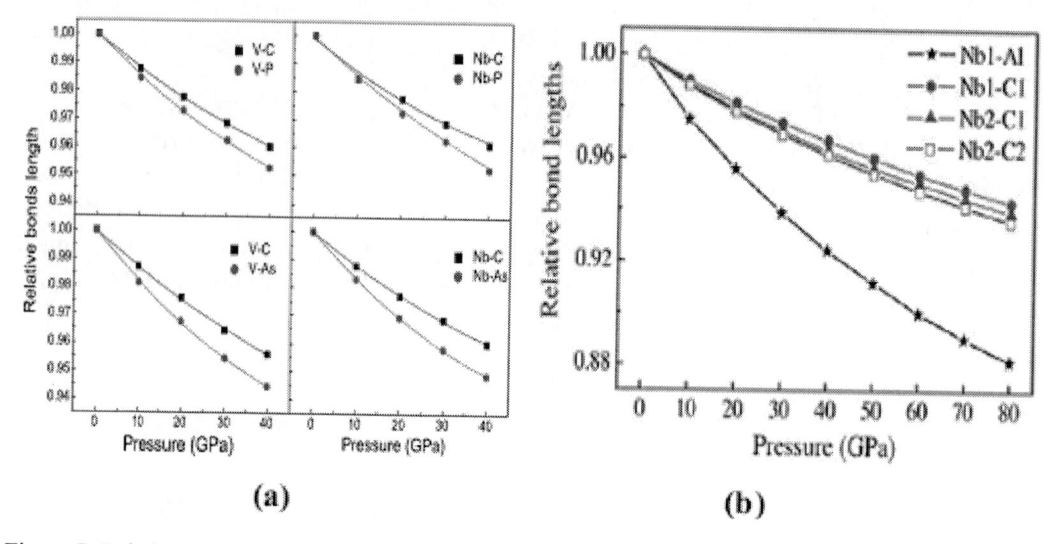

Figure 7. Relative changes in the bonds length of a-M_2AC; (M=Ti, Nb and A= P, AS)[8] [24], and b-Nb_4AlC_3[9] [75].

CONCLUSION

Up to now, MAX phases show the best attributed properties of both metals and ceramics. The structural parameters are well established, most of these compounds are stable against increasing hydrostatic pressure (≈ 50 GPa), the bulk modulus ranging from 127 and 261 GPa, which remains lower than that of their corresponding binary MX. MAX phases do not melt, but decompose peritectically at high temperature (350- 2300 °C). The good conductivity of these compounds is attributed to *d* states of the transition metal atoms. The difference in electronegativity between M, A and C generates a puzzle of covalent, ionic and metallic bonds, where M–C bonds are stiffer than M–A bonds. Despite the important of experimental results, many of them remain still unclear, and sometimes contrast with theoretical studies. Besides, experiments show that MAX phases properties are much affected by the presence of impurities, vacancies, elaboration methods and environment conditions.

8 Reprinted from figure 4 in Ref.[24] with permission of ELSEVIER (2010).
9 Reprinted from figure 7 in Ref.[75] with permission of ELSEVIER (2010).

ACKNOWLEDGMENTS

We gratefully acknowledge the following people for permission to reproduce figures from their publications:

Elsevier, American Physical Society (APS), American Institute of Physics (AIP), John Wiley and Sons and Pr. Bouchaib MANOUN.

REFERENCES

[1] Barsoum, M. W. *Prog. Solid St. Chem.* 2000, *28*, 201-281.
[2] Nowotny, V. H. Prog. *Solid. St. Chem.* 1970, *2*, 27-70.
[3] Barsoum, M. W. *Physical Properties of the MAX Phases. Encyclopaedia of Materials: Science and Technology*, Elsevier, Amsterdam, 2006, pp 1-11.
[4] El-Raghy, T.; Zavaliangos, A.; Barsoum, M. W.; Kalidindini, S. R. *J. Am. Ceram. Soc.* 1997, *80*, 513- 516.
[5] Barsoum, M. W.; Radovic, M. *Mechanical Properties of the MAX Phases. Encyclopaedia of Materials: Science and Technology*, Elsevier, Amsterdam, 2004, pp1-16.
[6] Hug, G. *Phys. Rev. B. 74,* 2006, 184113-184119.
[7] Zhou, Y.; Sun, Z.; Wang, X.; Chen, S. *J. Phys.: Condens. Matter.* 2001, *13,* 10001-10010.
[8] Etzkorn, J.; Ade, M.; Hillebrecht, H. *Inorg. Chem.* 2007, *46*, 1410-1418.
[9] Wang, J.; Wang, J.; Zhou, Y.; Lin, Z.; Hu, C. *Scripta Mater.* 2008, *58*, 1043-1046.
[10] Manoun, B.; Leaffer, O.D.; Gupta, S.; Hoffman, E.N.; Saxena, S.K.; Spanier, J.E.; Barsoum. M.W. *Solid State Comm.* 2009, *149*, 1978-1983.
[11] Manoun, B.; Liermann, H. P.; Gulve, R. P.; Saxena, S. K.; Ganguly. A.; Barsoum, M. W.; Zha, C. S. *Appl. Phys. Lett.* 2004, *84,* 2799-2800.
[12] Pietzka, M. A.; Schuster, J. C. *J. Am. Ceram. Soc.* 1996, *79*, 2321- 2330.
[13] Manoun, B.; Gulve, R. P.; Saxena, S. K.; Gupta, S. ; Barsoum, M.W.; Zha, C. S. *Phys. Rev. B.* 2006, *73*, 024110- 024116.
[14] Manoun, B.; Zhang, F. X.; Saxena, S. K.; El-Raghy, T.; Barsoum, M. W. *J. Phys. Chem. Solids* 2006, *67*, 2091- 2094.
[15] Zhang, H.; Wu, X.; Nickel, K. G.; Chen, J.; Presser, V. *J. Appl. Phys.* 2009, *106*, 013519- 13523.
[16] Lin, Z.; Zhuo, M.; Zhou, Y.; Li, M.; Wang. J. *J. Am. Ceram. Soc.* 2006, *89,* 3765- 3769.
[17] Manoun, B.; Saxena, S. K.; El- Raghy, T.; Barsoum, M. W. *Appl. Phys. Lett.* 2006, *88*, 201902- 201904.
[18] Manoun, B.; Saxena, S. K.; Barsoum, M. W. *Appl. Phys. Lett.* 2005, *86,*101906-101908.
[19] Hu, C. F.; Li, F. Z.; Zhang, J.; Wang, J. M.; Wang, J. Y.; Zhou, Y. C. *Scr. Mater.* 2007,*57*, 893-896.
[20] Du, Y. L.; Sun, Z. M.; Hashimoto, H.; Tian, W. B. *Phys. Status Solidi B* 2009, *246*, 1039–1043.

[21] Etzkorn, J.; Ade, M.; Hillebrecht, H. *Inorg. Chem.* 2007 ,*46,* 7646- 7653.

[22] A. Onodera, H. Hirano, T. Yuasa, N. F. Gao, and Y. Miyamoto, *Appl. Phys. Lett.* 1999, *74*, 3782- 3784

[23] Magnuson, M.; Mattesini, M.; Wilhelmsson, O.; Emmerlich, J.; Palmquist, J. –P.; Li, S.; Ahuja, R.; Hultman, L.; Erikson, O.; Jansson, U. *Phys. Rev. B.* 2006, *74*, 205102-205108.

[24] Medkour, Y.; Roumili, A. ; Boudissa, M. ; Maouche, D. ; Louail, L. ; Gamoura, A. *Comp. Mater. Sci.* 2010 , *48,*174–178.

[25] Kulkarni, S. R. ; Vennila, R. S. ; Phatak, N. A. ; Saxena, S. K. ; Zha, C. S. ; El-Raghy, T.; Barsoum, M. W.; Luo, W.; Ahuja, R. *J. Alloys and Comp.* 2006, *448*, L1- L4.

[26] Kulkarni, S.K.; Phatak, N.A.; Saxena, S.K.; Fei, Y.; Hu, J. *J. Phys.: Condens. Matter.* 2008, *20,* 135211-135216.

[27] Bouhemadou, A.; Khenata, R. *Phys. Lett.* A 2008, *372*, 6448- 6452.

[28] Bouhemadou, A.; Khenata, R. *J. Appl. Phys.* 2007 ,*102*, 043528- 043533.

[29] Phatak, N. A.; Kulkarni, S. R.; Drozd, V.; Saxena, S K.; Deng, L.; Fei, Y.; Hu, J.; Luo, W.; Ahuja, R. *J. Alloys and Comp.* 2008, *463,* 220–225.

[30] Sun, Z.; Music, D.; Ahuja, R.; Li, S.; Schneider, J. M. *Phys. Rev. B* 2004 ,*70,* 092102-092104.

[31] Bouhemadou, A. *Solid State Sci.* 2009, *11,* 1875–1881.

[32] Etzkorn, J.; Ade, M.; Kotzott, D.; Kleczek, M.; Hillebrecht, H.; *J. Solid State Chem.* 2009, *182,* 995- 1002.

[33] Bai, Y.; He, X.; Li, Y.; Zhu, C.; Li, M. *Solid State Comm.* 2009 ,*149,* 2156- 2159.

[34] Phataka, N. A.; Saxena, S. K.; Fei , Y.; Hu, J. *J. Alloys Comp.* 2009 ,*474,* 174–179.

[35] Manoun, B.; Amini, S.; Gupta, S.; Saxena, S. K.; Barsoum, M. W. *J. Phys: Condens. Matter* 2007, *19*, 456218- 456225.

[36] Manoun, B.; Yang, H.; Saxena, S. K.; Ganguly, A.; Barsoum, M. W.; El Bali, B.; Liu, Z. X.; Lachkar, M. *J. Alloys Comp.* 2007,*433,* 265–268.

[37] Högberg, H.; Eklund, P.; Emmerlich, J.; Birch, J.; Hultman, L. *J. Mater. Res.* 2005, *20*, 779- 783.

[38] Kumar, R. S.; Rekhi, S.; Cornelius, A. L.; Barsoum, M.W. *Appl. Phys. Lett.* 2005, *86*, 111904- 111906.

[39] Bai, Y.; He, X.; Li, M.; Sun, Y.; Zhu, C.; Li, Y. *Solid State Sci.* 2010, *12,* 144- 147.

[40] Bouhemadou, A. *Modern Phys. Lett. B*, 2008, *22*, 2063-2076.

[41] Medkour, Y.; Bouhemadou, A.; Roumili, A. *Solid. State Comm.* 2008, *148*, 459-463.

[42] Bouhemadou, A. *Physica B* 2008, *403*, 2707- 2713.

[43] Roumili, A.; Medkour, Y.; Maouche, D. *Inter. J. Modern Phys. B* 2009, *23,* 5155-5161.

[44] Dubois, S.; Cabioc'h, T.; Chartier, P.; Gauthier, V.; Jaouen, M. *J. Am. Ceram. Soc.* 2007, *90*, 2642- 2644.

[45] Kanoun, M. B.; Jaouen, M. *J. Phys: Condens. Matter.* 2008, *20*, 085211- 085215.

[46] Warner, J. A.; Patil, S. K. R.; Khare, S. V.; Masiulaniec, K. C. *Appl. Phys. Lett.* 2006, *88*, 101911- 101913.

[47] Schreiber, E.; Anderson, O. L.; Soga, N. *Elastic Constants and Their Measurement.* McGraw- Hill, New York, 1973.

[48] Green, D. J. *An introduction to the mechanical properties of ceramic*, Cambridge university press, 1998.

[49] Grossman, J. C.; Mizel, A.; Coté, M.; Cohen, M. L.; Louie, S. G. *Phys. Rev. B.* 1999, 60, 6343- 6347.
[50] Liermann, H. P.; Snigh, A. K.; Manoun, B.; Saxena, S. K.; Prakapenka, V. B.; Shen, G. *Int. J. Refract. Met. Hard Mater.* 2004, 22. 129-132.
[51] Liermann, H. P.; Snigh, A. K.; Somayazulu, M.; Saxena, S. K. *Int. J. Refract. Met. Hard Mater.* 2007, 25. 386- 391.
[52] Hettinger, J. D.; Lofland, S. E.; Finkel, P.; Meehan, T.; Palma, J.; Harrell, K.; Gupta, S.; Ganguly, A.; El-Raghy, T.; Barsoum, M. W. *Phys. Rev B.* 2005, *72*, 115120- 115125.
[53] Manoun, B.; Zhang, F.; Saxena, S. K.; Gupta, S.; Barsoum, M. W. *J. Phys.: Condens. Matter.* 2007, *19*, 246215- 246223.
[54] Emmerlish, J.; Music, D.; Houben, A.; Dronskowski, R.; Schneider, J. M. *Phys. Rev. B.* 2007, *76*. 224111- 224117.
[55] Liao, T.; Wang, J.; Zhou, Y. *J. Mater. Res.* 2009, *24*, 556- 564 .
[56] Liermann, H. P.; Singh, A. K.; Manoun, B.; Saxena, S. K.; Zha, C. S.; *Int. J. Refract. Met. Hard Mater.* 2005, *23*, 109- 114.
[57] Farber, L.; Levin, I.; Barsoum, M. W.; El-Raghy, T.; Tzenov, T.; *J. Appl. Phys.* 1999, *86*, 2540- 25430.
[58] Wang, Z.; Zha, C.S.; Barsoum, M.W. *Appl. Phys. Lett.* 2004, 85, 3453–3455.
[59] Barsoum, M. W.; El-Raghy. T.; *Am. Sci.* 2001, *89*, 334- 343.
[60] Sun, Z. M.; Zhang, Z. F.; Hashimoto, H.; Abe, T. *Mater. Trans.* 2002, *42,* 432-435.
[61] Yoo, H. I.; Barsoum, M. W.; El-Raghy. T. *Nature* 2000, *407*, 581-582.
[62] Sun, Z.; Zhou, J.; Music, D.; Ahuja, R.; Schneider, J.M.; *Scripta Mater.* 2006, *54*, 105–107.
[63] Williams, W. S. *Prog. Solid St. Chem.* 1971, *6*, 57- 118.
[64] Barsoum, M. W.; El-Raghy, T.; Porter W. D.; Wang, H.; Ho, J. C.; Chakraborty, S. *J. Appl. Phys.* 2000, *88*, 6313- 6316.
[65] Euklund, P.; Beckers, M.; Jansson, U.; Hogberg, H.; Hultman, L. *Thin Solid Films* 2010, *518*, 1851- 1878.
[66] Barsoum, M. W.; Yoo, H.-I.; Polushina, I. K.; Rud', V. Yu.; Rud', Yu. V; El-Raghy, T. *Phys. Rev B* 2000, *62*, 10194-10198.
[67] Lofland, S. E.; Hettinger, J. D.; Meehan, T.; Bryan, A.; Finkel, P.; Gupta, S.; Barsoum, M. W.; Hug, G. *Phys. Rev. B* 2006, *74*, 174501- 174505.
[68] Chaput, L.; Hug, G.; Pécheur, P.; Scherrer, H.; *Phys. Rev. B.* 2007, *75*, 035107-035111.
[69] Y. C. Zhou, H. Y. Dong, X. H. Wang and S. Q. Chen, *J. Phys.: Condens. Matter.* 2000, *12,* 9617- 9627.
[70] Scabarozi, T.H.; Eklund, P.; Emmerlich, J.; Högberg, H.; Meehan, T.; Finkel, P.; Barsoum, M.W.; Hettinger, J.D.; Hultman, L.; Lofland, S.E.; *Solid State Comm.* 2008, *146*, 498- 501.
[71] Zhou, Y.; Sun, Z. *Phys. Rev. B.* 2000, *61*, 12570- 12573.
[72] Haddad, N.; Garcia- Laurel, E.; Hultman, L.; Barsoum, M. W.; Hug, G.; *J. Appl. Phys.* 2008, *104*, 023531- 023540.
[73] Alexander L. Ivanovsky, Dmitry L. Novikov, Gennady P. Shveikin. Mendeleev Comm. 1995, 5, 90-91.
[74] Wang, J. Y.; Zhou, Y. C. *J. Phys.: Condens. Matter.* 2003, *15*, 1983- 1991.
[75] Wang, J.; Wang. J.; Zhou, Y.; Hu, C. *Acta Mater.* 2008, *56*, 1511-1518.
[76] Li, C.; Wang, B.; Li, Y.; Wang, R. *J. Phys. D: Appl. Phys.* 2009, *42*, 065407- 065411.

In: MAX Phases: Microstructure, Properties and Applications ISBN 978-1-61324-182-0
Editors: It-Meng (Jim) Low and Yanchun Zhou

Chapter 8

SUPERCONDUCTIVITY IN THE M_2AX NANOLAMINATES COMPOUNDS

Antonio Jefferson da Silva Machado and Ausdinir Danilo Bortolozo
University of São Paulo – EEL – USP, Materials Science Department,
Pólo Urbo Industrial Gleba AI 6 – Lorena – São Paulo – Brazil

ABSTRACT

In this chapter, we show that the two-dimensional nature (2D) of the M_2AX lamellar structure compounds is favorable for the existence of superconductivity in this new class of the compounds. In spite of there being numerous theoretical and experimental studies devoted to M_2AX phases, the physical properties of only 6 compounds were reported as having superconducting behavior. We discuss experimental and theoretical results which suggest enormous perspectives to explore the possibility of other high T_c superconductors with lamellar structure. It is also shown that nitride compound (Ti_2InN) possesses more than twice superconducting critical temperature of the carbide compound (Ti_2InC). Our results suggest that other nitride compounds may also exhibit high superconducting critical temperature which may represent a new class of high temperature superconducting.

1. INTRODUCTION

Superconductivity was discovered by H. Kammerlingh Onnes in 1911 [1]. The first superconductor material discovered by Onnes was Mercury, which exhibits zero resistivity below 4.1 K. This discovery was followed by the observation of other metals which exhibit zero resistivity below a certain critical temperature. Since then, many other pure elements such as Nb, Sn, In, were discovered as superconductors. Also, many alloys exhibit superconductivity, many of which are Nb based, such as Nb_3Sn, Nb_3Ge, etc. The fact that the resistance is zero has been demonstrated by sustaining currents in superconducting lead rings for many years with no measurable reduction. An induced current in an ordinary metal ring would decay rapidly from the dissipation of ordinary resistance, but superconducting rings have exhibited a decay constant of over a billion years.

Superconductivity is a fascinating and challenging field of physics. Scientists and engineers throughout the world have been striving to develop an understanding of this remarkable phenomenon for many years. Superconductivity is being applied to many diverse areas such as: medicine, theoretical and experimental science, a new kind public transportation, electronic devices, as well as many other areas. In ordinary conductors such as copper and silver, at lower temperature the resistivity is limited by impurities and other defects. Even near absolute zero, a real sample of copper shows some electrical resistance. In a superconductor however, despite these imperfections, the resistance drops abruptly to zero when the material is cooled below a certain temperature which is called superconducting critical temperature (T_c). One of the properties of a superconductor is that it will exclude magnetic fields, a phenomenon called the Meissner effect. The properties of superconductors were modeled successfully by the efforts of John Bardeen, Leon Cooper, and Robert Schrieffer in what is commonly called the BCS theory [2]. A key conceptual element in this theory is the pairing of electrons close to the Fermi level into Cooper pairs through interaction with the crystal lattice. This pairing result from a slight attraction between the electrons related to lattice vibrations; the coupling to the lattice is called a phonon interaction. Pairs of electrons can behave very differently from single electrons, which are fermions and must obey the Pauli exclusion principle. The pairs of electrons act more like bosons, which can condense into the same energy level. The electron pairs have a slightly lower energy and leave an energy gap above them on the order of .001 eV, which inhibits the kind of collision interactions which lead to ordinary resistivity. For temperatures such that the thermal energy is less than the band gap, the material exhibits zero resistivity. Bardeen, Cooper, and Schrieffer received the Nobel Prize in 1972 for the development of the theory of superconductivity.

In 1986, it was discovered that some cuprates-perovskite ceramic of the La-Ba-Cu-O system, possess superconducting critical temperature above 30 K [3]. It was shortly found that replacing the Lanthanum per Yttrium, i.e. making YBCO, the superconducting critical temperature goes up to 92 K [4]. A new era in the study of superconductivity began with the discovery of these materials, renewing interest in the topic because the current theory (BCS), could not explain them. From a practical perspective, 90 K is easy to reach with the readily available liquid nitrogen (boiling point 77 K). This means more experimentation and more commercial applications are feasible, especially if materials with even higher critical temperatures can be discovered. Like ferromagnetism and atomic spectral lines, superconductivity is a quantum mechanical phenomenon. It cannot be understood simply as the idealization of "perfect conductivity" in classical physics.

High-temperature superconductors can operate at liquid nitrogen temperatures (77 K). Unique and exciting opportunities exist today for students to explore and experiment with this new and important technological field of physics.

In this chapter we discuss superconductivity in MAX nanolaminates compounds, which can represent a new class of superconducting materials. We show evidence which gives support to the prospect of a new high critical superconducting temperature.

1.1. General Aspects about *MAX* Nanolaminate Compounds

The MAX phases, which were introduced by Barsoum et al. [5] as MAX, are a group of ternary carbides/nitrides with $M_{n+1}AX_n$ chemistry where M is an early transition element, A is an A-group element, X is carbon and/or nitrogen, and n is an integer 1, 2, or 3. Henceforth, based on the stoichiometry, these phases will be referred to as 211, 213, and 413 for n=1,2,3 respectively. Fig 1 shows where in the periodic table the MAX-phase can be found. There are about 50 different thermodynamically stable MAX-compounds. The MAX phases are readily machinable and possess a lamellar structure. Several different structures exist within the group depending on the value of n, thereby forming different unit cells (Figure 2) [6]. The carbon spacing in the 211 structure is the most evenly spaced, followed by the 312 and 413 structures. The ratio of the unit cell height to its width, c/a, for a particular structure gives an estimate of the uniformity of the carbon spacing. A material with a lower ratio, has more uniformly distributed carbon than a material with a higher ratio. The nominator for all MAX-compounds is the nanolaminates structure with a matrix of MX slabs interleaved by single layer of A-element. As can be seen in Figure 2 all the composition has a hexagonal unit cell and they belong to space group $D_{6h}^{4} - P6_3/mmc$.

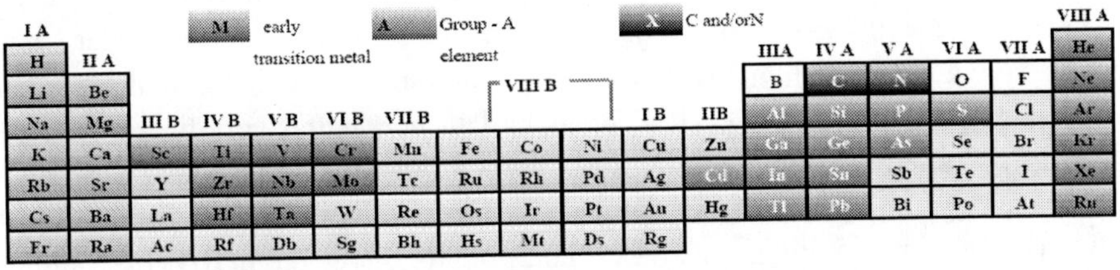

M early transition metal; A Group - A element; X C and/orN

I A	II A	III B	IV B	V B	VI B	VII B	VIII B			I B	IIB	IIIA	IV A	V A	VI A	VII A	VIII A
H																	He
Li	Be											B	C	N	O	F	Ne
Na	Mg											Al	Si	P	S	Cl	Ar
K	Ca	Sc	Ti	V	Cr	Mn	Fe	Co	Ni	Cu	Zn	Ga	Ge	As	Se	Br	Kr
Rb	Sr	Y	Zr	Nb	Mo	Tc	Ru	Rh	Pd	Ag	Cd	In	Sn	Sb	Te	I	Xe
Cs	Ba	La	Hf	Ta	W	Re	Os	Ir	Pt	Au	Hg	Tl	Pb	Bi	Po	At	Ru
Fr	Ra	Ac	Rf	Db	Sg	Bh	Hs	Mt	Ds	Rg							

Figure 1. MAX phases are made by metal transition elements, A-group (IIIA or IVA) and C or N.

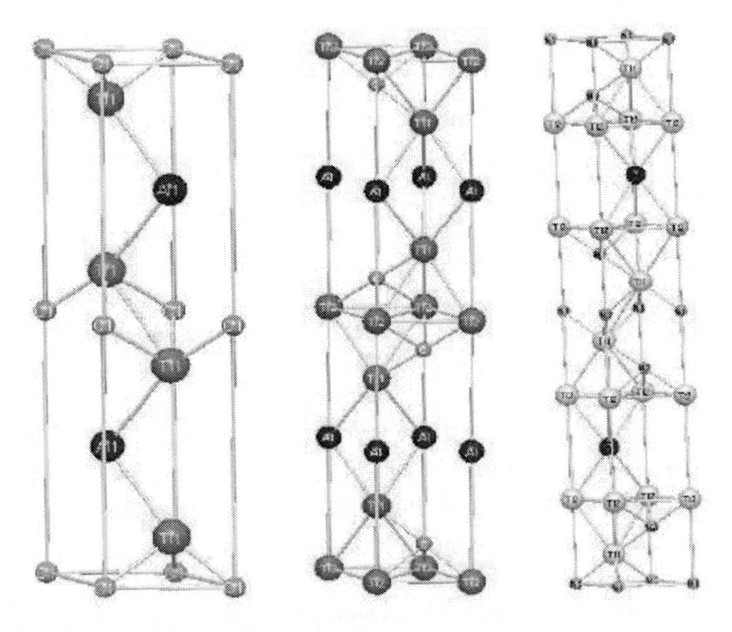

Figure 2. MAX unit cells of 211, 213 and 314 family.

Despite the numerous theoretical and experimental studies devoted to M_2AX phases, the physical properties of the majority of the 240 elemental possible combinations have not been effectively studied. On the other hand, an examination of published data suggests that much work has been done with Al as an A element and C as X element in MAX-series. However, individual properties vary from phase to phase depending on M, A and X elements. It is worthwhile to know the effect on physical properties by replacing Al by Ga and C by N in M_2AlC.

The advances in synthesis of anisotropic materials 2D (two-dimensional) are responsible for development of the modern superconductivity concepts. Among such systems, the greatest interest has been shown in magnesium diboride MgB_2, which presents the surprising superconducting critical temperature of 39 K [7]. In this material, as well as in related boride phases [8], the superconducting transition is caused by interaction between the electrons of boron atoms, which form planar graphite – like networks in the crystal and optical phonons. Another good example of superconductivity in layered compounds is the family of the borocarbides [9]. This family is composed of alternating molecular layers of a rare-earth (RE) carbide and transition metal (M) boride, which obey the following sequence: M_2B_2-REC-M_2B_2-RE…., where B and C represents boron and carbon atoms respectively. The superconducting critical temperature can reach 23 K and the superconductivity in these crystals is due to electron-phonon interaction in boride layers. Within this context, the MAX phases have great similarity with other layered compounds. In particular the M_2AX compounds have a hexagonal lattice (space group P63/mmc) in which the lattice is composed of molecular M_2X layers where each X atom has an octahedral coordination (XM_6). Each M_2X layer is separated from neighboring ones by A atom which obeys the sequence X-M-A-M-X. Alternatively, this structure can be seen as a rocksalt-like MX separated by A group elements. The lamellar natures of these compounds give them 2D (two-dimensional) characteristics where the electric and thermal properties are anisotropies. Indeed, bands structure calculations show that the electrical conductivity in the basal plane (a lattice parameter) is a magnitude order higher than the perpendicular axis (c parameter) [10-11]. These interesting features are promising from the superconductivity view point and so far Mo_2GaC [12], Nb_2SC [13], Nb_2SnC [14], Nb_2InC [15], Nb_2AsC [16] and Ti_2InC [17] are the superconducting materials known. For some of them, the features of their electronic band structure were elucidated through the use of density functional calculations.

1.2. Superconductivity Mechanisms

Most superconductors discovered so far have a niobium base such as Nb_2SC [13], Nb_2SnC [14], Nb_2InC [15] and Nb_2AsC [16]. Discussions about the possible superconducting mechanism would be interesting in order to compare the electronic properties of two compounds with the same base. Recently J. R. Shein et.al [18], published his results about calculations of the electronic properties of the Nb_2SC and Nb_2SnC superconductors. In this article, the authors discuss a comparison of the band structure of these two materials. Some authors have suggested that the superconductivity in Nb_2SC arises due to pairing of electrons of sulfur atoms arranged in planar networks [17]. At the same time, according to the conventional BCS theory (Bardeen-Cooper-Schrieffer) [19] the superconducting critical temperature (Tc) has the $Tc \sim 1/M^{1/2}$ dependence. However, the Nb_2SnC compound displays the higher superconducting critical temperature (Tc ~ 7.8 K) than Nb_2SC (Tc < 5.0 K). Considering that the sulfur atom is lighter than the tin atom, the substitution of tin atoms for sulfur atoms in the nanolaminates network should cause a sensitive increase at the superconducting critical temperature. Therefore, the experimental results show that the superconducting critical temperature for Nb_2SnC is higher than Nb_2SC, which strongly suggests that the pairing mechanism in these phases should have a different nature. Perhaps the major role in the superconductivity of nanolaminates is played by the states of carbide molecular layers. These layers in nanolaminates compound are composed of CNb_6 octahedra, such as occur in NbC (niobium monocarbide – Tc ~ 11 K [20]). But a change in the composition of these layers (for example due to carbon nonstoichiometry) provokes an abrupt change of the superconducting critical temperature of nanolaminates compounds. According to the McMillan formula, this gives for equation 1:

$$\lambda = N(EF)\langle I2\rangle/M\langle\omega 2\rangle \tag{1}$$

where $N(E_F)$ is the density of electronic states near the Fermi level (E_F) and $\langle I2\rangle$ and $\langle\omega 2\rangle$ are the averaged squares of the electron-phonon and phonon-phonon matrix elements respectively. Then, the λ and Tc values for both Nb_2SnC and Nb_2SC nanolaminates compounds should correlate with $N(E_F)$ values in the Nb-C layers because in both there are identical niobium carbide layers. Using this approach, the band structure was calculated through the density functional calculations [18]. These results for both compounds (Nb_2SnC and Nb_2SC) revealed that indeed the density of states near the Fermi level increase from Nb_2SC to Nb_2SnC. These authors suggest that the decisive role in the superconducting of the nanolaminates compounds is played by electron-phonon coupling involving the states of molecular layers of niobium carbide. Thus, these results arise as a consequence from the 2D (quasi-two-dimensional) characteristics of these compounds.

Another compound which displays relatively high critical temperature is the Nb_2InC (T_c ~ 7.5 K) compound [15]. Superconducting behavior was recently reported, and the data presented in this article is more comprehensive. Besides the conventional resistivity and magnetic measurements, the heat capacity in low temperature measurements is shown. The normal-state specific heat measured at $\mu_0 H = 0$ T was fitted to $Cp = \gamma T + \beta T^3$, where the first, and second, terms represent the electronic and the lattice contributions, respectively. The linear fit to these data gives: $\gamma \sim 12.6$ mJ mol^{-1} K^{-2} and $\beta \sim 0.54$ mJ mol^{-1} K^{-4} [15]. These values are higher than in the isostructural materials Ti_4AlN_3 ($\gamma = 8.12$ mJ mol^{-1} K^{-2}) [21], and Ti_3SiC_2 ($\gamma = 5.21$ mJ mol^{-1} K^{-2}) [22], suggesting relatively high density states at Fermi level. The Debye model connects the β coefficient and Debye temperature (Θ_D) through equation 2 [18] .

$$\theta_D = \left(\frac{12\pi^4}{5\beta_m} nR \right)^{\frac{1}{3}} \tag{2}$$

where $R = 8.314$ J mol^{-1} K^{-1}, $\beta_m = 4\beta$, and $n = 4$ for Nb_2InC. With this data it was possible to estimate $\Theta_D \sim 154$ K. With this value the electron–phonon coupling constant (λ_{ep}) can be estimated from another form of the McMillian's relation (equation 3) [23].

$$\lambda_{ep} = \frac{1.04 + \mu^* \ln\left(\frac{\theta_D}{1.45T_c}\right)}{(1 - 0.62\mu^*)\ln\left(\frac{\theta_D}{1.45T_c}\right) - 1.04} \qquad (3)$$

where μ^* is a Coulomb repulsion constant. A typical value for μ^* is 0.10. Taking μ^* in the range 0.05–0.2, is found $\lambda_{ep} \sim 0.8$–1.2, which implies that Nb_2InC is a moderately strong coupled superconductor. However, it is important to emphasize that all data available in the literature are in reference to the polycrystalline materials. So, due to 2D behavior of these materials, the values can represent an average because any anisotropic factors are being used. But independent of these questions, all these articles show unambiguously that these layered configurations are favorable for the existence of superconductivity in this intriguing new class of the materials. In order to clarify the 2D behavior of these materials, a comparison between two compounds that have the same M and A atoms, but different X elements will be discussed in the next section.

2. Superconductivity in the Ti_2InX (X = C or N)

The Ti_2InC compound was first reported by Jeitschko *et al.* and recently synthesized in bulk form [24-25]. The presence of covalent-bonded TiC octahedral, built in 2D oriented packets connected by In metallic layers is also presented in this compound such as mentioned for Nb_2InC. Recently Bortolozo *et al.* [17] reported that this is also a superconductor material with critical temperature close to 3.1 K. The band structure calculations also reveal 2D – like behavior of the quasi-flat electronics bands along the *c* axis [26].

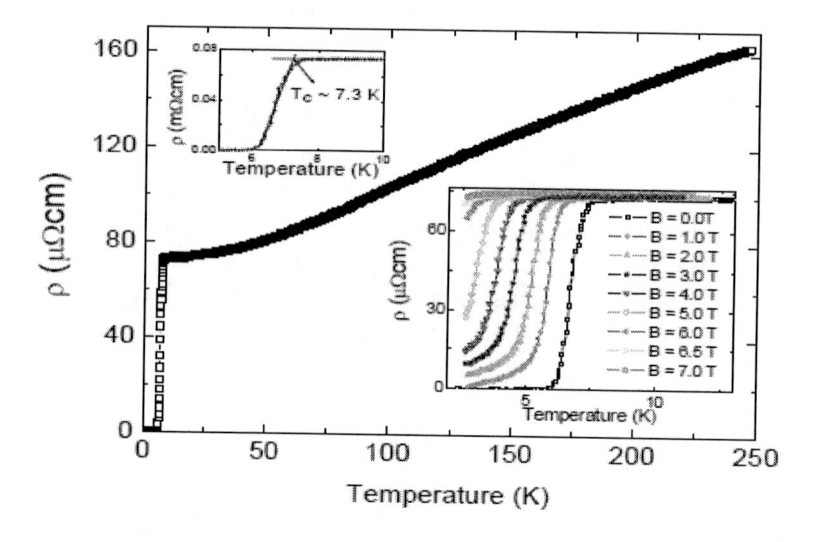

Figure 3. Resistivity as a function of temperature showing a clear superconducting transition near 7.1 K for Ti_2InN compound. The top inset shows the details from transition and the T_c criteria definition. The bottom inset displays the dependence of the ρ(T,B) displaying a progressive shift of T_c with increasing B extracted from reference [29].

The near-Fermi bands involved in the formation of superconducting state are formed mainly by Ti *3d* states. On the other hand, in 1963 Jeitschko *et al.* discovered the Ti_2InN compound [27], which crystallizes in the same prototype structure of carbides (Cr_2AlC). In this structure the basic structural component is an octahedron of six Ti atoms with an N atom instead of carbon residing in the center [23]. The electronic band structure shows that the interactions in the TiN octahedra are weaker than those in the Ti_6N octahedra in agreement with the general trend known for binary carbides and nitrides [28]. The electronic structure calculations show that Ti_2InC has 3.67 states/(eV cell) at Fermi level (E_F), while the Ti_2InN has 4.02 states/(eV cell) at E_F [28]. It is clear that the nitrogen atoms change the electronic structure of these compounds and increase the state density at Fermi level. This characteristic of band structure could probably affect the transport properties if compared with the close relative Ti_2InC. Indeed Bortolozo *et al.* [29] studied the electrical transport of the Ti_2InN compound and they found a superconducting critical temperature close to 7.3 K. The resistivity as a function of temperature, which clearly displays a superconducting transition, is shown in the Figure 3. The inset (top) shows the detail of the transition while the inset on the bottom displays the dependence of the $\rho(T, B)$ indicating a progressive shift of T_c with increasing B.

The $\rho(T, B)$ dependence suggests that the upper critical field (H_{c2} or B_{c2}) is $\mu_o H_{c2} \sim 10.8$ T which can be obtained through the relation given by equation (4) and (5) [30].

$$\mu_o H_{c2}(0) = 0.693\left[-\left(\frac{dB_{c2}}{dT}\right)_{T=Tc}\right]T_c \tag{4}$$

$$\mu_o H_{c2}(T) = \mu_o H_{c2}(0)\left[1-\left(\frac{T}{T_c}\right)^2\right] \tag{5}$$

In the equation (4) the $H_{c2}(0)$ value can be estimated through the slop of the curve estimated from data obtained in $\rho(T, B)$ dependence. Actually, the dependence of $\mu_o H_{c2}$ as a function of temperature is a phase diagram which defines the upper critical field [$\mu_o H_{c2}(0)$] as a thermodynamic critical parameter. The fitting of the dates can be made with the auxiliary of the equation (5). The Cooper pairs, which are responsible for zero resistance into a superconducting state, are directly related with the upper critical field through equation 6.

$$H_{c2}(0) = \frac{\Phi_0}{2\pi\mu_0\xi^2} \tag{6}$$

where $H_{c2}(0)$ is the upper critical field at zero temperature, Φ_o represents the quantum magnetic flux (2.07×10^{-15} Tm^2), μ_o is the vacuum permeability and $\xi_{(0)}$ signifies the Ginzburg-Landau coherence length, which is related to the distance between two electrons that form the Cooper-pair. So, the value estimated from equation 6 yields to a coherence length $\xi_{(0)} \sim 55$ Å at zero temperature. Independently of the discussion about critical parameters calculated for this compound in literature, the most important discussion is the sensitive increase of the superconducting critical temperature. This result represents the first evidence of the superconductivity in a nitride of this family of materials. In comparison with

Ti_2InC, the nitride compound possesses more than twice the superconducting critical temperature. This represents clear evidence of the 2D characteristic of this class of materials. The nature of the covalent bond for the two materials (Ti_2InC and Ti_2InN) is undoubtedly different, such as shown in the electronic band calculations for both materials; therefore superconductivity appears to be occurring in the M_2X layer. Considering that many nitrides of the early transition metal are superconductors (see Table 1), it is possible to make a simple speculation. From table 1 the TiN superconducts at 5.6 K and Ti_2InN posses more than twice the superconducting critical temperature of the correspondent carbide (Ti_2InC). If it is true that the nitride layers are responsible for the improvement of T_c in the Ti_2InX, it is reasonable to admit that a nitride like Nb_2AN (remember that A means an element of the IVB group) could have the critical temperature higher than the corresponding carbide compound. This speculation is possible because NbN is superconductor at 16.6 K.

Table 1. Some nitrides of the early transition metal with its respective superconducting critical temperature

Compound	Superconducting critical temperature (T_c) optimized	Reference
NbN	16.6 K	31
MoN	14.8 K	31
TaN	14 K	31
ZrN	10.7 K	31
HfN	8.8 K	31
VN	8.5 K	31
TiN	5.6 K	31

These results are promising because they open enormous perspectives to explore the possibility of other high T_c superconductors with lamellar structure.

3. Other Superconducting Compounds of This Family

There are about 50 carbides reported in this family which posses M_2AX composition, but only 6 compounds were reported as superconductor. However it is possible to find two more superconducting compounds which belong to this family [29] which are not yet reported in the literature. These compounds are summarized in Table 2.

Table 2. Superconducting compounds found by the work of Bortolozo *et al.* [29]. The * symbols indicate that the material was not yet reported in the literature

Compound	T_c (K)	Reference
Ti_2InC	3.1	17
Ti_2InN	7.5	29*
Nb_2SnC	7.8	14
Nb_2InC	7.5	11
Ti_2GeC	9.8	29*
Nb_2GeC	9.0	29*

In particular, it will be necessary to discuss the Ti_2GeC compound which displays the highest superconducting critical temperature among the compounds of this intriguing family of materials. There is no superconducting relationship with any binary phase in the Ti – Ge system. But the critical temperature of Ti_2GeC is high enough if compared with other carbide compounds found so far. Figure 4 shows the electrical behavior of this material which can be observed as a sharp superconducting transition close to 9.8 K. The inset of this figure (bottom) shows the R(B,T) versus T, which suggests the upper critical field at zero temperature of about 8.1 T. Through equation 5, the Ginzburg-Landau coherence length is shown to be about 64 Å at zero temperature.

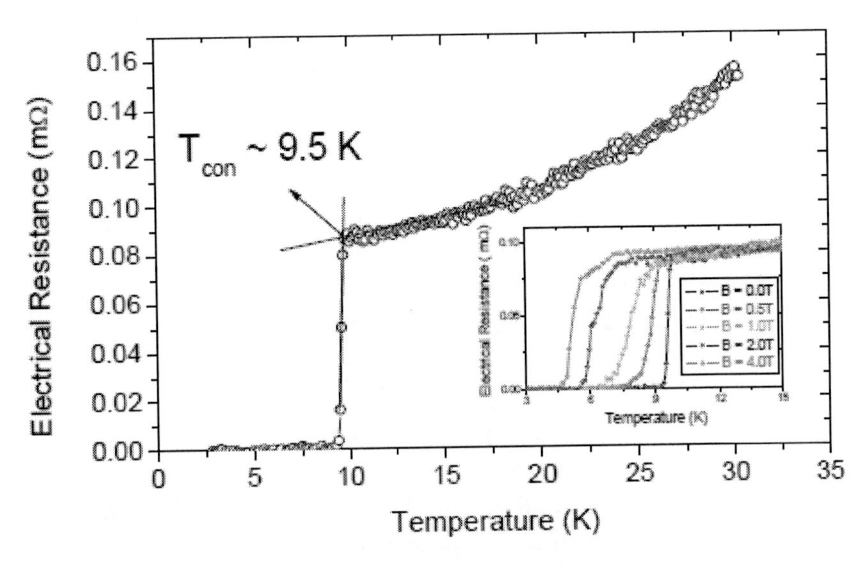

Figure 4. Electrical resistance as a function of temperature for Ti_2GeC compound. Inset shows the electrical resistance dependence with applied magnetic field [29].

These results are promising because other compounds of this family can exhibit superconducting behavior in many critical temperatures different than those presented in this chapter. Once again it is important to emphasize that the possibility of the existence of nitrides of these compounds like Ti_2GeN can represent the possibility of the existence of a new high T_c superconductor family. Another important aspect is the synthesis process used to produce these compounds. For example, Nb_2SnC is a superconductor when produced by the method reported in the reference [15]. The choice of the synthesis route is fundamental for the quality of the superconducting phase. For this reason, we will discuss the importance of the method of synthesis for superconductivity in these materials.

4. Synthesis Method Versus Superconducting Behavior

The thermodynamic equilibrium of the phases in the M-A-X system is extremely complicated for some compounds. Transformations such as peritectic or other kinds of transformations are often present in most compounds of this family. This fact does not permit the synthesis of good quality (single phase), for example, by the arc-melting method. Sometimes a small superconducting signal has been observed in magnetization measurements (low superconducting volume fraction) in samples prepared by conventional powder metallurgy. For example, we can mention the Nb_2SnC compound; when it is prepared by powder metallurgy, it displays a low superconductor fraction. However, if this compound is submitted to high pressure during the synthesis in low temperature (500°C), the superconducting signal sensitively increases. This fact suggests an improvement in the superconducting volume. As a consequence of high pressure synthesis, the superconducting transition is very sharp in the resistivity measurements. The simplest to synthesize is Ti_2InC compound which has good quality even in conventional powder metallurgy. But the best results are obtained when the synthesis is realized in temperature lower than 800°C. At temperature higher than 800°C, secondary phases like In and Ti_2C are observed which produce poor quality material. The success in Ti_2InN was possible through heat treatment in a tubular furnace under 130 bar argon pressure at 900°C for 10 h. Finally, Ti_2GeC often displays segregation of Ti_2C and Ge, when this compound is synthesized through conventional powder metallurgy, suggesting that the best route could be the high pressure synthesis. Besides, most of the superconductor compounds reported was submitted to the quenching even when made by powder metallurgy method. Perhaps, some defect structure could be generated during each kind of the heat treatment, but to clarify this suspect neutron diffraction experiments are necessary. Therefore, in each material synthesis, the superconducting behavior is extremely dependent on the heat treatment method used.

Conclusions

We have shown that the two-dimensional nature of the M_2AX lamellar structure compounds is favorable for the existence of superconductivity. The results about nitride compound of Ti_2InN composition possess more than twice the critical temperature of the carbide correspondent. This suggests that other nitride compounds of this family may exhibit high superconducting critical temperature (T_c) and may represent a new class of high temperature superconductor.

Acknowledgment

The authors are grateful to the journalist Frank Braun for the corrections and some suggestions in the text.

References

[1] Onnes, K.H. *Comm. Phys. Lab.* Univ. Leiden. 1911, 122 and 124.

[2] Bardeen, J.; Cooper, L.N.; Schrieffer, J.R. *Phys. Rev.* 1957, 108, 1175-1204.

[3] Berdnorz, J.G.; Muller. K.A. *Z. Phys.* 1986, B64 (2), 189-193.

[4] Wu, K.; Ashburn, J. R.; Torng, C.J.;. Hor, P.H; Meng, R.L.; Gao, L.; Huang, Z.J.; Wang Y.Q.; Chu C.W. *Phys. Rev. Lett.* 1987, 58, 908-910.

[5] Barsoum, M.W. *Prog. Solid State Chem.* 2000, 28, 201-281.

[6] Barsoum, M.W.; El-Raghy, T. *American Scientist*. 2001, 89, 334-343.
[7] Nagamatsu, J.; Nakagawa, N.; Muranaka, T.; Zenitani, Y.; Akimitsu. J. *Nature*. 2001, 410, 63-64.
[8] Ivanovskii, A.L. *Fiz. Tverd. Tela*. 2003, 45, 1742–1769.
[9] Cava, R.J., Takagi, H., Zandbergen, H.W., Krajewski, J.J., Peck, W.F., Siegrist, T., Barlogg, B., Vandover, R.B., Fedler, R.J., Mizuhashi, K., Lee, J.O., Eisaki, H., and Uchida, S. *Nature*. 1994, 367,6460, 252–253.
[10] Zhou Y.; Sun, Z. *Phys. Rev. B*. 2000, 61, 12570-12573.
[11] Zhou Y.; Sun, Z.*J. App. Phys*. 1999, 86, 1430-1432.
[12] Toth, L.E.; Jeitschiko, W.; Yen, M. *J. Less-Common Metals*. 1966, 10, 29-32.
[13] Sakamaki, K.; Wada, W.; Nozaki, H.; Onuki, Y.; Kawai, M. *Solid State Comm*. 1999, 112, 323-327.
[14] Bortolozo, A.D.; Santanna, O.H. ; Da Luz, M.S.; Dos Santos, C.A.M.; Machado, A.J. S.; Pereira, A.S.; Trentin, K.S. *Solid State Communications*. 2006, 139, 57-59.
[15] Bortolozo, A.D.; Fisk, Z.; Santanna, O.H..; Dos Santos, C.A.M.; Machado, A.J.S. *Physica C*. 2009, 469, 256-258.
[16] Lofland, S.E.; Hettinger, J.D.; Meehan, T.; Bryan, A.; Finkel, P.; Gupta, S.; Barsoum, M.W.; Hug, G. *Physical Review B*. 2006, 74, 174501/1-174501/5.
[17] Bortolozo, A.D.; Santanna, O.H.; Dos Santos, C.A.M.; Machado, A.J.S. *Solid State Communications*. 2007, 144, 419-421.
[18] Shein, I.R.; Bamburov, V.G.; Ivanoskii, A.L. *Doklady Physical Chemistry*. 2006, 411, 317-321.
[19] Bardeen, J.; Cooper, L.N.; Schrieffer, J.R. *Phys. Rev*. 1957, 108, 1175-1204.
[20] Matthias, B.T. *Physical Review*. 1953, 92, 874-876.
[21] Ho, J.C.; Hamdeh, H. H.; Barsoum, M.W.; El-Raghy, T. *J. Appl. Phys*. 1999, 86, 3609-3611.
[22] Drulis, M.K.; Drulis, H.; Hackemer A.E.; Gangul, A.; El-Raghy, T.; Barsoum, M.W. *J. Alloys and Compounds*. 2007, 433, 59-62.
[23] McMillian, W.L. *Phys. Rev*. 1967, 167, 331-344.
[24] Jeitschko, W.; Nowotny, H.; Benesovsky, F. *Monatsh. Chem*. 1963, 94, 1201-1205.
[25] Barsoum, M.W.; Golczewski, J.; Seifert, H.J.; Aldinger, F. *J. Alloy Compounds*. 2002, 340, 173-179.
[26] He, X.; Bai, Y.; Li, Y.; Zhu, C.; Li, M. *Solid State Communications*. 2009, 149, 564-566.
[27] Jeitschko, W.; Nowotny H.; Benesovsky, F. *Monatsh. Chem*. 1964, 95, 178-179.
[28] Ivanovskii, A.L.; Sabiryanov, R.F.; Skazkin, A.N.; Zhukovskii, V.M.; Shveikin, G.P.; *Inorganic Materials*. 2000, 36, 28-31.
[29] Bortolozo, A.D. *Superconductivity in the lamellar carbide of the M_2AX family*. 2009, Brazil. Thesis (Doctoral in Materials Engineering) – Universidade de São Paulo, Escola de Engenharia de Lorena, Lorena, São Paulo.
[30] Werthamer, N.R.; Helfand, E.; Hohenberg, P.C. *Phys. Rev*. 1966, 147, 295-302.
[31] Roberts, BW. *J. Phys. Chem. Ref. Data*. 1976, 5, 581-821.

In: MAX Phases: Microstructure, Properties and Applications ISBN 978-1-61324-182-0
Editors: It-Meng (Jim) Low and Yanchun Zhou

Chapter 9

STRUCTURAL AND MICROSTRUCTURAL IRRADIATION-INDUCED DAMAGE IN TITANIUM SILICON CARBIDE

Jean-Christophe Nappé[1] and Fabienne Audubert[2]
[1]École Nationale Supérieure des Mines, SPIN/PMMC, LPMG UMR CNRS 5148
158 cours Fauriel, 42023 Saint-Étienne cedex 2, France
[2]CEA, DEN, DEC/SPUA/LTEC, Cadarache, 13108 St Paul lez Durance, France

ABSTRACT

This chapter presents an overview of the behaviour of the Ti_3SiC_2 MAX phase under ionic irradiation, from both structural and microstructural points of view. Thus, it appears that Ti_3SiC_2 is relatively resistant towards irradiations: even for a large number of displacements per atom, this material remains crystallized. Nevertheless, a loss of its nanolamellar structure and an anisotropic variation of its lattice parameters, have been observed. Moreover, the swelling induced by the irradiation of the ternary compound appears as anisotropic, but seems relatively low compared to other materials. Concerning the effect of electronic interactions, they seem to enhance the oxidation of Ti_3SiC_2. Eventually, the presence of an oxide layer would cause the formation of some reliefs that seem not to have ever been observed in other materials.

1. INTRODUCTION

Because of their remarkable thermo-mechanical properties, MAX phases, and more particularly Ti_3SiC_2, seem to have aroused the interest of the nuclear industry. However, before considering this material as an element of either a high-temperature fission reactor such as the Gas-cooled Fast Reactor or a fusion facility like Iter, their behaviour under neutron irradiation has to be studied. That is why from a few years some research is led to understand how Ti_3SiC_2 would react in nuclear facilities.

In this Chapter, a non-exhaustive list of the major both structural and microstructural damage induced by irradiation is established as a function of the interactions, which may occur in reactor. Irradiations also cause some changes of the thermo-mechanical properties – for instance, variation of the hardness [1-7] or of the thermal conductivity [1, 8-12] – but these phenomena, which are generally the consequence of structural damage, are not discussed here.

2. Introduction to Ion-Mater Interaction

Neutron irradiations would have been ideal to understand the behaviour of Ti_3SiC_2 under irradiation for use in nuclear facilities. However, such irradiations are rather long, and the irradiated materials require characterization in specialized laboratories. Also, if a few Ti_3SiC_2 samples have been neutron irradiated [13], they do not seem to be characterized yet. Nevertheless, in order to obtain results quickly and so to know whether a material is interesting for the aimed application, the neutron irradiations are often simulated by ion irradiations [14-21].

For a better understanding of this chapter, some bases of the ion-matter interactions, and of the simulation by ion irradiations, are presented in what follows. More extensive complements may be found elsewhere [22, 23].

When ion penetrates a material, it interacts with atoms of the target by transferring energy *via* two kinds of interactions. The nuclear interactions are elastic collisions between the ion and the nucleus of the target atoms. These interactions are dominant for low-speed projectile (ion energy below 10 keV amu^{-1}). In nuclear reactors, they are usually induced by neutrons, which transfer their energy to atoms by elastic collisions, but also by the α-decay reactions (both α particles and recoil atoms).

The electronic interactions are inelastic collisions between the ion and the electron cloud of the target atoms, producing mainly electronic excitations and ionizations. These interactions are dominant for high-speed projectile (ion energy above 0.1 MeV amu^{-1}). In nuclear reactors, they are induced by fission products, the energy of which is near 1 MeV amu^{-1}. Ions inducing mainly electronic interactions are usually called “swift heavy ions”.

The stopping power S is defined by the ion kinetic-energy loss per unit distance along the path inside the material (usually expressed in keV nm^{-1}). Two kinds of stopping power are generally defined, each associated with interactions previously defined: the stopping power is hence the sum of the nuclear stopping power and the electronic one.

From these considerations, the impact of different interactions occurring in reactor can be simulated in a material, depending on the ion energy used to irradiate it: low energy ions may simulate the neutron damage, whereas swift heavy ions of energy near 1 MeV amu^{-1} would help to simulate the impact of fission products.

Figure 1 shows an example of the damage produced by ions of different energies in Ti_3SiC_2 [24]; it was obtained with the TRIM-2008 code [25]. In this Figure, one can see the evolution of both the electronic stopping power and the number of displacements per atom[1] (dpa) of the target caused by four irradiations in Ti_3SiC_2. Thus, the nuclear interactions induced by neutrons have been simulated over about 760 nm by irradiating Ti_3SiC_2 with 4 MeV Au ions (Figure 1a), while the damage caused by fission products has been simulated with both 74 MeV Kr and 92 MeV Xe ions (Figure 1b). Figure 1b shows that fission products generate not only electronic interactions, but also nuclear shocks, which grow along the ion path. In view to know the effect of electronic interactions alone, irradiations with high energy ions may be carried out (Figure 1c, 930 MeV Xe).

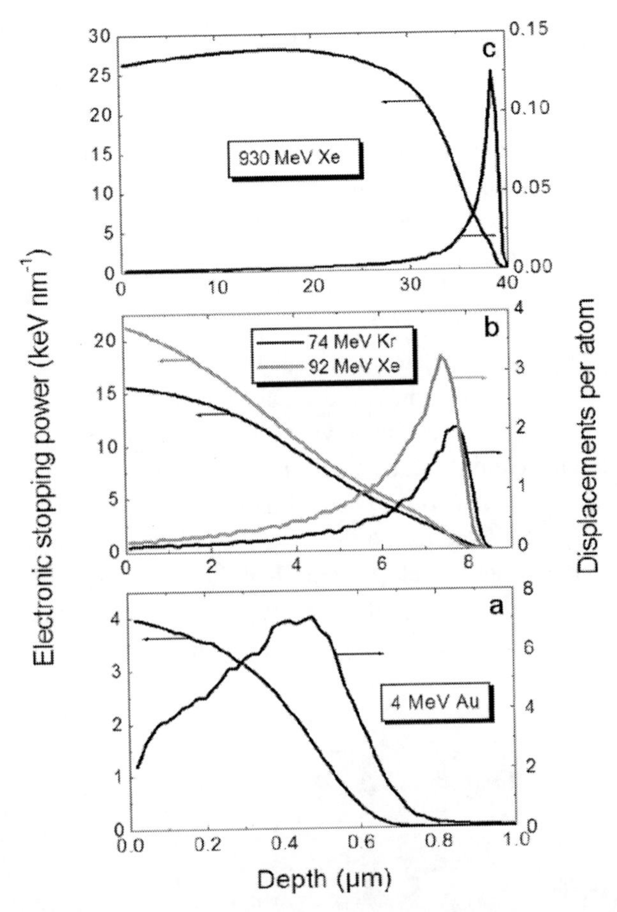

Figure 1. Electronic stopping power and number of displacements per atom induced by different kinds of ion irradiation as a function of the depth in Ti_3SiC_2; dpa are for a fluence of 10^{15} cm^{-2} (a-b) and $4.5x10^{13}$ cm^{-2} (c).

The damage of materials in the reactor is not only due to these two interactions. Indeed, the fission products (often noble gas atoms) and the α particles (helium nuclei) can lead to gas production in materials in which nuclear reactions occur. Therefore, many studies exist about the impact of this gas production in different materials [19, 26-32]. For these studies, the authors simulate the particle at the origin of this phenomenon by irradiating samples with

1 The number of dpa is a parameter allowing both the prediction of the damage caused by nuclear interactions during an irradiation, and the comparison of different irradiations regardless of whether they be ionic, neutron or electronic.

ions of noble gas to high fluences[2] (above 10^{16} cm^{-2}) in order to allow the agglomeration of the implanted ions, and the formation of nano-bubbles. However, as far as we know, no such study has yet been carried out on Ti_3SiC_2.

3. Effects of Ion Irradiation on Ti_3SiC_2 Structure

3.1. Electronic Interaction Effects

The electronic interactions are known to cause many phenomena involving electrons, ions and atoms in the solid: it follows a change in the solid that may be local or not [33]. The defects induced by such interactions depend on the nature of the solid, and notably the electronic structure of the material [34]. Thus, electronic collisions induce structural defects in both insulators and semiconductors, while energy is dissipated as thermal energy in metals, without any defect creation [35]. In the case of ceramics, these interactions mainly lead to the formation of generally-amorphous linear tracks, so-called latent tracks, along the ion path [36]. To represent the formation of latent tracks, different models have been developed, the most common of which are the Coulomb explosion [37-39] and the thermal spike [40-44] models. The overlapping of these latent tracks, obtained for fluences above 10^{12}-10^{13} cm^{-2} [16, 17, 45], leads to the amorphization of the irradiated material.

Therefore, since Ti_3SiC_2 is a ceramic with a metallic character, electronic interactions could induce either the formation of latent tracks (and consequently the amorphization of the material), or no structural defect as is the case for metals.

The effect of electronic interactions on the Ti_3SiC_2 structure can be observed in materials irradiated with high energy ions such as 930 MeV Xe ions [24]; indeed, for such a high energy irradiation nuclear shocks are negligible (Figure 1c). Thus, inside a sample containing no native defect (Figure 2a), it has been observed that Ti_3SiC_2 is not damaged by such irradiation, whatever the fluence reached (Figure 2b-c).

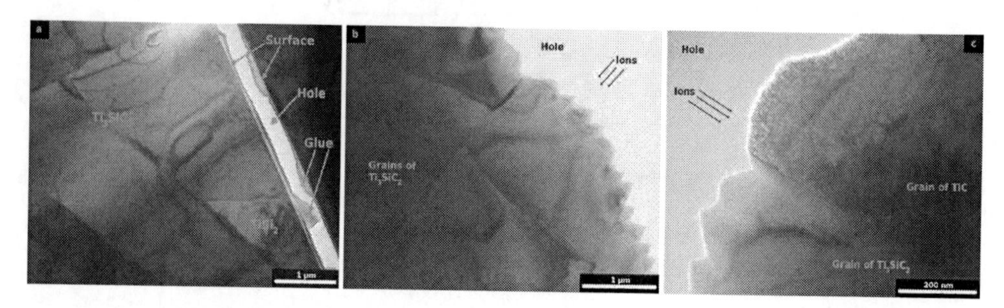

Figure 2. Cross-sectional transmission electron micrographs of (a) a virgin sample, and samples irradiated at room temperature with 930 MeV Xe ions to (b) 10^{11} cm^{-2} and (c) $4.5x10^{13}$ cm^{-2}.

In addition, high-resolution investigations have shown that the typical nanolamellar structure of Ti_3SiC_2, which may be observed by transmission electron microscopy when the basal planes are oriented parallel to the electron beam, remains intact to that observed before irradiation (Figure 3) [24].

[2] The fluence is the number of particles that impact a target per unit area (usually expressed in cm^{-2}).

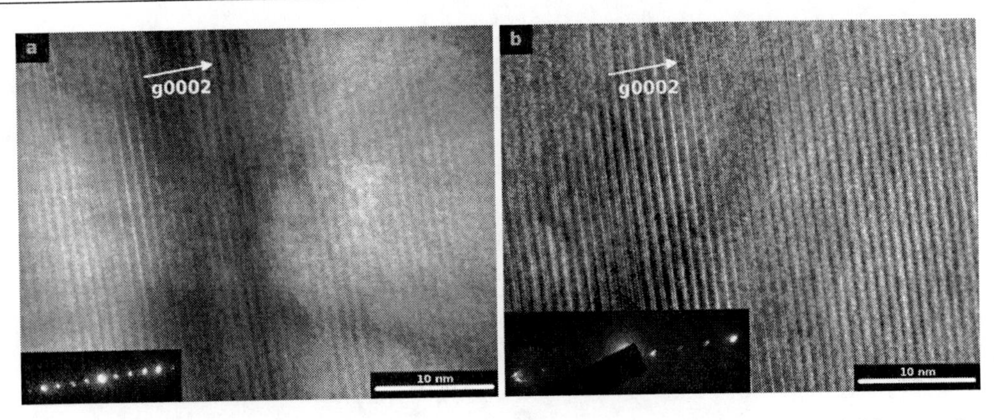

Figure 3. High resolution transmission electron micrographs of (a) a virgin sample, and (b) a sample irradiated at room temperature with 930 MeV Xe ions to $4.5x10^{13}$ cm^{-2}.

The fact that neither defect nor latent track is observable inside the samples irradiated with high energy ions leads to the conclusion that Ti_3SiC_2 is insensitive to electronic interactions. This result is not proper to this material since most metals [46, 47], but also some ceramics such as SiC [48-50], have this characteristic.

However, even in metals [51-54], it has been shown that latent tracks may be formed for electronic stopping powers higher than a threshold value Se_{th}, which depends on material. Also, if such a threshold exists in Ti_3SiC_2, it would be greater than 28 keV nm^{-1}, the maximum electronic stopping power hitherto reached in this material [24].

3.2. Nuclear Interaction Effects

In a nuclear collision, if the energy transmitted to the target atom exceeds a certain threshold value, called atomic displacement threshold energy E_d, the atom is ejected from its site and becomes a projectile; a Frenkel pair is hence created. This atom is then called primary atom and can eject other atoms. All the primary, secondary, etc. displacements form a cascade of displacements, which leads to regions containing defects [55]. Depending on the target material, this defect accumulation may cause the amorphization of the structure. This amorphization occurs beyond a dose, called critical amorphization dose, depending on the material, but also on both the mass and the energy of the incident particle [56-61].

In a nuclear reactor, the nuclear interactions are mainly induced by neutron elastic collisions. As previously explained (Section 2), the simulation of such interactions is usually performed with low energy ions. However, since the electronic interactions are not harmful to Ti_3SiC_2 (Section 3.1), irradiations with higher energy ions (e.g. with ions of about 1 MeV amu^{-1}) would induce damage only attributable to nuclear shocks; such a damage increases along the ion path (Figure 1b) [24]. Also, in this Section, to compare the nuclear damage induced by irradiations with different energy ions, the results will be as often as possible presented in terms of number of displacements per atom (dpa) induced by these irradiations.

3.2.1. Defects and Disorder

Figure 4 shows cross-sectional transmission electron micrographs of a Ti_3SiC_2 sample irradiated at room temperature with 92 MeV Xe [24].

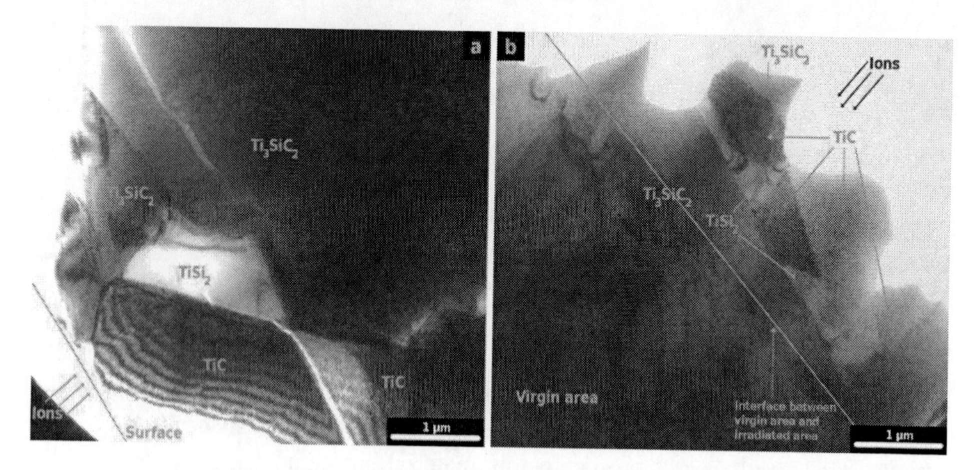

Figure 4. Cross-sectional transmission electron micrographs of Ti_3SiC_2 grains irradiated at room temperature with 92 MeV Xe ions to 10^{15} cm^{-2}, (a) at the beginning and (b) at the end of the ion range.

In a material containing no defects before irradiation, similar to that shown in Figure 2a, the nuclear shocks create many defects that appear as black dots (Figure 4a). These defects are too small to be identified by transmission electron microscopy; they may be clusters of Frenkel pairs or dislocation loops [62]. Along the 92 MeV Xe ion path, an increase of the concentration of irradiation induced defects has been observed, confirming first that they are induced by nuclear shocks, and second that electronic interactions do not damage Ti_3SiC_2 (Figure 1b). Figure 4b shows the end of the ion course, where the damage is ten times higher than at the beginning of the course. A Ti_3SiC_2 partly irradiated grain can be observed: compared to the virgin area, the irradiated area of the grain seems amorphous. To verify this observation, high-resolution transmission electron microscopy has been carried out (Figure 5).

This Figure shows the evolution of the Ti_3SiC_2 structure as a function of the number of dpa [24]. These high-resolution micrographs were obtained after orientation of the Ti_3SiC_2 basal planes parallel to the electron beam, in order to appreciate the nanolamellar structure of this ternary compound (Figure 5a); this structure can also be observed in the diffraction pattern, through three diffraction spots of low intensity between two more intense ones.

Through Figures 5b-d, it may be noted that when the number of dpa increases, the nanolamellar structure disappears, as well in the micrographs as in the diffraction patterns [24, 62]. However, the close-packed hexagonal stack is still observable, indicating that Ti_3SiC_2 does not become amorphous even for 7 dpa; for comparison, the critical amorphization dose of silicon carbide at room temperature ranges between 0.1 and 0.5 dpa [56, 63-70]. The cause of this nanolamellar structure disappearance has apparently not been determined yet. However, considering the crystal structure of Ti_3SiC_2 [71-73], it seems that the most likely hypothesis to explain this nuclear-shock caused damage would be the substitution of titanium atoms by silicon atoms (and/or vice versa), inducing the creation of defects such as Si_{Ti} (and/or Ti_{Si}) [24, 62].

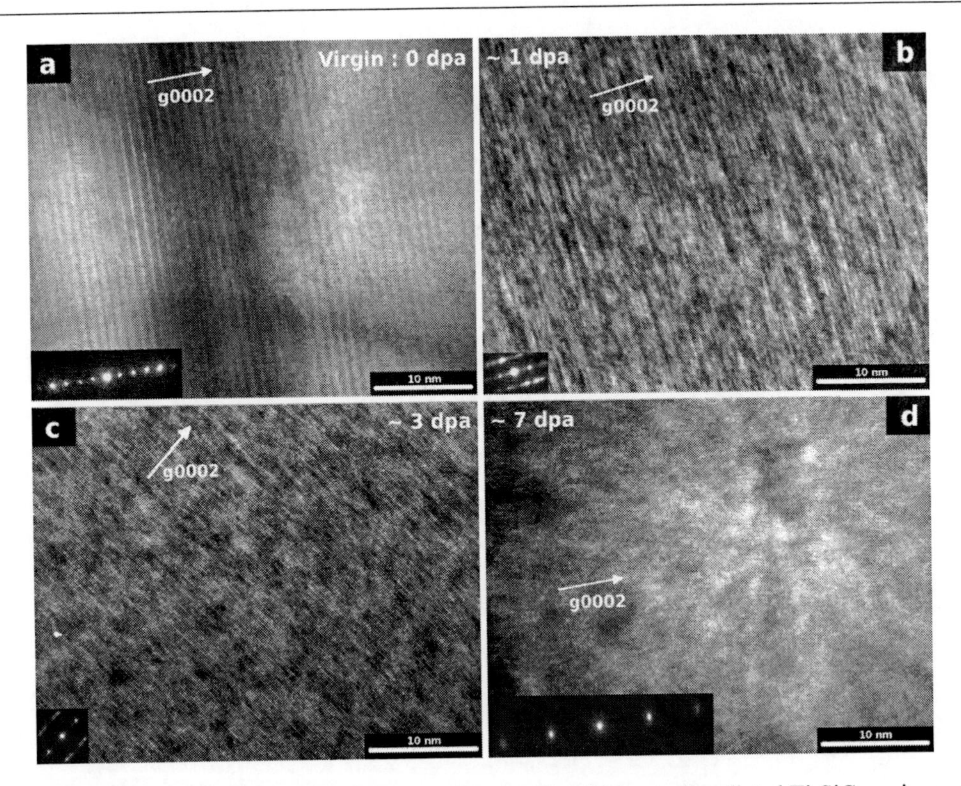

Figure 5. High resolution transmission electron micrographs of (a) an unirradiated Ti_3SiC_2 grain, and of grains irradiated at room temperature to (b) 1 dpa, (c) 3 dpa and (d) 7 dpa.

When the irradiation temperature increases, it has been observed that both the concentration of defects decreases, and the Ti_3SiC_2 nanolamellar structure is less damaged (Figure 6a), to appear as intact for an irradiation performed at 950 °C generating 7 dpa (Figure 6b). The creation of irradiation defects being an athermal phenomenon, this damage decrease is attributed to a defect annealing by temperature [24]. Indeed, the higher the temperature, the higher the diffusion of the species constituting the material, and consequently the higher the probability of recombination of Frenkel pairs.

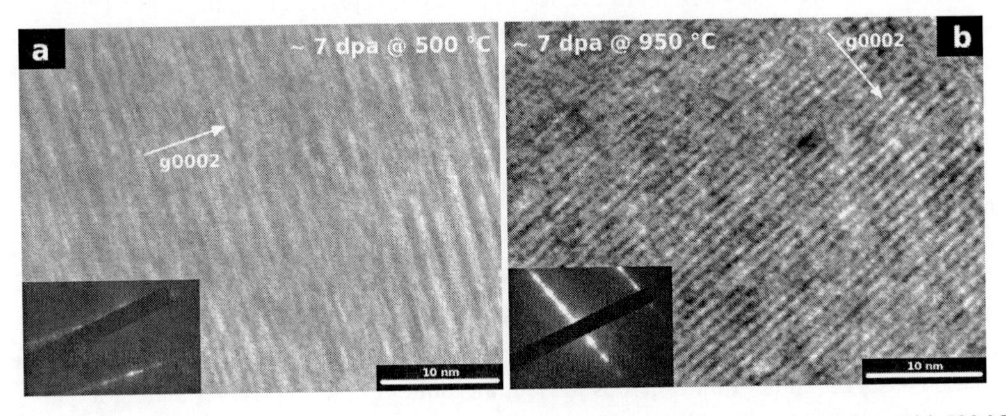

Figure 6. High resolution transmission electron micrographs of Ti_3SiC_2 irradiated to 7 dpa at (a) 500 °C and (b) 950 °C.

3.2.2. Change in Lattice Parameters

The creation of defects by nuclear shocks usually leads to other structural damage, the most often reported of which is the variation of lattice parameters [17, 74-77]. Figure 7 shows different diffraction patterns obtained by low incidence X-ray diffraction on samples irradiated with both low energy (Figure 7a) and medium energy (Figure 7b) ions; incidence angle utilized are 1° for low energy irradiations and 3° for medium energy ones, corresponding to an estimated analyzed thickness of 230 and 760 nm respectively [24]. Despite the low incidence used, it has been shown that the diffractograms of samples irradiated with 4 MeV Au ions present a contribution of both the irradiated area and the virgin one located at about 1 micron [24].

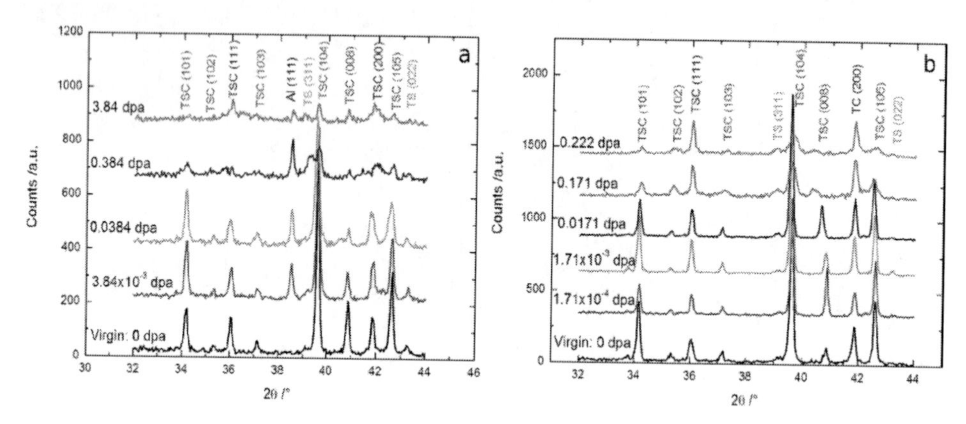

Figure 7. Low-incidence X-ray diffraction patterns of a commercial Ti_3SiC_2 irradiated at room temperature with (a) 4 MeV Au ions and (b) 92 MeV Xe ions; TSC, TC and TS stand for Ti_3SiC_2, $TiC_{0.92}$ and $TiSi_2$ phases respectively.

From these diffractograms, one can see a shift of the Ti_3SiC_2 (008) line towards lower diffraction angles as a function of the dose applied in the samples. This reflects a change in lattice parameters of the ternary compound. Figure 8 shows the evolution of both a and c parameters of Ti_3SiC_2 as a function of the number of dpa, as well as the variation in unit volume [24].

Thus, it has been shown that the Ti_3SiC_2 lattice both expands along the c axis (Figure 8b), and shrinks slightly along the a axis (Figure 8a). These leads to an increase in the unit volume up to a critical dose beyond which the unit volume decreases (Figure 8c); this critical dose has been estimated at 0.1 dpa for irradiations performed at room temperature. It is interesting to note that these results obtained on a sample of Ti_3SiC_2 differ slightly from that obtained by X-ray diffraction in Bragg-Brentano configuration on $Ti_3Si_{0.90}Al_{0.10}C_2$ samples irradiated under similar conditions [78]: indeed, if the expansion of the lattice along c axis was confirmed on these latter samples (same order of magnitude), the shrinkage along a has not been observed, resulting in an increase of the unit volume with the dose.

Eventually, an increase in irradiation temperature reducing the damage caused by nuclear shocks (Section 3.2.1), the changes in lattice parameters are not large enough to observe variations as a function of dpa such as those noticeable in Figure 8. Nevertheless, it has been shown that the defect annealing allows both the parameters to be less modified, and the critical dose (beyond which the unit volume decreases) to increase with increasing temperature [24].

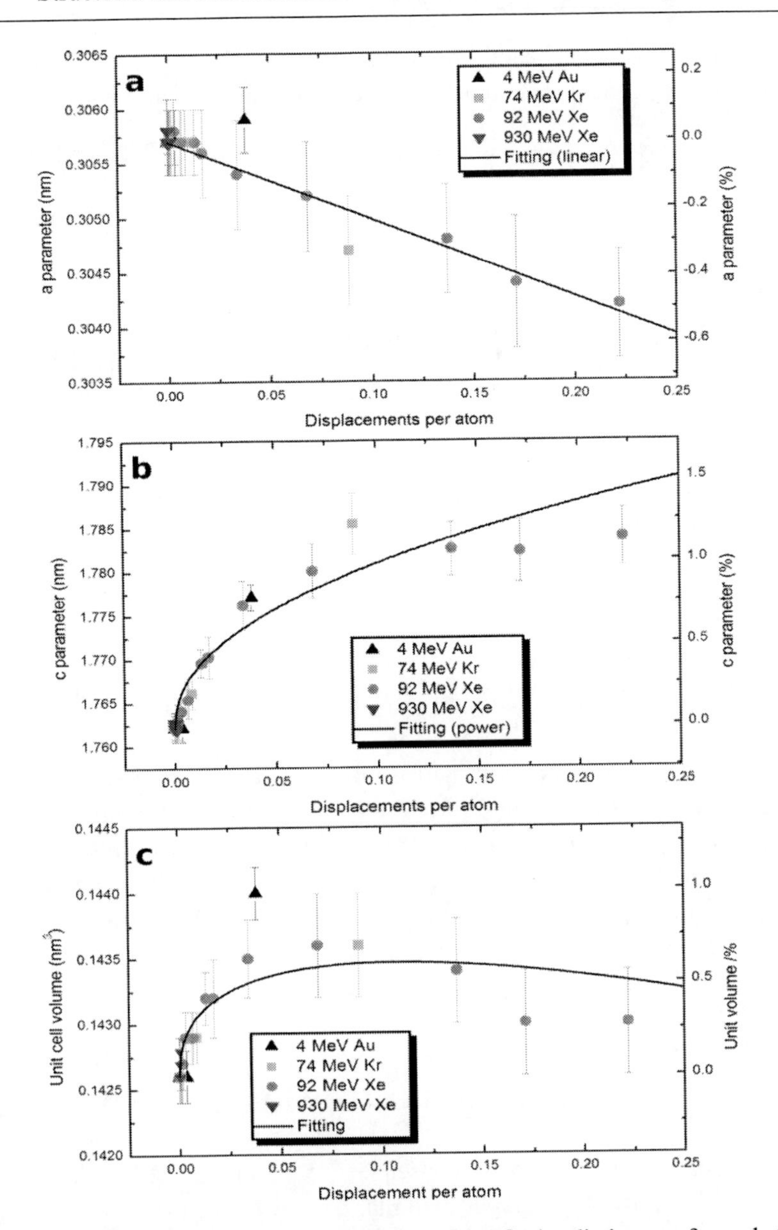

Figure 8. Change in the lattice parameters as a function of dpa for irradiations performed at room temperature; the fitting curve of the unit cell volume data was obtained from those of a and c parameters.

3.2.3. Change in Microstrain Yield

The previously observed Ti_3SiC_2 structural changes may induce microdistorsions in the crystal network. To verify this, a study of X-ray diffraction line broadening by Williamson and Hall method has been performed [24, 78]. Thus, Figure 9 shows the evolution of the microstrain as a function of both 74 MeV Kr and 92 MeV Xe ion fluence [24], considering that the crystallite size of the ternary compound (a few microns before irradiation [79]) is not affected by irradiations, and therefore does not induce any line broadening. For this Figure, the microstrain yields were evaluated from the broadening of the Ti_3SiC_2 (104) line alone, but the results seems in agreement with others obtained considering many lines [78].

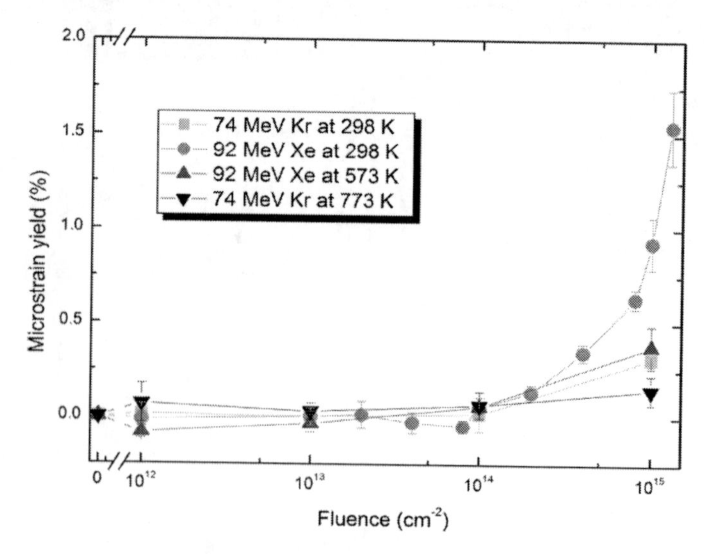

Figure 9. Variation of the microstrain yield as a function of the ion fluence for some medium energy irradiations.

In Figure 9, it can be observed an increase of the microstrain yield as a function of the fluence. This increase is smaller for 74 MeV Kr ion irradiations than for 92 MeV Xe ions, which induce more damage (Figure 1b). Moreover, since no microstrain could be measured for some samples irradiated with high energy ions (930 MeV Xe [80]), it can be conclude that the microstrains are induced by nuclear shocks. Finally, an increase of the irradiation temperature is still beneficial for the Ti_3SiC_2 damage: the higher the irradiation temperature, the fewer the microdistorsions in the crystal network (Figure 9).

4. Swelling Induced by Irradiation

One of the major microstructural consequences of the structural damage caused by irradiation is the swelling of the target material. The swelling may have different origins: creation of point defects (Frenkel pairs), agglomeration of point defects (dislocation loops, cavities), change in lattice parameters, phase transformation, amorphization, etc. It is hence often difficult to know the exact origin of the swelling.

Previously, it was shown that electronic interactions do not produce amorphization of Ti_3SiC_2 (Section 3.1). Thus, it seems unlikely that such interactions induce any swelling of the material. On the contrary, nuclear shocks induce both the creation of small clusters of defects, and the variation of lattice parameters. Therefore it seems necessary to verify whether these structural changes engender any swelling, and if the need arises to evaluate it.

Depending on the type of irradiation, two methods are commonly used to evaluate the swelling of an irradiated material. For neutron irradiations, the swelling concerns the whole material, and is three dimensional. In such cases, samples volume or density measurements, before and after irradiation, allow the estimation of the volume swelling [81-85]. However, as previously explained (Section 2), the characterization of neutron irradiated Ti_3SiC_2 specimens does not seem to be achieved yet.

For ion irradiations, the swelling only affects the thickness irradiated. Since the damage caused by ion irradiations is not homogeneous along the ion path (Figure 1), it is usually difficult to estimate the thickness which induces the swelling: some consider the ion path [86], while others the total thickness affected by irradiation [87]. In addition, ion irradiation mainly generates a linear swelling, parallel to the ion beam [88-90]. Thus, the volume swelling can be estimated by the linear swelling. Therefore, for such irradiations, measurements of the height between a virgin and an irradiated area on partly irradiated samples are usually performed to determine the volume swelling.

4.1. Effect of Nuclear Interactions

Height difference measurements have been carried out by atomic force microscopy on grains of Ti_3SiC_2 partly irradiated at room temperature with 4 MeV Au ions to 10^{15} cm^{-2} (Figure 10) [79, 91]. Performing several sections as those shown in Figure 10, and assuming a damaged thickness of 760 nm (Figure 1a), the swelling was estimated to 2.2 ± 0.8 % for an average damage of 4.3 dpa.

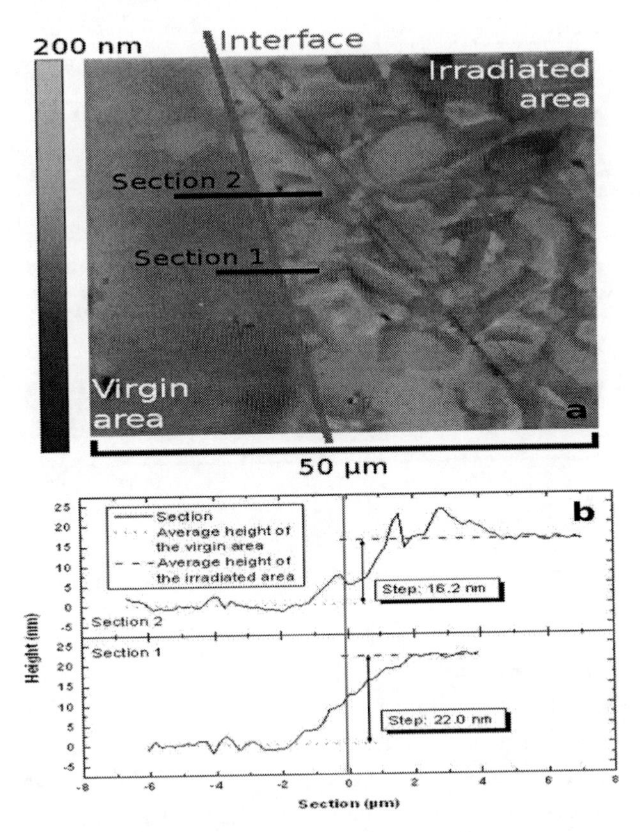

Figure 10. Swelling induced by an irradiation performed at room temperature with 4 MeV Au ions to 10^{15} cm^{-2}: (a) atomic force micrograph of the interface between the irradiated and the virgin areas, and (b) profiles of the sections indicated in (a).

This swelling estimated for an irradiation performed at room temperature appears to be relatively low compared to that of silicon carbide, which varies between 10 and 20 % for damage above a few tenths of dpa [2, 3, 14, 87, 92, 93].

4.1.1. Effect of Ion Fluence and Irradiation Temperature

According to the literature, it appears that no other swelling measurement has been performed on this material, irradiated with ions generating only nuclear shocks. Indeed, as shown in Figure 11, the microstructure revealing associated to the swelling of Ti_3SiC_2 is minor when nuclear damage is lower and/or when the irradiation temperature is higher [79]. Therefore, the difference of step height between the irradiated and the virgin parts of the samples irradiated at doses below 4.3 dpa and/or temperatures above room temperature is not important enough to be measurable, and swelling can be considered as negligible [79].

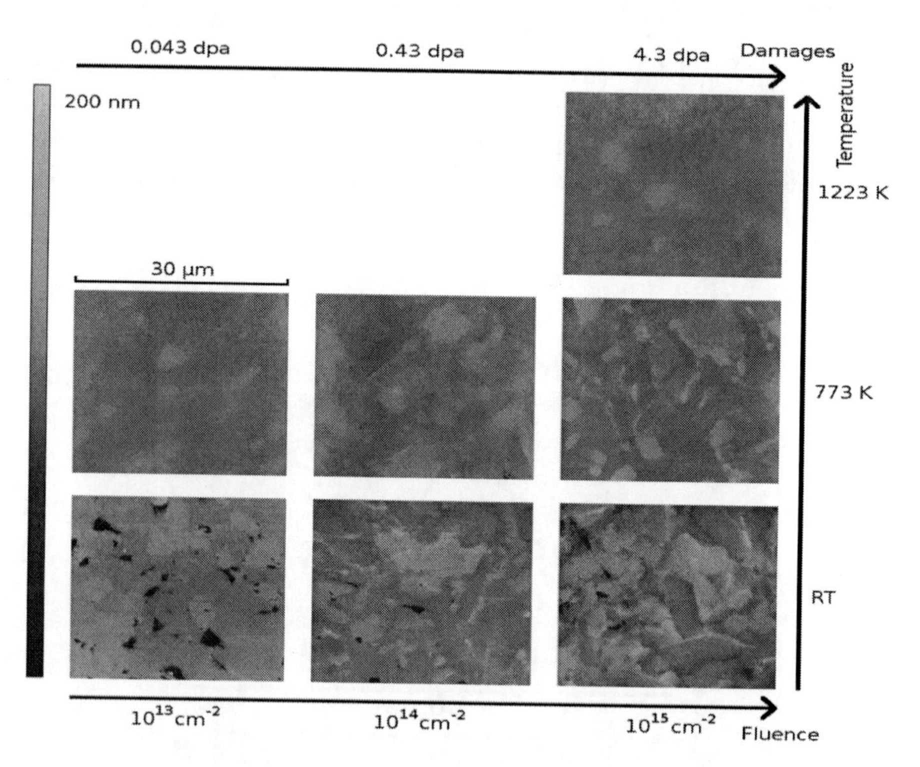

Figure 11. Evolution of the microstructure of areas irradiated with 4 MeV Au ions as a function of both the temperature and the fluence (or average dpa); RT stands for room temperature.

4.1.2. The Microstructure Revealing

The nuclear-shock caused swelling induces a microstructure revealing. The comparison of electron back-scatter diffraction micrographs performed on a virgin sample with atomic force micrographs of irradiated samples highlights that the revealed grains are of the same shape and same size as the crystallites of the sample before irradiation (Figure 12) [79]. Thus, the swelling would not be the same from a crystallite to the other, suggesting that Ti_3SiC_2 exhibits anisotropic swelling. Unfortunately, due to the loss of the Ti_3SiC_2 crystallinity under the impact of nuclear shocks, electron back-scatter diffraction analyzes could not be performed on samples irradiated to high doses [79].

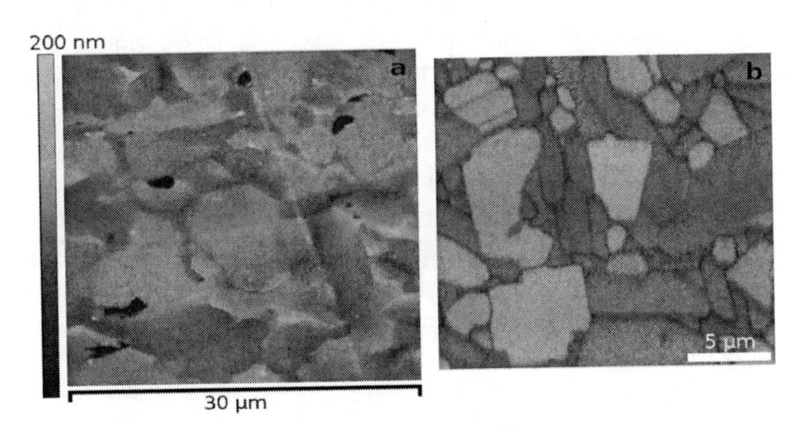

Figure 12. Comparison of the microstructure of (a) a Ti_3SiC_2 sample irradiated at room temperature with 4 MeV Au ions to 10^{15} cm^{-2} observed by atomic force microscopy, and (b) a virgin sample from the same origin observed by electron back-scatter diffraction.

Anisotropic swelling is not proper to Ti_3SiC_2. Quite the contrary, it has often been highlighted in materials owning anisotropic crystal structure, such as rhombohedral α-Al_2O_3 [82, 94, 95] and B_4C [96], hexagonal AlN [85, 94], and body-centered tetragonal $ZrSiO_4$ [97]. Furthermore, due to different swelling from a crystallite to the other, such a swelling induces high stresses in the irradiated area, causing fractures and microcracks at grain boundaries of polycrystalline materials [82, 85]. Such microcracks were observed by scanning electron microscopy on a Ti_3SiC_2 sample irradiated at room temperature with 4 MeV Au ions to 10^{15} cm^{-2} (average of 4.3 dpa) [79, 98], as shown in Figure 13. This latter observation confirms the hypothesis that Ti_3SiC_2 exhibits an anisotropic swelling under the effect of nuclear interactions.

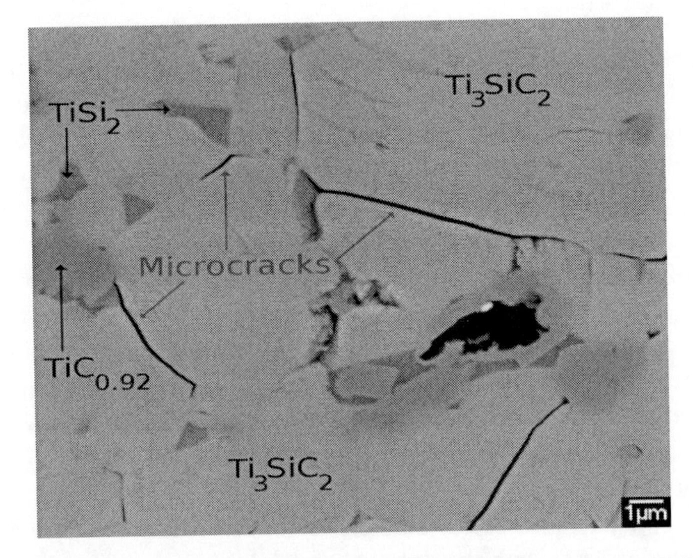

Figure 13. Back-scattered electron micrograph of the formation of microcracks on a Ti_3SiC_2 sample irradiated at room temperature with 4 MeV Au ions to 10^{15} cm^{-2}.

4.2. Effect of Electronic Interactions

It seems that no step height measurement could have been performed on samples irradiated with swift heavy ions inducing only electronic interactions. Indeed, the low fluences reached to date (up to 4.5×10^{13} cm^{-2} of 930 MeV Xe ions) would not allow the observation of a difference by atomic force microscopy between the virgin and irradiated parts of the samples [80]. Also, to determine the effect of electronic interactions, measurements of step heights on samples irradiated with swift heavy ions of lower energy (92 MeV Xe) have been attempted; such an irradiation causes a damage that is both nuclear and electronic (Figure 1b).

Figure 14 shows an atomic force micrograph carried out at the interface between a virgin area and an area irradiated at room temperature with 92 MeV Xe ions to 10^{15} cm^{-2} [80].

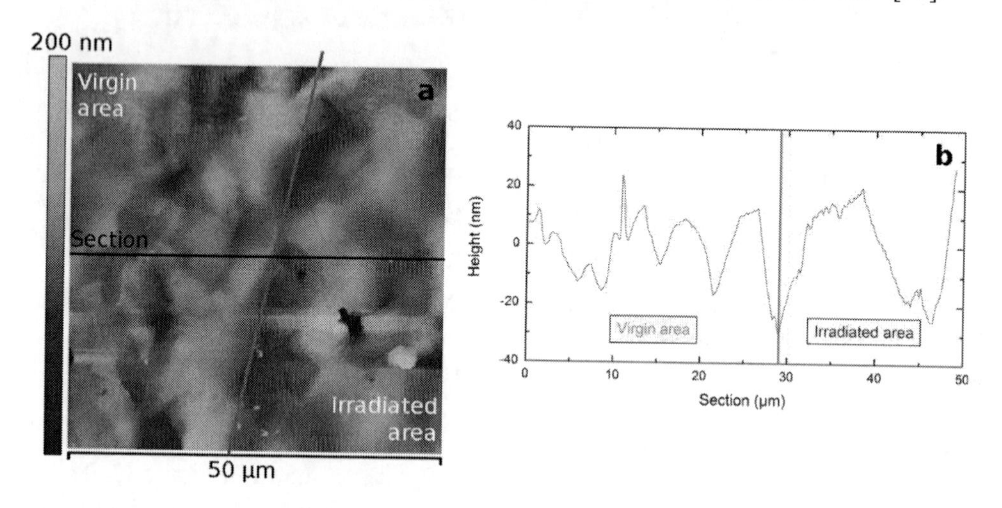

Figure 14. (a) Atomic force micrograph of a sample irradiated at room temperature with 92 MeV Xe ions to 10^{15} cm^{-2} achieved at the interface between the irradiated and the virgin areas, and (b) profiles of the sections indicated in (a).

In this micrograph, no obvious difference between the virgin and the irradiated parts can be observed. Thus, in agreement with the structural damage previously shown (Section 3.1), it appears that the electronic interactions do not induce any swelling of Ti_3SiC_2, for the to-date performed irradiations.

4.3. Model of Swelling Exhibited by Ti_3SiC_2

The swelling of a material strongly depends on its structural behaviour under irradiation. Thereof, many models have been developed to explain this phenomenon, depending on the irradiated material. For example, concerning silicon carbide, three swelling regimes have been distinguished [99], depending on the irradiation temperature:

- At low temperatures, SiC is amorphous for few tenths of a dpa [56, 63-70]. So, the swelling increases with the dose to reach a saturation (independent of temperature) for a dose slightly higher than the critical amorphization dose. This is the "amorphization regime";

- Beyond a certain temperature, called critical amorphization temperature, SiC does not become amorphous anymore, whatever the applied dose; this temperature would range between 400 and 650 K [57, 64, 100-104].Therefore, the swelling increases with the dose owing to the creation and agglomeration of defects. It reaches a saturation, significantly lower than that of the amorphization regime (no amorphous phase), which decreases when increasing the irradiation temperature. This last tendency is generally attributed to a recombination of the Frenkel pairs by the temperature. This is the "saturatable regime";
- When the irradiation temperature is much higher, beyond 1200-1300 K [105, 99, 106, 107], the created vacancies are mobile enough to create vacancy clusters that can form cavities in the material. Thus, for doses of several dpa [108], these cavities become the major cause of the swelling, which increases not only with the irradiation temperature, but also with the dose without reaching saturation. This is the "non-saturatable regime".

If this model has also been described for other ceramics like oxides [82] or nitrides [94], it does not seem applicable to metals. Indeed, according to literature, it appears that steels [109-113], but also various alloys [83, 84], do not become amorphous during high dpa irradiations. Also, the swelling of metals does not saturate and increases with the dose due to the agglomeration of point defects into extended defects, whatever the irradiation temperature. Moreover, the evolution of the swelling of metals as a function of the irradiation temperature is variable depending on the considered material: it may increase with temperature [84], decrease [83, 84] or show a maximum for a given temperature [112].

Taking into account the results obtained concerning the swelling of Ti_3SiC_2 under the effect of nuclear interactions, let us try to determine whether this material follows a swelling model typical of ceramics or of metals. First, consider that Ti_3SiC_2 follows the same swelling model as silicon carbide [99]. The results obtained previously (Section 4.1) do not rule out this hypothesis. Instead, they suggest that Ti_3SiC_2 is in the "saturatable regime" for the temperature range studied to date (from room temperature to 950 °C). This assumption can be deduced from the following four points:

- Ti_3SiC_2 is not amorphous and contains many small defect clusters [24, 62] for the highest dose reached (up to a mean of 4.3 dpa, with a maximum of about 7 dpa), while SiC, which is in the "amorphization regime" at room temperature, is amorphous from few tenths of a dpa [2, 3, 14, 87, 92, 93];
- No extended defect, such as cavities, was observed by transmission electron microscopy, even for the highest irradiation temperature, suggesting that Ti_3SiC_2 is not in the "insaturatable regime";
- The swelling both increases with the dose and decreases when the irradiation temperature increases, variations typical of the "saturatable regime";
- Considering the maximum underwent dose, the measured swelling is relatively low (2.2 ± 0.8 % for 4.3 dpa), the same order of magnitude as that of other materials irradiated in the "saturatable regime": Al_2O_3 irradiated under similar conditions [88]

or SiC [99], MgO [82], AlN [85, 94], and Si_3N_4 [114] irradiated at higher temperature.

Therefore, the critical amorphization temperature of Ti_3SiC_2 would lower than room temperature, explaining that no amorphous phase has been observed by transmission electron microscopy in the temperature range hitherto investigated.

However, if the results seem to fit with the typical swelling model of ceramics, they cannot rule out the hypothesis that the swelling of Ti_3SiC_2 is similar to that of metal. Indeed, as previously explained, metals do not seem to become amorphous during irradiations with high dpa in the temperature range studied up to now for Ti_3SiC_2 [83, 84, 109-113], inducing a gradual increase of swelling with dose as observed in the ternary compound. Moreover, the decrease of the swelling observed in Ti_3SiC_2 with increasing temperature has ever been noticed in some metals [83, 84].

Thus, for irradiation conditions hitherto investigated, Ti_3SiC_2 does not become amorphous, and its swelling does not seem to have reached saturation, making it impossible to know if the swelling of the ternary compound is similar to that of ceramics or of metals. To find out it, irradiations at other temperatures (especially at lower temperatures to favor an eventual amorphization), and higher doses (creation of extended defects or amorphization) should be carried out.

5. HILL FORMATION

For irradiations performed with swift heavy ions, a hill formation phenomenon was observed on Ti_3SiC_2 surface [62, 98, 115, 116]. Figure 16 illustrates this phenomenon on a commercial polycrystalline sample whose secondary phases are TiC and $TiSi_2$ [115]; the phase identification was achieved by energy-dispersive X-ray spectroscopy analysis coupled with a scanning electron microscope.

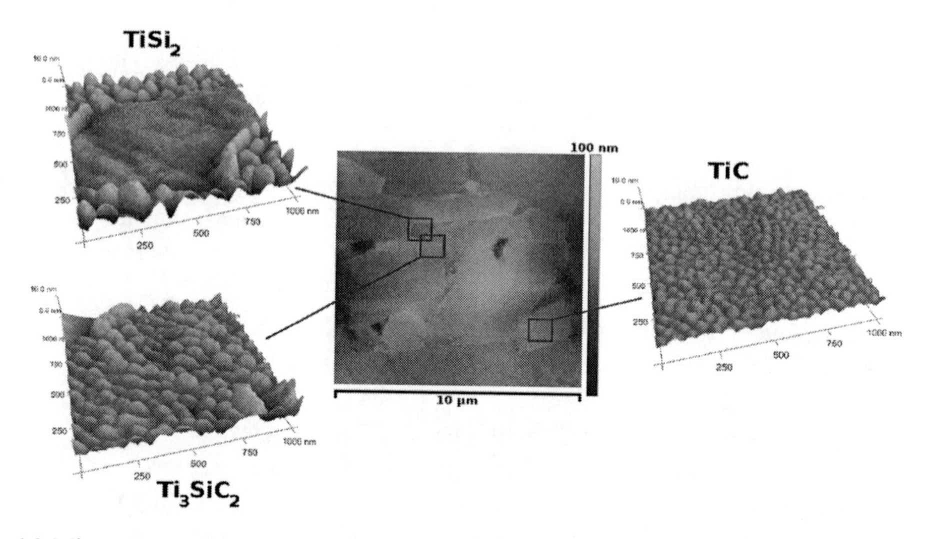

Figure 16. Microstructure by atomic force microscopy of a sample irradiated at room temperature with 92 MeV Xe ions to 10^{15} cm^{-2}; hills are observable on both Ti_3SiC_2 and TiC grains, but not on $TiSi_2$ ones.

This section proposes to understand this phenomenon, which does not seem to have been observed in other materials in the conditions of irradiation used for Ti_3SiC_2 [98].

5.1. Structure and Composition of the Hills

To know the hill composition, X-ray photo-electron spectrometry analyses were performed before and after irradiation [115]. Figure 17 shows both the Ti 2p and the Si 2p peaks of the virgin sample. In Figure 17a one can see both the Ti $2p_{1/2}$ and the Ti $2p_{3/2}$ characteristic peaks of Ti-C bond in Ti_3SiC_2, and in Figure 17b one another Ti_3SiC_2 characteristic peak around 99 eV ever mentioned in earlier studies [72] but not identified then. Based on both the Si-Ti bond length [72] and some work concerning the growth of $TiSi_2$ on Si(111) wafer [117], this peak may be attributed to the silicide Si-Ti bond [115].

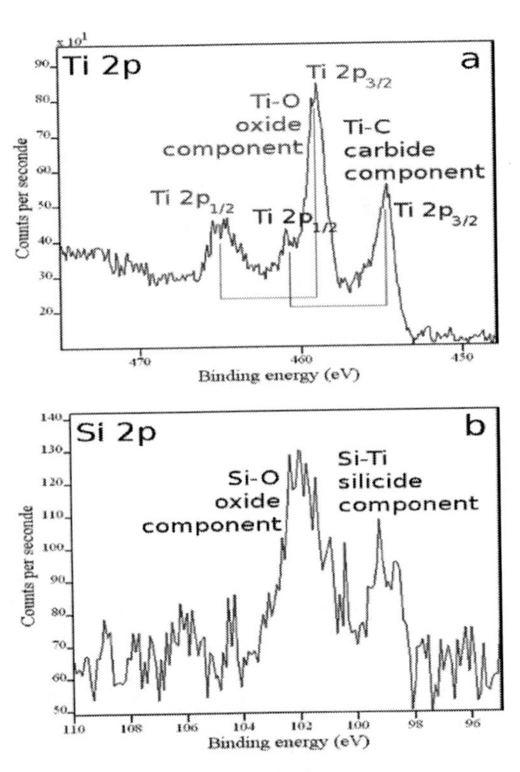

Figure 17. X-ray photo-electron spectra of the virgin sample: (a) Ti 2p peaks, and (b) Si 2p peaks.

In addition to these Ti_3SiC_2 characteristic peaks, these spectra show the presence of oxygen in the analyzed thickness. Oxygen is present as both titanium and silicon oxides, the two oxides formed on the surface of Ti_3SiC_2 during its oxidation in temperature [118-120]. A cross-sectional transmission electron microscopy observation of the surface (Figure 18) has confirmed the presence of an amorphous oxide layer, with uniform thickness estimated at about 3 nm [115] ; the thickness analyzed by X-ray photo-electron spectrometry was estimated to 5 nm. This layer has certainly formed after polishing operations carried out before irradiation, which would activate the surface of the material, making it highly reactive toward the oxygen-rich environment.

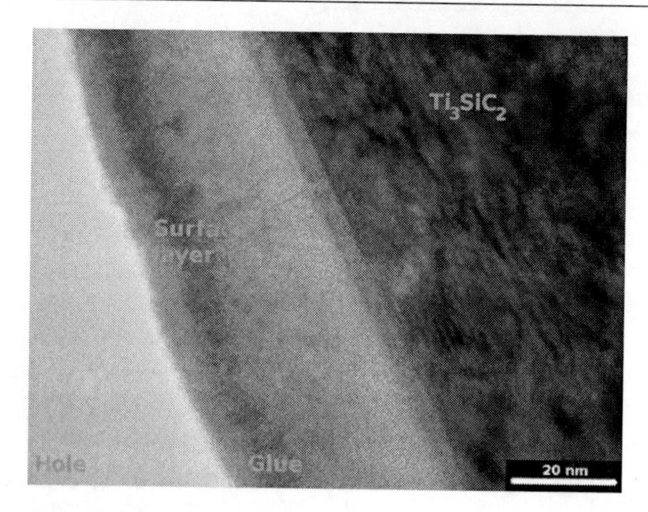

Figure 18. Observation by cross-sectional transmission electron micrograph of an amorphous oxide layer on the surface of the virgin sample.

After an irradiation at room temperature with 92 MeV Xe ions to 10^{15} cm^{-2} (irradiation conditions allowing the obtaining of the Figure 16), it was observed that the X-ray photo-electron spectrometry characteristic peaks of Ti_3SiC_2 disappeared, as shown in Figure 19: only the peaks related to the oxide layer remain [115].

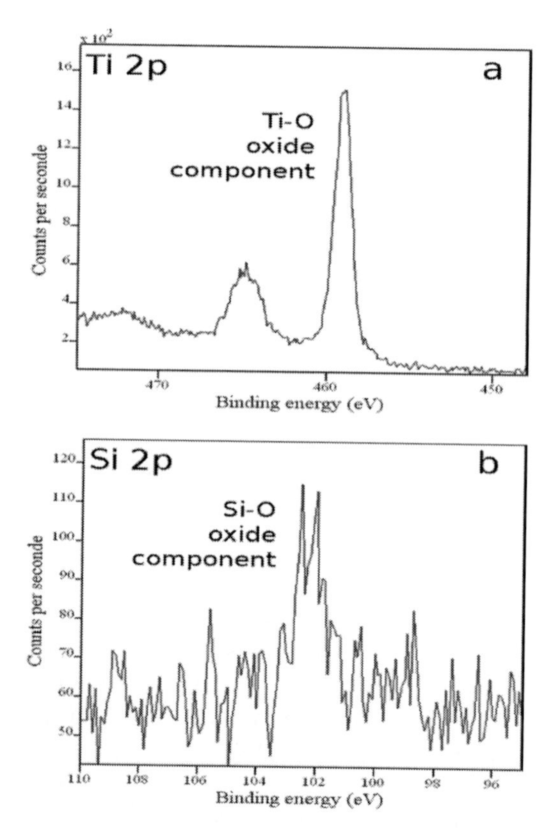

Figure 19. X-ray photo-electron spectra of sample irradiated at room temperature with 92 MeV Xe ions to 10^{15} cm^{-2}: (a) Ti 2p peaks, and (b) Si 2p peaks.

This result is consistent with the observations of the surface of the same sample by cross-sectional transmission electron microscopy (Figure 20). Indeed, one can observe [115]:

- That the thickness of the oxide layer is larger than the one on the non-irradiated sample;
- That the hills are present on the oxide layer, without any change in the roughness of the crystalline part under the oxide layer;
- That the hills are amorphous.

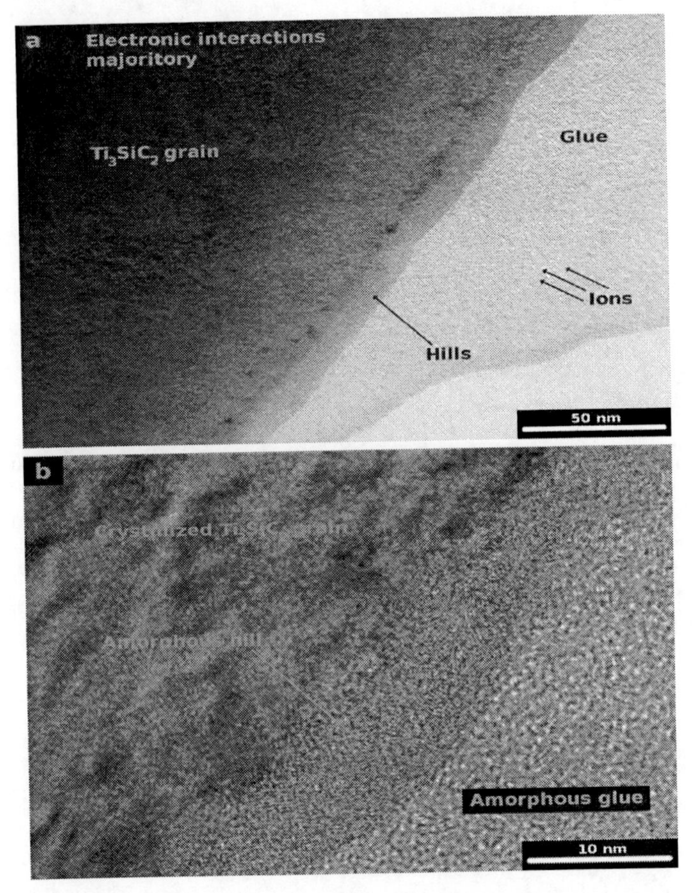

Figure 20. Cross-sectional transmission electron micrographs of the hills on a sample irradiated at room temperature with 92 MeV Xe ions to 10^{15} cm^{-2}: hills are amorphous.

These observations indicate that the hills are composed of both titanium and silicon oxides, and that their formation is certainly due to the presence of the native oxide layer, which probably increases during irradiation [115].

5.2. Effect of 92 Mev Xe Ion Fluence

Figure 21 shows the evolution of the Ti_3SiC_2 surface as a function of the 92 MeV Xe ion fluence for irradiations performed at room temperature. In agreement with this study achieved by atomic force microscopy, the hills appear for a fluence of $4x10^{14}$ cm^{-2}. Subsequently, increasing the fluence, the hills begin to grow laterally without much variation in height, then, from 10^{15} cm^{-2} their height increases at quasi-constant diameter [115].

Figure 21. Evolution of the hill microstructure by atomic force microscopy as a function of the fluence in 92 MeV Xe ions.

To complete this study, the oxide proportion inside the thickness analyzed by X-ray photo-electron spectrometry (about 5 nm) has been estimated (Figure 22) [115].

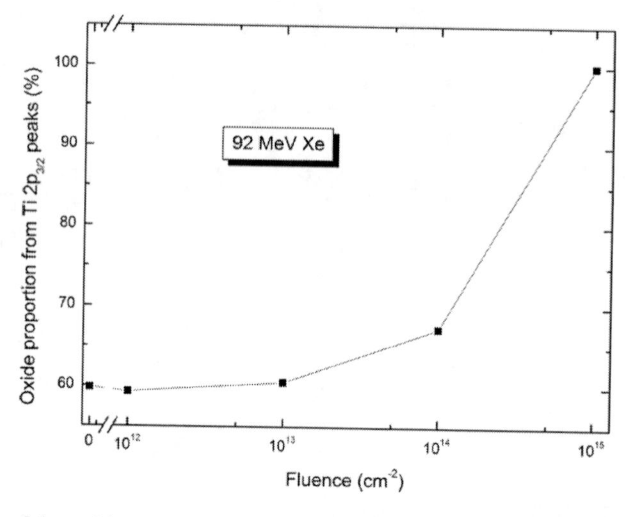

Figure 22. Evolution of the oxide proportion in the surface of samples irradiated with 92 MeV Xe ions as a function of the fluence; the oxide proportion is estimated with the Ti $2p_{3/2}$ X-ray photo-electron spectrometry peaks.

It hence appears that the thickness of the oxide layer remains stable up to 10^{13} cm^{-2} before slightly increasing between 10^{13} and 10^{14} cm^{-2}. Eventually, for a fluence of 10^{15} cm^{-2}, the thickness of the oxide layer measures more than 5 nm, reaching 100% of the thickness analyzed by X-ray photo-electron spectrometry. The thickness of the oxide layer would so increase before the formation of hills at about $4x10^{14}$ cm^{-2}.

Note that few data exist concerning the effect of the temperature on the hill formation, but it seems that for an irradiation performed at 300 °C with 92 MeV Xe ions to 10^{15} cm^{-2} [121], the hills are bigger in diameter compared with those observed for irradiations performed at room temperature, without being really different in height (Figure 23). This could be due to a phenomenon close to the Ostwald ripening observed during sintering.

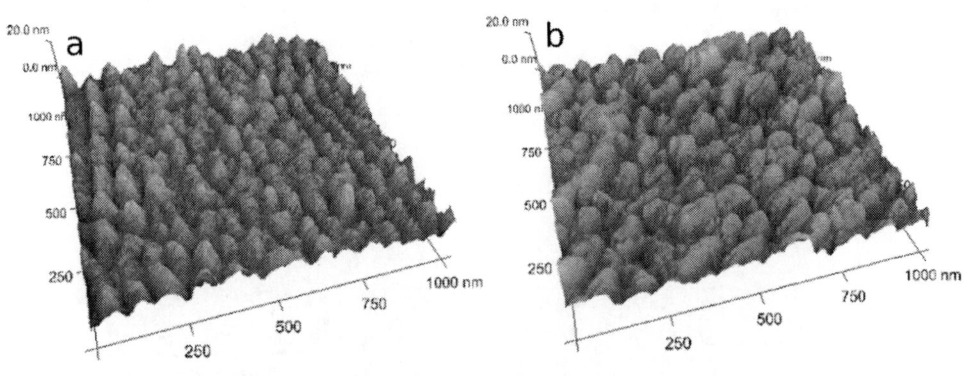

Figure 23. Atomic force micrographs of the hill microstructure for samples irradiated at (a) room temperature, and (b) 300 °C with 92 MeV Xe ions to 10^{15} cm^{-2}.

5.3. Effect of Stopping Power

First, as shown in Figure 11, no hill was observed for high fluences on samples irradiated with ions engendering only nuclear interactions [79, 98, 115]. It hence seems that this phenomenon is induced by the strong electronic interactions caused at the beginning of the swift heavy ion path (Figure 1). Figure 24 shows the surface state of samples irradiated at room temperature with some other swift heavy ions. One can observe that no hill was formed on these samples [115].

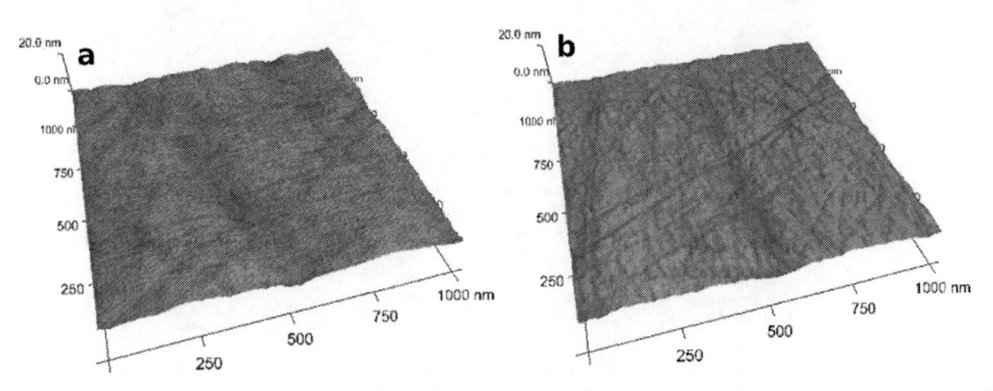

Figure 24. Atomic force micrographs of Ti_3SiC_2 grains irradiated at room temperature with (a) 74 MeV Kr ions to 10^{15} cm^{-2}, and (b) 930 MeV Xe ions to $4.5x10^{13}$ cm^{-2}.

Concerning the sample irradiated with 74 MeV Kr ions (Figure 24a), the electronic stopping power seems too low for the hill formation (in the surface, about 15.6 and 21.2 keV nm^{-1} for 74 MeV Kr and 92 MeV Xe respectively). Therefore, this phenomenon could take place either beyond a threshold electronic stopping power (see Section 3.1), or beyond a critical fluence ϕ_c, which decreases with increasing electronic stopping power. Both of these hypotheses could explain the lack of hills on the sample irradiated with 930 MeV Xe ions (Figure 24b): this irradiation presents an electronic stopping power above the one of 92 MeV Xe ions (about 26.3 keV nm^{-1} in the surface), but its reached maximum fluence is relatively low (4.5x 10^{13} cm^{-2}).

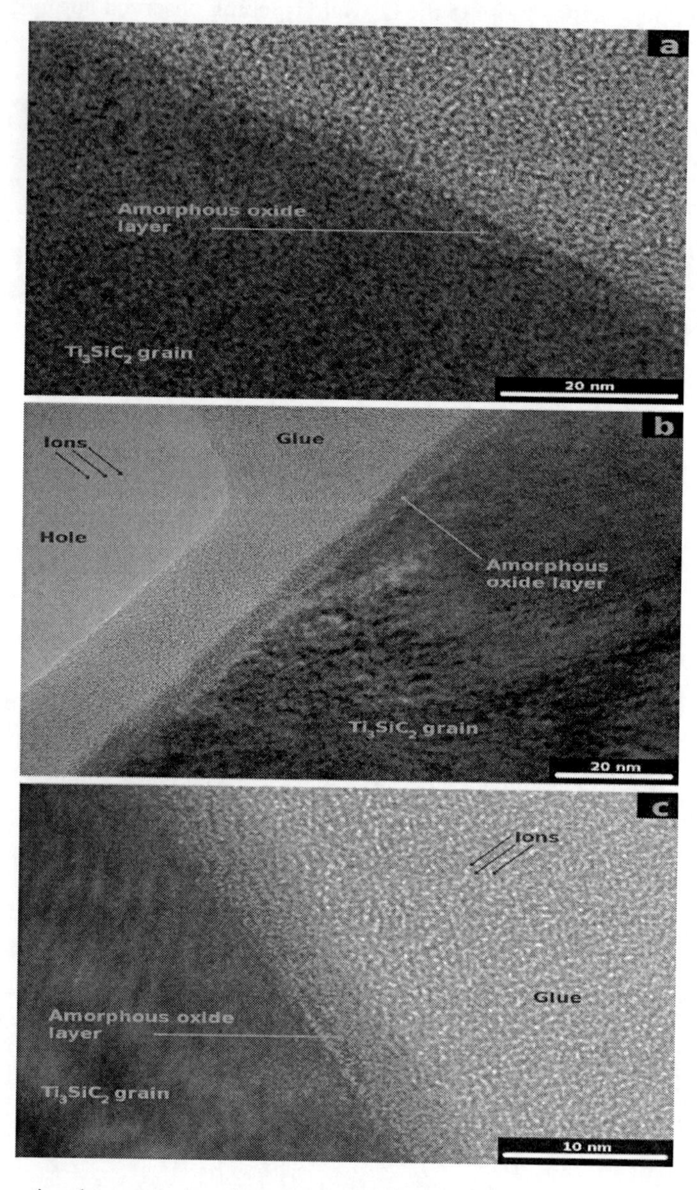

Figure 25. Cross-sectional transmission electron micrographs of the surface of samples irradiated at room temperature with (a) 4 MeV Au and (b) 74 MeV Kr ions both to 10^{15} cm^{-2}, and with (c) 930 MeV Xe ions to 4.5x10^{13} cm^{-2}.

The surface of samples irradiated at room temperature with 4 MeV Au, 74 MeV Kr, and 930 MeV Xe ions to high fluences has also been observed by cross-sectional transmission electron microscopy (Figure 25) [115]. These micrographs confirm the absence of hills observed by atomic force microscopy (Figure 24).

In addition, these micrographs allow the observation of an increase in the thickness of the native oxide layer for irradiations carried out with swift heavy ions (Figure 25b-c), while for the irradiation inducing only nuclear interactions no variation of the thickness was observed (Figure 25a). Thus, as was previously mentioned (Section 5.2), the growth of the oxide layer occurs before the formation of hills.

This growth was also evaluated by X-ray photo-electron spectrometry [115], estimating the proportion of oxide in the analyzed thickness for the irradiations performed at room temperature to 10^{15} cm^{-2}, as a function of the electronic stopping power (Figure 26).

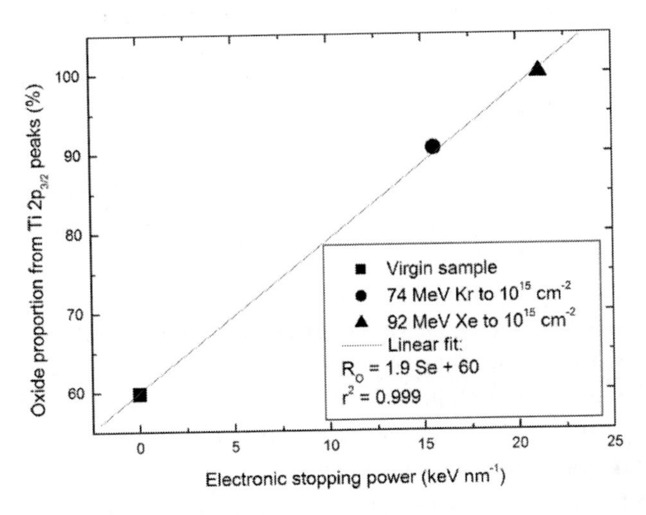

Figure 26. Evolution, as a function of the electronic stopping power, of the oxide proportion in the surface of samples irradiated to 10^{15} cm^{-2}; the oxide proportion is estimated with the Ti $2p_{3/2}$ X-ray photo-electron spectrometry peaks.

From Figure 26, it is interesting to note that the thickness of the oxide layer seems to be directly related to the electronic stopping power, confirming that, as for the hill formation, the oxide layer growth is induced by electronic interactions.

5.4. Origin of the Phenomena

As was previously explained (Section 3.1), the electronic interactions lead to many phenomena having for origin the interactions between electrons, ions and atoms of the target material. Among them, we can for instance cite:

- The desorption of volatile species such as H_2 and N_2 molecules [122];
- The oxidation of the material [18, 123-127];
- The electronic sputtering [33, 128-131];
- The mixing of atoms at the interfaces [132-134];

- The phase transformation [135-143];
- The phase separation such as Si and SiO_2 in SiO_x layers [144].

Most of these phenomena are generally driven by the presence of latent tracks in the material, appearing for electronic stopping powers above a threshold Se_{th}.

Moreover, such tracks may induce the formation of reliefs similar to those observed on Ti_3SiC_2, so-called hillocks [20, 145-151]. These hillocks, observed by atomic force microscopy, are usually about a few nanometers in height. However, such hillocks are only observed for low fluences (below the overlapping the latent tracks occurring between 10^{12} and 10^{13} cm^{-2}) and their density increases with the fluence. This is not in agreement with the formation of hills on Ti_3SiC_2 that begins at about $4x10^{14}$ cm^{-2}, and the density of which decreases when increasing the fluence.

According to Section 3.1, the electronic interactions do not produce latent tracks in Ti_3SiC_2, and the threshold electronic stopping power in the ternary compound would be greater than 28 keV nm^{-1}. On the contrary, it was shown that this threshold would be around 6 keV nm^{-1} in titanium oxide [152], and would be less than 5 keV nm^{-1} in silicon oxide [128, 153, 154]. Therefore, the surface phenomena observed in this fifth Section (both oxide layer growth and hill formation) were obtained for swift heavy ion irradiation, the electronic stopping power of which exceeds the threshold electronic stopping power of the oxide layer.

5.4.1. Increase of the Oxide Layer Thickness

The increase of the native oxide-layer thickness under the effect of electronic interactions could be the result of either a swelling or an oxidation [115]. However, even if some swelling has ever been observed in amorphous phases [155-157], the thickness of the Ti_3SiC_2 native oxide layer almost doubles, which would correspond to a volume swelling close to 100 %. Also, such a swelling value seems to have never been reported, all the more so no extended defect such as cavities was observed in this layer. Therefore, the hypothesis of an oxidation seems the most likely.

Oxidation induced by electronic interactions has ever been observed in zirconium irradiated with some fission products [125] or some swift heavy ions [18, 126], but also in some iron targets [123, 124]. More particularly for the iron targets [123], it was shown that for certain electronic stopping power conditions, the electronic interactions lead to an oxidation at the substrate/oxide interface. This occurs when the electronic stopping power is both higher than the threshold Se_{th} for latent track formation in the oxide and lower than Se_{th} in the substrate (the iron target). These conditions induce an increase of the diffusion of the adsorbed oxygen atoms along the tracks all the way to the substrate/oxide interface.

This adsorbed oxygen would come from the residual oxygen in the irradiation chambers [18, 124, 126], despite the low pressure existing there (between 10^{-6} and 10^{-7} mbar [115]).

5.4.2. Formation of the Hills

For the studied irradiation conditions, the hill formation phenomenon does not seem to have been observed in other materials than Ti_3SiC_2. Actually, in addition to the previously mentioned “hillocks” that appear for low fluences during irradiations inducing the formation of latent tracks, we can mention other similar reliefs observed on materials irradiated with ions of lower energy (few hundred of keV) to higher fluences (above 10^{16} cm^{-2}); they are then rather called “ripples” [31, 32, 148, 158, 159]. Nevertheless, this phenomenon differs from that observed on Ti_3SiC_2, first because of the irradiation conditions, and second because the crystalline structure, being generally under an amorphous layer, presents also ripples [31, 32].

However, it seems that the swift heavy ion irradiation of amorphous silica leads to a plastic deformation attributable to the ion hammering effect [155], also called Klaumünzer effect [160, 161]. The model relative to this effect, which seems proper to amorphous materials submitted to electronic interactions [162-166], predicts an anisotropic growth of the irradiated phase, to wit both an expansion perpendicular, and a shrinkage parallel to the ion beam (Figure 27). Moreover, it was observed that the characteristics of this anisotropic growth are in agreement with the hill formation as a function of the irradiation parameters [115]. Thus, considering this model, a hypothesis to explain the hill formation would be that the ion hammering of the oxide layer induces some strong stresses, which are opposed to the oxidation phenomenon. This stress opposition in the oxide layer could then be accommodated by the formation of hills on the surface by a mechanism still unknown.

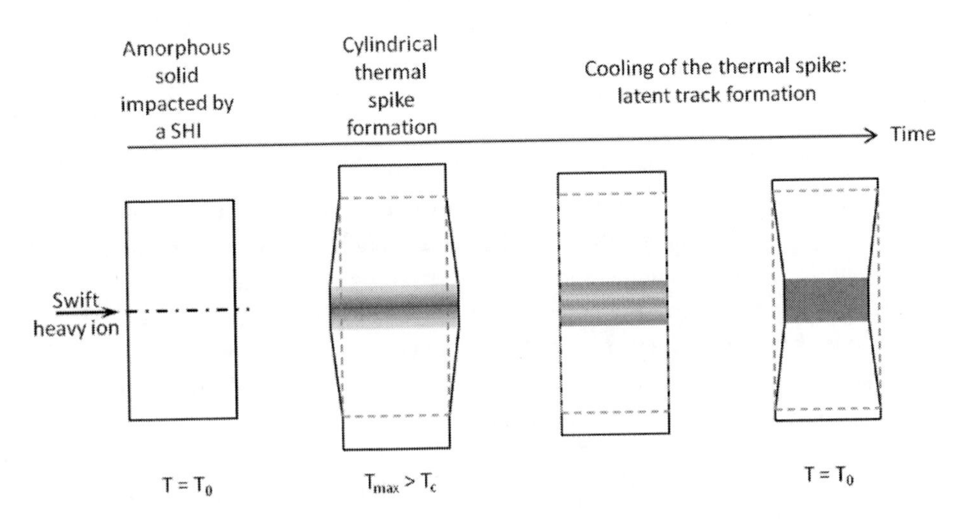

Figure 27. Schematic illustration of the anisotropic growth in an amorphous solid impacted by a swift heavy ion; T_c is the critical temperature of thermal spike formation.

According to both the Klaumünzer effect model and the results obtained [115], the accommodation of the stresses would take place after an incubation fluence, which decreases when the electronic stopping power increases. This incubation fluence would be of about $4x10^{14}$ cm^{-2} for irradiations carried out at room temperature with 92 MeV Xe ions (about 21.2 keV nm^{-1}), and above $4.5x10^{13}$ cm^{-2} for irradiations with 930 MeV Xe ions (about 26.3 keV nm^{-1}).

CONCLUSIONS

In this chapter, we attempted to present an overview of the works led to understand the behaviour of Ti_3SiC_2 under irradiation. Such research could in particular interest the nuclear industry, but also other industries involving ionic or electronic irradiations.

Even though some neutron irradiations have been carried out, the studies hitherto conducted to understand the behaviour of Ti_3SiC_2 under irradiation seem to have been achieved only with ion irradiations; however, the presentation of results as a function of dpa provides an estimate of the damage underwent by this material under the effect of nuclear interactions, whatever their origin.

Throughout this chapter, it appeared that the ternary compound reacts differently depending on whether it is subjected to nuclear shocks or electronic interactions. Thus, it seems clear that the Ti_3SiC_2 structure is not damaged by electronic interactions, notably explaining the absence of swelling regardless of the reached fluence. This result allowed the conclusion that if there exist a threshold electronic stopping power of latent track formation in Ti_3SiC_2, this is greater than 28 keV nm^{-1}. However, from a microstructural point of view, the presence of a small amount of oxygen in the irradiation chamber leads to the ternary-compound oxidation, enhanced by the electronic excitations. Finally, probably through an accommodation of stresses caused partly by the oxidation and partly by an ion hammering effect, the presence of an oxide layer on the surface can induce the formation of hills on Ti_3SiC_2 by a yet unknown mechanism.

Conversely to the electronic excitations, nuclear interactions are harmful for the Ti_3SiC_2 structure. Moreover, the anisotropic structure of this material seems to behave in anisotropic way towards these interactions. Actually, in addition to the creation of many defects, which lead to an important disorder, the nuclear shocks cause the disappearance of the nanolamellar structure of the ternary compound, without however affecting the basal plane staking. Furthermore, the Ti_3SiC_2 hexagonal lattice deforms: it highly expands along the c axis, while shrinking along the a axis, inducing some relatively strong microstrains in the network. From a microstructural point of view, the nuclear interactions are the origin of anisotropic swelling, which induce both a microstructure that differs depending on the crystallite orientation, and some microcracks at Ti_3SiC_2 grain boundaries. Eventually, in general, irradiations performed at high temperatures reduce the overall damage caused by nuclear shocks, which have been observed for room temperature irradiations. More particularly, for irradiations achieved at 950 °C to high dpa, the Ti_3SiC_2 structural damage is almost nonexistent. This result reflects the positive effect of temperature, which can anneal the defects created by irradiation.

These first studies led on the behaviour of Ti_3SiC_2 under irradiation allow to grasp the damage of this material in nuclear reactors. Nevertheless, to complete the knowledge, it seems important to study the evolution of the properties of this ternary compound under irradiation, and more particularly the thermo-mechanical properties. Indeed, it has been shown in many metals that the creation of defect clusters induce a decrease of the dislocation mobility, and therefore an increase of the hardness of the irradiated material [1, 4, 167-169].

REFERENCES

[1] Yano, T.; Iseki, T. *J. Nucl. Mater.* 1991, *179*, 387-390.
[2] Snead, L.L.; Hay, J.C. *J. Nucl. Mater.* 1999, *273*, 213-220.
[3] Weber, W.J.; Yu, N.; Wang, L.M.; Hess, N.J. *J. Nucl. Mater.* 1997, *244*, 258-265.
[4] Hartmann, T.; Wang, L.M.; Weber, W.J.; Yu, N.; Sickafus, K.E.; Mitchell, J.N.; Wetteland, C.J.; Nastasi, M.A.; Hollander, M.G.; Baker, N.P.; Evans, C.R.; Tesmer,

J.R.; Maggiore, C.J. *Nucl. Instrum. Methods Phys. Res. Sect. B: Beam Interact. Mater. Atoms* 1998, *141*, 398-403.

[5] Yano, T.; Miyazaki, H.; Akiyoshi, M.; Iseki, T. *J. Nucl. Mater.* 1998, *253*, 78-86.

[6] Kim, C.S.; Iseki, T.; Yano, T.; Tezuka, M. *J. At. Energy Soc. Jpn.* 1992, *34*, 335-341.

[7] Manika, I.; Maniks, J.; Schwartz, K.; Trautmann, C. *Nucl. Instrum. Methods Phys. Res. Sect. B: Beam Interact. Mater. Atoms* 2002, *196*, 299-307.

[8] Hollenberg, G.W.; Henager, C.H.; Youngblood, G.E.; Trimble, D.J.; Simonson, S.A.; Newsome, G.A.; Lewis, E. *J. Nucl. Mater.* 1995, *219*, 70-86.

[9] Gomes, S.; David, L.; Roger, J.P.; Carlot, G.; Fournier, D.; Valot, C.; Raynaud, M. *Eur. Phys. J.: Spec. Top.* 2008, *153*, 87-90.

[10] David, L.; Gomes, S.; Carlot, G.; Roger, J.P.; Fournier, D.; Valot, C.; Raynaud, M. *J. Phys. D: Appl. Phys.* 2008, *41*, 035502.

[11] Akiyoshi, M.; Takagi, I.; Yano, T.; Akasaka, N.; Tachi, Y. *Fusion Eng. Des.* 2006, *81*, 321-325.

[12] Price, R.J. *J. Nucl. Mater.* 1973, *46*, 268-272.

[13] Audubert, F.; Abrivard, G.; Tallaron, C. In *Proceedings of the 30th International Conference and Exposition on Advanced Ceramics and Composites*; Cocoa Beach, FL, US, 2006.

[14] Snead, L.L.; Zinkle, S.J.; Hay, J.C.; Osborne, M.C. *Nucl. Instrum. Methods Phys. Res. Sect. B: Beam Interact. Mater. Atoms* 1998, *141*, 123-132.

[15] Gosset, D.; Dollé, M.; Simeone, D.; Baldinozzi, G.; Thomé, L. *Nucl. Instrum. Methods Phys. Res. Sect. B: Beam Interact. Mater. Atoms* 2008, *266*, 2801-2805.

[16] Beauvy, M.; Dalmasso, C.; Thiriet-Dodane, C.; Simeone, D.; Gosset, D. *Nucl. Instrum. Methods Phys. Res. Sect. B: Beam Interact. Mater. Atoms* 2006, *242*, 557-561.

[17] Sattonnay, G.; Moll, S.; Thomé, L.; Legros, C.; Herbst-Ghysel, M.; Garrido, F.; Costantini, J.-M.; Trautmann, C. *Nucl. Instrum. Methods Phys. Res. Sect. B: Beam Interact. Mater. Atoms* 2008, *266*, 3043-3047.

[18] Bérerd, N.; Moncoffre, N.; Chevarier, A.; Jaffrézic, H.; Faust, H.; Balanzat, E. *Nucl. Instrum. Methods Phys. Res. Sect. B: Beam Interact. Mater. Atoms* 2006, *249*, 513-516.

[19] Chernov, I.I.; Kalashnikov, A.N.; Kalin, B.A.; Binyukova, S.Y. *J. Nucl. Mater.* 2003, *323*, 341-345.

[20] Skuratov, V.A.; Efimov, A.E.; Havancsak, K. *Nucl. Instrum. Methods Phys. Res. Sect. B: Beam Interact. Mater. Atoms* 2006, *250*, 245-249.

[21] Yu, J.; Zhao, X.; Zhang, W.; Yang, W.; Chu, F. *J. Nucl. Mater.* 1997, *251*, 150-156.

[22] Lehmann, C. *Interaction of radiation with solids and elementary defect production*; Series on Defects in Crystalline solids; Elsevier-North Holland; Amsterdam, NL, 1977; Vol. 10.

[23] Nastasi, M.; Mayer, J.; Hirvonen, J.K. *Ion-Solid Interactions: Fundamentals and Applications*; Cambridge Solid State Science Series; Cambridge University Press; Cambridge, UK, 1996.

[24] Nappé, J.C.; Monnet, I.; Grosseau, Ph.; Audubert, F.; Guilhot, B.; Beauvy, M.; Thomé, L.; Benabdesselam, M. *J. Nucl. Mater.* 2011, *409*, 53-61.

[25] Ziegler, J.F. <http://www.srim.org/>.

[26] Binyukova, S.Y.; Chernov, I.I.; Kalin, B.A.; Kalashnikov, A.N.; Timofeev, A.A. *At. Energy* 2002, *93*, 569-577.

[27] Jiao, Z.; Ham, N.; Was, G.S. *J. Nucl. Mater.* 2007, *367-370*, 440-445.

[28] Ono, K.; Arakawa, K.; Hojou, K. *J. Nucl. Mater.* 2002, *307-311*, 1507-1512.
[29] Oliviero, E.; Tromas, C.; Pailloux, F.; Declémy, A.; Beaufort, M.F.; Blanchard, C.; Barbot, J.F. *Mater. Sci. Eng. B* 2003, *102*, 289-292.
[30] Sasajima, N.; Matsui, T.; Hojou, K.; Furuno, S.; Otsu, H.; Izui, K.; Muromura, T. *Nucl. Instrum. Methods Phys. Res. Sect. B: Beam Interact. Mater. Atoms* 1998, *141*, 487-493.
[31] Chini, T.K.; Okuyama, F.; Tanemura, M.; Nordlund, K. *Physical Review B* 2003, *67*, 205403.
[32] Chini, T.K.; Datta, D.P.; Luchhesi, U.; Mücklich, A. *Surf. Coat. Technol.* 2009, *203*, 2690-2693.
[33] Arnoldbik, W.M.; Tomozeiu, N.; Habraken, F.H.P.M. *Nucl. Instrum. Methods Phys. Res. Sect. B: Beam Interact. Mater. Atoms* 2003, *203*, 151-157.
[34] Rotaru, C. *SiO_2 sur silicium: comportement sous irradiation avec des ions lourds (SiO_2 on silicon: behavior under heavy ion irradiation)*. PhD thesis in Materials Science, University of Caen, FR, 2004.
[35] Lehmann, C. *Interaction of radiation with solids and elementary defect production*; Series on Defects in Crystalline solids; Elsevier-North Holland; Amsterdam, 1977; Vol. 10.
[36] Dalmasso, C. *Caractérisation des défauts ponctuels induits par l'irradiation dans les céramiques : exemple des oxydes α-Al_2O_3, MgO et $MgAl_2O_4$ (Characterization of point defects induced by irradiation in ceramics: example of the α-Al_2O_3, MgO and $MgAl_2O_4$ oxides)*. PhD thesis in Physic, University of Nice-Sophia Antipolis, FR, 2005.
[37] Fleischer, R.L.; Price, P.B.; Walker, R.M. *J. Appl. Phys.* 1965, *36*, 3645-3652.
[38] Remillieux, J. *Nucl. Instrum. Methods* 1980, *170*, 31-40.
[39] Cooney, P.J.; Faibis, A.; Kanter, E.P.; Koenig, W.; Maor, D.; Zabransky, B.J. *Nucl. Instrum. Methods Phys. Res. Sect. B: Beam Interact. Mater. Atoms* 1986, *13*, 160-166.
[40] Desauer, F. *Z. Phys.* 1923, *12*, 38-47.
[41] Seitz, F.; Kohler, J.S. *Solid State Phys.: Adv. Res. App.* 1956, *2*, 305-448.
[42] Toulemonde, M.; Dufour, C.; Paumier, E. *Phys. Rev. B* 1992, *46*, 14362-14369.
[43] Toulemonde, M.; Paumier, E.; Dufour, C. *Radiat. Eff. Defects Solids* 1993, *126*, 201-206.
[44] Szenes, G. *Phys. Rev. B: Condens. Matter* 1995, *51*, 8026-8029.
[45] Quentin, A.; Monnet, I.; Gosset, D.; Lefrançois, B.; Bouffard, S. *Nucl. Instrum. Methods Phys. Res. Sect. B: Beam Interact. Mater. Atoms* 2009, *267*, 980-982.
[46] Dunlop, A.; Legrand, P.; Lesueur, D.; Lorenzelli, N.; Morillo, J.; Barbu, A.; Bouffard, S. *Europhys. Lett.* 1991, *15*, 765-770.
[47] Ossi, P. M.; Pastorelli, R. *Nucl. Instrum. Methods Phys. Res. Sect. B: Beam Interact. Mater. Atoms* 1997, *122*, 566-570.
[48] Benyagoub, A.; Audren, A.; Thome, L.; Garrido, F. *Appl. Phys. Lett.* 2006, *89*, 241914.
[49] Audren, A.; Monnet, I.; Gosset, D.; Leconte, Y.; Portier, X.; Thomé, L.; Garrido, F.; Benyagoub, A.; Levalois, M.; Herlin-Boime, N.; Reynaud, C. *Nucl. Instrum. Methods Phys. Res. Sect. B: Beam Interact. Mater. Atoms* 2009, *267*, 976-979.
[50] Intarasiri, S.; Yu, L.D.; Singkarat, S.; Hallen, A.; Lu, J.; Ottosson, M.; Jensen, J.; Possnert, G. *J. Appl. Phys.* 2007, *101*, 084311.
[51] Ossi, P.M.; Pastorelli, R. *Nucl. Instrum. Methods Phys. Res. Sect. B: Beam Interact. Mater. Atoms* 1997, *122*, 566-570.
[52] Ossi, P.M. *Philos. Mag. B: Phys. Condens. Matter Stat.* 1997, *76*, 541-548.

[53] Nozières, J.P.; Ghidini, M.; Dempsey, N.M.; Gervais, B.; Givord, D.; Suran, G.; Coey, J.M.D. *Nucl. Instrum. Methods Phys. Res. Sect. B: Beam Interact. Mater. Atoms* 1998, *146*, 250-259.
[54] Gupta, A. *Vac.* 2000, *58*, 16-32.
[55] Audren, A. *Effets d'irradiation et diffusion des produits de fission (césium et iode) dans le carbure de silicium (Irradiation effects and diffusion of fission products (cesium and iodine) in silicon carbide)*. PhD thesis in Materials Science, University of Caen, FR, 2007.
[56] Edmond, J.A.; Davis, R.F.; Withrow, S.P.; More, K.L. *J. Mater. Res.* 1988, *3*, 321-328.
[57] Weber, W.J.; Wang, L.M.; Yu, N.; Hess, N.J. *Mater. Sci. Eng. A* 1998, *253*, 62-70.
[58] Grimaldi, M.G.; Calcagno, L.; Musumeci, P.; Frangis, N.; VanLanduyt, J. *J. Appl. Phys.* 1997, *81*, 7181-7185.
[59] Bolse, W.; Conrad, J.; Harbsmeier, F.; Borowski, M.; Rodle, T. *Mater. Sci. Forum* 1997, *248-249*, 319-325.
[60] Wendler, E.; Heft, A.; Wesch, W.; Peiter, G.; Dunken, H.H. *Nucl. Instrum. Methods Phys. Res. Sect. B: Beam Interact. Mater. Atoms* 1997, *127-128*, 341-346.
[61] Musumeci, P.; Calcagno, L.; Grimaldi, M.G.; Foti, G. *Nucl. Instrum. Methods Phys. Res. Sect. B: Beam Interact. Mater. Atoms* 1996, *116*, 327-331.
[62] Le Flem, M.; Liu, X.; Doriot, S.; Cozzika, T.; Monnet, I. *Int. J. Appl. Ceram. Technol.* 2010, *7*, 766-775.
[63] Weber, W.J.; Jiang, W.; Thevuthasan, S. *Nucl. Instrum. Methods Phys. Res. Sect. B: Beam Interact. Mater. Atoms* 2001, *175-177*, 26-30.
[64] Zinkle, S.J.; Snead, L.L. *Nucl. Instrum. Methods Phys. Res. Sect. B: Beam Interact. Mater. Atoms* 1996, *116*, 92-101.
[65] Hart, R.R.; Dunlap, H.L.; Marsh, O.J. *Radiat. Eff. Defects Solids* 1971, *9*, 261-266.
[66] White, C.W.; McHargue, C.J.; Sklad, P.S.; Boatner, L.A.; Farlow, G.C. *Mater. Sci. Rep.* 1989, *4*, 41-146.
[67] Williams, J.M.; McHargue, C.J.; Appleton, B.R. *Nucl. Instrum. Methods Phys. Res.* 1983, *209-210*, 317-323.
[68] Spitznagel, J.A.; Wood, S.; Choyke, W.J.; Doyle, N.J.; Bradshaw, J.; Fishman, S.G. *Nucl. Instrum. Methods Phys. Res. Sect. B: Beam Interact. Mater. Atoms* 1986, *16*, 237-243.
[69] Chechenin, N.G.; Bourdelle, K.K.; Suvorov, A.V.; Kastilio-Vitloch, A.X. *Nucl. Instrum. Methods Phys. Res. Sect. B: Beam Interact. Mater. Atoms* 1992, *65*, 341-344.
[70] Horton, L.L.; Bentley, J.; Romana, L.; Perez, A.; McHargue, C.J.; McCallum, J.C. *Nucl. Instrum. Methods Phys. Res. Sect. B: Beam Interact. Mater. Atoms* 1992, *65*, 345-351.
[71] Jeitschko, W.; Nowotny, H. *Monatsh. Chem.* 1967, *98*, 329-337.
[72] Kisi, E.H.; Crossley, J.A.A.; Myhra, S.; Barsoum, M.W. *J. Phys. Chem. Solids* 1998, *59*, 1437-1443.
[73] Barsoum, M.W.; El-Raghy, T. *J. Mater. Synth. Process.* 1997, *5*, 197-216.
[74] Kobayashi, N.; Kobayashi, H.; Sakamoto, I.; Hayashi, N. *J. Nucl. Mater.* 1991, *179-181*, 469-472.
[75] Scardi, P.; Kothari, D.C.; Guzman, L. *Thin Solid Films* 1991, *195*, 213-223.
[76] Ishikawa, N.; Iwase, A.; Chimi, Y.; Maeta, H.; Tsuru, K.; Michikami, O. *Phys. C: Supercond.* 1996, *259*, 54-60.

[77] Jiang, W.; Nachimuthu, P.; Weber, W.J.; Ginzbursky, L. *Appl. Phys. Lett.* 2007, *91*, 091918.

[78] Liu, X.; Le Flem, M.; Béchade, J.L.; Onimus, F.; Cozzika, T.; Monnet, I. *Nucl. Instrum. Methods Phys. Res. Sect. B: Beam Interact. Mater. Atoms* 2010, *268*, 506-512.

[79] Nappé, J.C.; Maurice, C.; Grosseau, Ph.; Audubert, F.; Guilhot, B.; Beauvy, M.; Thomé, L.; Benabdesselam, M. *J. Eur. Ceram. Soc.* 2011, *31*, 1503-1511.

[80] Nappé, J.C. *Évaluation du comportement sour irradiation de Ti_3SiC_2 : Étude de l'endommagement structural et microstructural (Assessment of the behaviour of Ti_3SiC_2 under irradiations: study of both the structural and microstructural spoiling).* PhD thesis in Chemical Engineering, École Nationale Supérieure des Mines de Saint-Étienne, FR, 2009.

[81] Newsome, G.; Snead, L.L.; Hinoki, T.; Katoh, Y.; Peters, D. *J. Nucl. Mater.* 2007, *371*, 76-89.

[82] Clinard Jr., F.W.; Hurley, G.F.; Hobbs, L.W. *J. Nucl. Mater.* 1982, *108-109*, 655-670.

[83] Mitchell, M.A.; Garner, F.A. *J. Nucl. Mater.* 1992, *187*, 103-108.

[84] Garner, F. A.; Gelles, D. S.; Takahashi, H.; Ohnuki, S.; Kinoshita, H.; Loomis, B. A. *J. Nucl. Mater.* 1992, *191-194*, 948-951.

[85] Yano, T.; Iseki, T. *J. Nucl. Mater.* 1993, *203*, 249-254.

[86] Terasawa, M.; Mitamura, T.; Liu, L.; Tsubakino, H.; Niibe, M. *Nucl. Instrum. Methods Phys. Res. Sect. B: Beam Interact. Mater. Atoms* 2002, *193*, 329-335.

[87] Harbsmeier, F.; Conrad, J.; Bolse, W. *Nucl. Instrum. Methods Phys. Res. Sect. B: Beam Interact. Mater. Atoms* 1998, *137*, 505-510.

[88] Zinkle, S.J.; Pells, G.P. *J. Nucl. Mater.* 1998, *253*, 120-132.

[89] Spitzer, W.G.; Hubler, G.K.; Kennedy, T.A. *Nucl. Instrum. Methods Phys. Res.* 1983, *209-210*, 309-312.

[90] Custer, J.S.; Thompson, M.O.; Jacobson, D.C.; Poate, J.M.; Roorda, S.; Sinke, W.C.; Spaepen, F. *Appl. Phys. Lett.* 1994, *64*, 437-439.

[91] Nappé, J.C.; Grosseau, Ph.; Audubert, F.; Beauvy, M.; Guilhot, B. In *Proceedings of the 11th International Conference and Exhibition of the European Ceramic Society*; Bucko, M.M.; Haberko, K. Z.; Pedzich; Zych, L.; Ed.; Krakow, PL, 2009; pp 1089-1092.

[92] Bolse, W.; Conrad, J.; Rödle, T.; Weber, T. *Surf. Coat. Technol.* 1995, *74-75*, 927-931.

[93] Weber, W.J.; Yu, N.; Wang, L.M. *J. Nucl. Mater.* 1998, *253*, 53-59.

[94] Yano, T.; Ichikawa, K.; Akiyoshi, M.; Tachi, Y. *J. Nucl. Mater.* 2000, *283*, 947-951.

[95] Youngman, R.A.; Mitchell, T.E.; Clinard, F.W.; Hurley, G.F. *J. Mater. Res.* 1991, *6*, 2178-2187.

[96] Gosset, D.; Simeone, D.; Quirion, D. *J. Phys. IV* 2000, *10*, Pr10-55-63.

[97] Pruneda, J.M.; Archer, T.D.; Artacho, E. *Phys. Rev. B* 2004, *70*, 104111.

[98] Nappé, J.C.; Grosseau, Ph.; Audubert, F.; Guilhot, B.; Beauvy, M.; Benabdesselam, M.; Monnet, I. *J. Nucl. Mater.* 2009, *385*, 304-307.

[99] Snead, L.L.; Nozawa, T.; Katoh, Y.; Byun, T.S.; Kondo, S.; Petti, D.A. *J. Nucl. Mater.* 2007, *371*, 329-377.

[100] Inui, H.; Mori, H.; Fujita, H. *Philos. Mag. B: Phys. Condens. Matter Stat.* 1990, *61*, 107-124.

[101] Weber, W.J.; Wang, L.M. Nucl. Instrum. *Methods Phys. Res. Sect. B: Beam Interact. Mater. Atoms* 1995, 106, 298-302.

[102] Weber, W.J.; Wang, L.M.; Yu, N. *Nucl. Instrum. Methods Phys. Res. Sect. B: Beam Interact. Mater. Atoms* 1996, 116, 322-326.
[103] Heft, A.; Wendler, E.; Bachmann, T.; Glaser, E.; Wesch, W. *Mater. Sci. Eng. B: Solid State Mater. Adv. Technol.* 1995, *29*, 142-146.
[104] Heft, A.; Wendler, E.; Heindl, J.; Bachmann, T.; Glaser, E.; Strunk, H.P.; Wesch, W. *Nucl. Instrum. Methods Phys. Res. Sect. B: Beam Interact. Mater. Atoms* 1996, *113*, 239-243.
[105] Snead, L.L.; Zinkle, S.J. *Mater. Res. Soc. Symp. Proc.* 1997, *439*, 595-606.
[106] Price, R.J. *J. Nucl. Mater.* 1973, *48*, 47-57.
[107] Senor, D.J.; Youngblood, G.E.; Greenwood, L.R.; Archer, D.V.; Alexander, D.L.; Chen, M.C.; Newsome, G.A. *J. Nucl. Mater.* 2003, *317*, 145-159.
[108] Katoh, Y.; Snead, L.L.; Henager Jr., C.H.; Hasegawa, A.; Kohyama, A.; Riccardi, B.; Hegeman, H. *J. Nucl. Mater.* 2007, *367-370*, 659-671.
[109] Zhang, C.H.; Jang, J.; Kim, M.C.; Cho, H.D.; Yang, Y.T.; Sun, Y.M. *J. Nucl. Mater.* 2008, *375*, 185-191.
[110] Neustroev, V.S.; Garner, F.A. *J. Nucl. Mater.* 2008, *378*, 327-332.
[111] Kozlov, A.V.; Portnykh, I.A. *J. Nucl. Mater.* 2009, *386-388*, 147-151.
[112] David, C.; Panigrahi, B.K.; Amarendra, G.; Abhaya, S.; Balaji, S.; Balamurugan, A.K.; Nair, K.G.M.; Viswanathan, B.; Sundar, C.S.; Raj, B. *Surf. Coat. Technol.* 2009, *203*, 2363-2366.
[113] Porollo, S.I.; Konobeev, Yu.V.; Garner, F.A. *J. Nucl. Mater.* 2009, *393*, 61-66.
[114] Clinard Jr., F.W.; Hurley, G.F.; Hobbs, L.W.; Rohr, D.L.; Youngman, R.A. *J. Nucl. Mater.* 1984, *123*, 1386-1392.
[115] Nappé, J.C.; Monnet, I.; Grosseau, Ph.; Audubert, F.; Guilhot, B.; Beauvy, M.; Benabdesselam, M. *Under review in Nucl. Instrum. Methods Phys. Res. Sect. B: Beam Interact. Mater. Atoms*.
[116] Liu, X.; Le Flem, M.; Béchade, J.L.; Monnet, I. *J. Nucl. Mater.* 2010, *401*, 149-153.
[117] Lee, S.M.; Ada, E.T.; Lee, H.; Kulik, J.; Rabalais, J.W. *Surf. Sci.* 2000, *453*, 159-170.
[118] Gao, N.F.; Miyamoto, Y.; Zhang, D. *Mater. Lett.* 2002, *55*, 61-66.
[119] Barsoum, M.W.; El-Raghy, T.; Ogbuji, L. *J. Electrochem. Soc.* 1997, *144*, 2508-2516.
[120] Sun, Z.; Zhou, Y.; Li, M. *Corros. Sci.* 2001, *43*, 1095-1109.
[121] Nappé, J.C.; Grosseau, Ph.; Guilhot, B.; Audubert, F.; Beauvy, M.; Benabdesselam, M. In *Mechanical Properties and Performance of Engineering Ceramics and Composites IV*; Ceramic Engineering and Science Proceedings; Singh, D.; Kriven, W.M.; Ed.; Hoboken, US, 2010; Vol. 30, pp 199-204.
[122] Marée, C.H.M.; Vredenberg, A.M.; Habraken, F.H.P.M. *Mater. Chem. Phys.* 1996, *46*, 198-205.
[123] Avasthi, D.K.; Assmann, W.; Tripathi, A.; Srivastava, S.K.; Ghosh, S.; Grüner, F.; Toulemonde, M. *Phys. Rev. B* 2003, *68*, 153106.
[124] Roller, T.; Bolse, W. *Phys. Rev. B* 2007, *75*, 054107.
[125] Bérerd, N.; Catalette, H.; Chevarier, A.; Chevarier, N.; Faust, H.; Moncoffre, N. *Surf. Coat. Technol.* 2002, *158-159*, 473-476.
[126] Bérerd, N.; Chevarier, A.; Moncoffre, N.; Jaffrézic, H.; Balanzat, E.; Catalette, H. *J. Appl. Phys.* 2005, *97*, 083528.
[127] Bérerd, N. *Effets d'irradiation sur l'oxydation du zirconium et la diffusion de l'uranium dans la zircone (irradiation effects in both the oxidation of zirconium and the diffusium*

of uranium in zirconia). PhD thesis in Nuclear Physic, Claude Bernard University, Lyon I, FR, 2003.

[128] Toulemonde, M.; Assmann, W.; Trautmann, C.; Gruner, F.; Mieskes, H.D.; Kucal, H.; Wang, Z.G. *Nucl. Instrum. Methods Phys. Res. Sect. B: Beam Interact. Mater. Atoms* 2003, *212*, 346-357.

[129] Johnson, R.E; Sundqvist, B.U.R. *Phys. Today* 1992, *45*, 28-36.

[130] Bringa, E.M.; Johnson, R.E. *Nucl. Instrum. Methods Phys. Res. Sect. B: Beam Interact. Mater. Atoms* 2001, *180*, 99-104.

[131] Assmann, W.; Toulemonde, M.; Trautmann, C. In *Sputtering by Particle Bombardment*; Topics in Applied Physics; 2007; Vol. 110, pp 401-450.

[132] Studer, F.; Hervieu, M.; Costantini, J.M.; Toulemonde, M. *Nucl. Instrum. Methods Phys. Res. Sect. B: Beam Interact. Mater. Atoms* 1997, *122*, 449-457.

[133] Schattat, B.; Bolse, W.; Klaumünzer, S.; Harbsmeier, F.; Jasenek, A. *Nucl. Instrum. Methods Phys. Res. Sect. B: Beam Interact. Mater. Atoms* 2002, *191*, 577-581.

[134] Schattat, B.; Bolse, W.; Klaumünzer, S.; Harbsmeier, F.; Jasenek, A. *Appl. Phys. A: Mater. Sci. Process.* 2003, *76*, 165-169.

[135] Dhara, S. *Crit. Rev. Solid State Mater. Sci.* 2007, *32*, 1-50.

[136] Sickafus, K.E.; Matzke, Hj.; Hartmann, Th.; Yasuda, K.; Valdez, J.A.; Chodak III, P.; Nastasi, M.; Verrall, R.A. *J. Nucl. Mater.* 1999, *274*, 66-77.

[137] Simeone, D.; Gosset, D.; Bechade, J.L.; Chevarier, A. *J. Nucl. Mater.* 2002, *300*, 27-38.

[138] Schuster, B.; Lang, M.; Klein, R.; Trautmann, C.; Neumann, R.; Benyagoub, A. *Nucl. Instrum. Methods Phys. Res. Sect. B: Beam Interact. Mater. Atoms* 2009, *267*, 964-968.

[139] Baldinozzi, G.; Simeone, D.; Gosset, D.; Surblé, S.; Mazérolles, L.; Thomé, L. *Nucl. Instrum. Methods Phys. Res. Sect. B: Beam Interact. Mater. Atoms* 2008, *266*, 2848-2853.

[140] Sprouster, D.J.; Giulian, R.; Schnohr, C.S.; Araujo, L.L.; Kluth, P.; Byrne, A.P.; Foran, G.J.; Johannessen, B.; Ridgway, M.C. *Phys. Rev. B* 2009, *80*, 115438.

[141] Chaudhari, P.S.; Bhave, T.M; Kanjilal, D.; Bhoraskar, S.V. *J. Appl. Phys.* 2003, *93*, 3486-3489.

[142] Thakurdesai, M.; Mahadkar, A.; Kanjilal, D.; Bhattacharyya, V. *Vac.* 2008, *82*, 639-644.

[143] Thakurdesai, M.; Mohanty, T.; Kanjilal, D.; Raychaudhuri, P.; Bhattacharyya, V. *Appl. Surf. Sci.* 2009, *255*, 8935-8940.

[144] Arnoldbik, W.M.; Knoesen, D.; Tomozeiu, N.; Habraken, F.H.P.M. *Nucl. Instrum. Methods Phys. Res. Sect. B: Beam Interact. Mater. Atoms* 2007, *258*, 199-204.

[145] Moll, S.; Thome, L.; Vincent, L.; Garrido, F.; Sattonnay, G.; Thome, T.; Jagielski, J.; Costantini, J.M. *J. Appl. Phys.* 2009, *105*, 023512.

[146] Jheeta, K.S.; Jain, D.C.; Kumar, R.; Garg, K.B. *Solid State Commun.* 2007, *144*, 460-465.

[147] Khalfaoui, N.; Gorlich, M.; Muller, C.; Schleberger, M.; Lebius, H. *Nucl. Instrum. Methods Phys. Res. Sect. B: Beam Interact. Mater. Atoms* 2006, *245*, 246-249.

[148] Lakhani, A.; Ganesan, V.; Elizabeth, S.; Bhat, H.L.; Singh, R.; Kanjilal, D. *Nucl. Instrum. Methods Phys. Res. Sect. B: Beam Interact. Mater. Atoms* 2006, *244*, 120-123.

[149] Singh, J.P.; Singh, R.; Mishra, N.C.; Ganesan, V.; Kanjilal, D. *Nucl. Instrum. Methods Phys. Res. Sect. B: Beam Interact. Mater. Atoms* 2001, *179*, 37-41.

[150] Skuratov, V.A.; Zagorski, D.L.; Efimov, A.E.; Kluev, V.A.; Toporov, Y.P.; Mchedlishvili, B.V. *Radiat. Meas.* 2001, *34*, 571-576.
[151] Thakurdesai, M.; Kanjilal, D.; Bhattacharyya, V. *Appl. Surf. Sci.* 2008, *254*, 4695-4700.
[152] Nomura, K.; Nakanishi, T.; Nagasawa, Y.; Ohki, Y.; Awazu, K.; Fujimaki, M.; Kobayashi, N.; Ishii, S.; Shima, K. *Phys. Rev. B* 2003, *68*, 064106.
[153] Meftah, A.; Brisard, F.; Costantini, J.M.; Dooryhee, E.; Hageali, M.; Hervieu, M.; Stoquert, J.P.; Studer, F.; Toulemonde, M. *Phys. Rev. B* 1994, *49*, 12457-12463.
[154] Jensen, J.; Razpet, A.; Skupinski, M.; Possnert, G. *Nucl. Instrum. Methods Phys. Res. Sect. B: Beam Interact. Mater. Atoms* 2006, *243*, 119-126.
[155] Hedler, A.; Klaumünzer, S.; Wesch, W. *Nucl. Instrum. Methods Phys. Res. Sect. B: Beam Interact. Mater. Atoms* 2006, *242*, 85-87.
[156] Wesch, W.; Schnohr, C.S.; Kluth, P.; Hussain, Z.S.; Araujo, L.L.; Giulian, R.; Sprouster, D.J.; Byrne, A.P.; Ridgway, MC *J. Phys. D: Appl. Phys.* 2009, *42*, 115402.
[157] Didyk, A.Y.; Hofman, A.; Savin, V.V.; Semina, V.K.; Hajewska, E.; Szteke, W.; Starosta, W. *Nukleonika* 2005, *50*, 149-152.
[158] Rusponi, S.; Costantini, G.; Boragno, C.; Valbusa, U. *Phys. Rev. Lett.* 1998, *81*, 2735-2738.
[159] Sarkar, S.; Van Daele, B.; Vandervorst, W. *N. J. Phys.* 2008, *10*, 083012.
[160] Trinkaus, H.; Ryazanov, A.I. *Phys. Rev. Lett.* 1995, *74*, 5072-5075.
[161] Trinkaus, H. *Nucl. Instrum. Methods Phys. Res. Sect. B: Beam Interact. Mater. Atoms* 1998, *146*, 204-216.
[162] Klaumünzer, S. *Radiat. Eff. Defects Solids* 1989, *110*, 79-83.
[163] Klaumünzer, S. *Int. J. Radiat. Appl.Instrum. Part D: Nucl. Tracks Radiat. Meas.* 1991, *19*, 91-96.
[164] Klaumünzer, S.; Schumacher, G. *Phys. Rev. Lett.* 1983, *51*, 1987-1990.
[165] Hou, M.D.; Klaumünzer, S.; Schumacher, G. *Phys. Rev. B* 1990, *41*, 1144-1157.
[166] Gutzmann, A.; Klaumünzer, S. *Nucl. Instrum. Methods Phys. Res. Sect. B: Beam Interact. Mater. Atoms* 1997, *127-128*, 12-17.
[167] Yang, Y.; Dickerson, C.A.; Swoboda, H.; Miller, B.; Allen, T.R. *J. Nucl. Mater.* 2008, *378*, 341-348.
[168] Fukushima, K.; Yamada, I. *Nucl. Instrum. Methods Phys. Res. Sect. B: Beam Interact. Mater. Atoms* 1996, *112*, 116-119.
[169] Bolse, W.; Peteves, S.D. *Nucl. Instrum. Methods Phys. Res. Sect. B: Beam Interact. Mater. Atoms* 1992, *68*, 331-341.

In: MAX Phases: Microstructure, Properties and Applications ISBN 978-1-61324-182-0
Editors: It-Meng (Jim) Low and Yanchun Zhou

Chapter 10

AB INITIO PREDICTION OF STRUCTURAL, ELECTRONIC AND MECHANICAL PROPERTIES OF Ti_3SiC_2

A. L. Ivanovskii, N. I. Medvedeva and A. N. Enyashin
Institute of Solid State Chemistry, Ural Branch of the Russian Academy of Sciences, Ekaterinburg, Russia

ABSTRACT

Layered ternary transition-metal carbides (so-called nano-laminates, or MAX phases) attract much attention of physicists and material scientists owing to their unique physical and chemical properties and their technological applications as high-temperature and ultrahigh-temperature structural materials. The *ab initio* approaches are very helpful in optimization and prediction of properties of these materials. Here, recent achievements in the theoretical studies of electronic band structure, chemical bonding and some properties of Ti_3SiC_2 as a basic phase of a broad family of nano-laminates are reviewed. Besides, the peculiarities of electronic properties of non-stoichiometric and doped Ti_3SiC_2 - based species are described, and Ti_3SiC_2 surface states and hypothetical Ti_3SiC_2 – based nanotubes are discussed.

1. INTRODUCTION

The MAX phases (known also as nano-laminates) are a wide group of ternary layered compounds with formal stoichiometry $M_{n+1}AX_n$ (n=1,2,3...), where M are transition d metals, A are p elements (such as Si, Ge, Al, S, Sn *etc.*) and X is carbon or nitrogen.

In recent years, the MAX phases have become materials of intense research owing to their remarkable mechanical properties, fully reversible plasticity, exceptional shock resistance, damage tolerance, negligible thermopower, high thermal conductivity and some others, which are very attractive for industrial applications of these systems as advanced ceramic materials [1-3].

Besides great efforts, which have been made in synthesis of new promising nano-laminates (in the form of both bulk materials and thin films) and in experimental characterizations of their properties, much activity was focused on the theoretical studies, which have shown powerful ability to understand and to predict the properties of these unusual materials.

Ti_3SiC_2 is a prototype of the M_3AX_2 phases (so-called 312 nano-laminates) and the most extensively studied member of this family. The theoretical studies on the band structure of Ti_3SiC_2 and related materials (such as Ti_3AlC_2, Zr_3SiC_2, V_3SiC_2, Ti_3AlN_2 *etc.*) by means of the *ab initio* calculations within the density-functional theory (DFT) approaches were started in the middle of 1990s by our group [4-9] and are continued till now, see [10-33].

In those works the following main topics were examined: (i) Detailed studies of the electronic band structure, chemical bonding and some properties (such as structural, optical properties, cohesive parameters *etc.*) of the basic phase Ti_3SiC_2 and some related compounds such as Ti_3AlC_2, Zr_3SiC_2, V_3SiC_2, Ti_3AlN_2, Ti_3GeC_2 *etc.*; (ii) The investigation on the phase stability of Ti_3SiC_2 and related phases at high pressures and polymorphism of Ti_3SiC_2; (iii) The influence of lattice defects and various impurities on the properties of bulk Ti_3SiC_2; (iv) The surface states for Ti_3SiC_2 and related compounds as well as some interfaces; (v) and simulations of the Ti_3SiC_2-based solid solutions such as $Ti_3SiC_{2-x}N_x$; $Ti_3Si_{1-x}Al_xC_2$ or $Ti_3Si_{1-x}Ge_xC_2$. The results obtained are discussed below in Sec. 1.

Besides, as Ti_3SiC_2 is of interest mostly as advanced ceramics, we will focus also on the mechanical properties of this material and discuss the cleavage characteristics, fracture properties under tensile stress and lattice resistance to sliding predicted from *ab initio* calculations, Sec. 2. Finally, the structural models and electronic properties of the recently proposed hypothetical Ti_3SiC_2 nanotubes are shortly discussed in Sec. 3.

1.1. Electronic Band Structure of Ti_3SiC_2 and Related Phases

Ti_3SiC_2 adopts a hexagonal structure (space group *P63/mmc*, *Z*= 2), see reviews [1-3]. There are two non-equivalent Ti atoms (denoted as Ti1 and Ti2), which are located at the Wyckoff positions $2a$= (0,;0;0) and $4f$ = (1/3;2/3;z_{Ti}), respectively. Here, two (per cell) Ti1 atoms have carbon atoms as nearest neighbors, while the other four (per cell) Ti2 atoms have both C and Si atoms as nearest neighbors. The silicon atoms are in $2b$ = (0;0;1/4) positions and carbon atoms are in positions $4f$ = (1/3;2/3;z_C). Here z_{Ti} and z_C are internal parameters. The experimentally determined lattice parameters are a = 3.068 Å and c = 17.669 Å and the internal coordinates are z_{Ti} = 0.135 and z_C = 0.567 [1-3]. The structure of Ti_3SiC_2 can be described as two edge-shared Ti_6C octahedron layers linked together by a two-dimensional atomic Si layer or, in other words, as formed from the hexagonal layers ...[Si-Ti2-C-Ti1-C-Ti2]... stacked in the repeated sequence as is shown in Figure 1.

To calculate the ground-state electronic properties of Ti_3SiC_2, it is necessary to obtain the stable atomic configuration of crystal. In numerous theoretical papers the equilibrium structure was determined by relaxation with respect to lattice parameters and internal parameters. In Table 1, the calculated lattice constants and internal parameters, as well as the bulk moduli B for Ti_3SiC_2 (and for some related materials) are listed, and the theoretical values are in reasonable agreement with the experimental data.

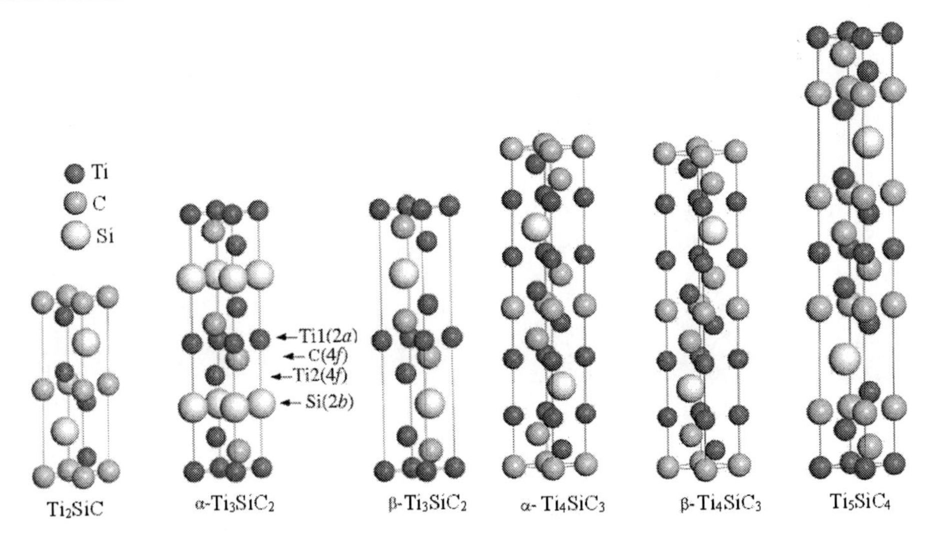

Figure 1. The crystal structures of $Ti_{n+1}SiC_n$ (n=1,2,3 and 4) [31]. The Wyckoff positions of various atoms and non-equivalent Ti atoms (Ti1 and Ti2, see text) are labeled for α-Ti_3SiC_2.

Table 1. Calculated lattice parameters (*a* and *c*, in Å), internal coordinates ($z_{Ti,C}$) and bulk modulus (*B*, in GPa) for Ti_3SiC_2 and some related materials in comparison with experiment

phase	a	c	c/a	z_{Ti}	z_C	B	References
Ti_3SiC_2	3.0563	17.6604	5.78	0.1366	0.5728	195	[23]
	3.0615	17.6094	5.75	-	-	184	[19]
	-	-	5.77	0.135	0.572	225	[11]
	3.0603	17.6707	5.77	0.1342	0.5720	-	[25]
	3.0705	17.6707	5.77	0.1370	0.5741	202	[14]
	3.076	17.713	5.76	-	-	-	[21]
	3.0665	17.671	5.78	-	-	-	[1]
Ti_3AlC_2	3.082	18.6523	6.05	0.1274	0.5693	-	[25]
	3.0634	18.5066	6.04	0.1286	0.5699	187	[23]
	3.0720	18.732	6.10	0.1290	0.5701	190	[15]
	3.0753	18.578	6.04	0.1280	0.5640	-	[24]
Ti_3GeC_2	3.0992	17.8948	5.77	0.1325	0.5718	-	[25]
	3.0823	17.7109	5.75	0.1361	0.5737	198	[14]
	3.07	17.76	5.79	-	-	-	[1]

* As calculated by CASTEP [14,15,19,23], FLMTO [11], VASP [21] and FLAPW-GGA [25] codes in comparison with experiments [1,24].

The calculated elastic constants (C_{ij}) for Ti_3SiC_2 [19] are positive and satisfy the well-known generalized criteria for mechanically stable hexagonal crystals: $C_{11} > 0$; $C_{44} > 0$; $C_{11} - C_{12} > 0$; $(C_{11} + 2C_{12})C_{33} - 2C_{13}^2 > 0$. The shear anisotropy ratio $A = C_{44}/C_{66}$, bulk modulus, shear modulus (G) and Young's modulus (Y) for Ti_3SiC_2 calculated from the elastic constants are presented in Table 2. It is seen that $B > G$; this implies that the parameter limiting the mechanical stability of this material is the shear modulus. According to the semi-empirical

Pugh's criteria (see, for example [49]) the material should behave in a ductile manner if *G/B* < 0.5, otherwise it should be brittle. For Ti_3SiC_2 *G/B* = 0.76, thus this phase is expected to behave as a brittle material, see also below.

Table 2. Calculated elastic constants (C_{ij}, in GPa) and shear anisotropy ratio *A*, bulk modulus, (*B*, in GPa), shear modulus (*G*, in GPa) and Young's modulus (*Y*, in GPa) for the α- and β- polymorphs of Ti_3SiC_2 [19]

polymorph	C_{11}	C_{12}	C_{13}	C_{33}	C_{44}	C_{66}	*A*	*B*	*G*
α	355	96	103	347	160	130	1.26	184	140
β	375	85	74	361	121	145	0.83	175	136

Let us discuss the electronic properties of Ti_3SiC_2. Figure 2(a) shows the calculated [7] band structure of Ti_3SiC_2 along the high-symmetry lines in the Brillouin zone (BZ) of the hexagonal lattice.

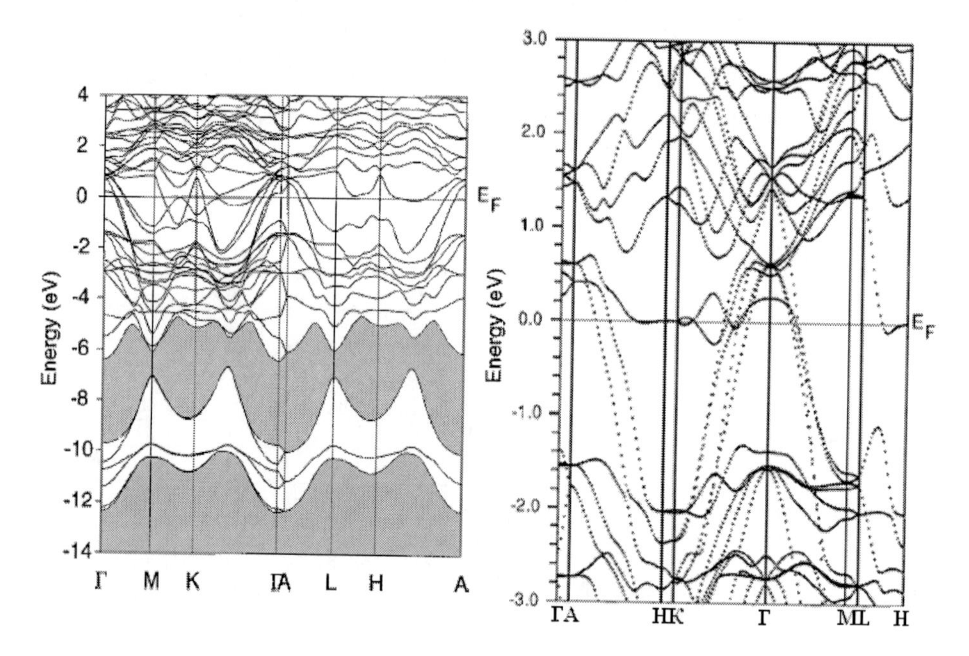

Figure 2. (a) Electronic band structure of Ti_3SiC_2 according to FLMTO calculations [7] and (b) the near-Fermi bands according to FLAPW calculations [25].

It is seen that the valence bands of Ti_3SiC_2 may be divided into two basic groups: a low-energy group composed of C, Si states mainly of s-symmetry (there is also a little contribution from Ti 3d states), and the next occupied bands containing predominantly valence Si 3p, C 2p and Ti 3d states. There is no direct overlap of C and Si s-bands (the four lower and the next two bands, Figure 2(a)). The energy dispersion of the Si bands is larger than that for the C bands owing to a more diffuse character of the 3s and 3p orbitals of silicon. The top valence bands are derived from the strongly hybridized Ti 3d, Si 3p, and C 2p states, see Figure 3, where the total and partial densities of states (DOSs) are depicted.

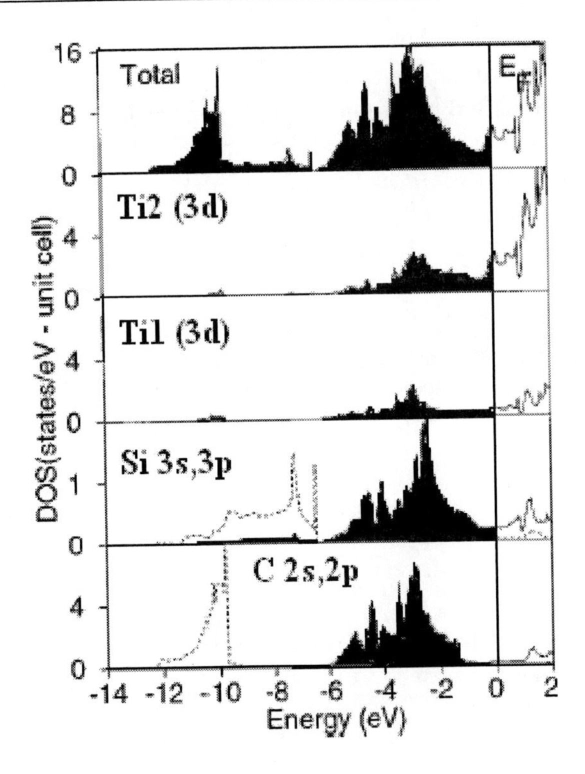

Figure 3. Total and partial densities of states of Ti_3SiC_2.

Ti_3SiC_2 exhibits a metallic behavior owing to bands crossing the Fermi level (E_F) along the K-Γ and Γ-M directions; this results in the non-zero DOS at the Fermi level $N(E_F) \sim 4.8$-4.9 states/eV per unit cell [7,19]. On the other hand, the band structure near and below the Fermi level is strongly anisotropic [23], which is evident from a very low k-dispersion along the Γ-A, H-K, M-L and L-H directions, Figure 2(b). It can be concluded that the conductivity for the single crystal Ti_3SiC_2 is also anisotropic.

The Brillouin zone and the Fermi surface (FS) for Ti_3SiC_2 [23] are depicted in Figure 4. The results [25] show a FS consisting of electron- and hole-like pockets at the center of BZ (around point Γ) as well as at the corners of BZ, which reflects the above quasi-two-dimensional-like bands dispersion for this crystal.

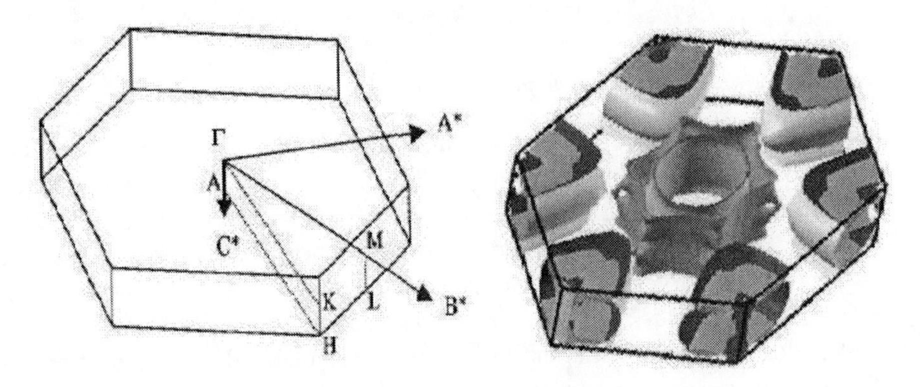

Figure 4. The Brillouin zone (BZ) and the Fermi surface (FS) of Ti_3SiC_2 [25].

To describe the inter-atomic bonding for Ti_3SiC_2, we begin with a simple ionic picture, which considers the standard oxidation numbers of atoms: Ti^{4+}, C^{4-} and Si^{4-}. However, the values of the total and partial electronic charges inside the muffin-tin (MT) spheres as calculated by the FLMTO method [7] reveal that the main effects of inter-atomic bonding are due to the participation of metal *d* and metalloid *p* states: the occupation of the outer *s,p* titanium states is low. Indeed, the DOSs distributions (Figure 4) demonstrate strong hybridization effects for Ti 3d, Si 3p, and C 2p states, i.e. the covalent Ti-C and Ti-Si bonding. It was shown that the inter-atomic interaction in the layers formed by Ti_6C octahedra is of a combined covalent-ionic-metallic type owing to hybridization of C 2p and Ti 3d states, partial charge transfer in the Ti1,2→C,Si direction (the estimated atomic effective charges are: Ti1 = + 0.641e, Ti2 = + 0.614e, Si = -0.304 e and C = -0.384e) and *dd-π* overlapping of Ti states [7,10-12,19]. The inter-atomic interaction in the Si atom layers is determined mainly by the hybridization of Si 3p–Si 3p states; the interaction with the layers made up of Ti_6C octahedron is relatively low. Thus, the chemical bonding in Ti_3SiC_2 is strongly anisotropic as shown in Figure 5. Let us note also a difference in the distribution of charge density in the Ti1 and Ti2 layers, which implies that Ti1 and Ti2 play different roles in the chemical bonding. In the Ti2–C–Ti1–C–Ti2–Si chains the inter-atomic distance between Ti1,2 and C is 2.13 Å, and that between Ti2 and Si is 2.67 Å. The bonding between Ti and carbon is stronger than that between Ti2 and silicon.

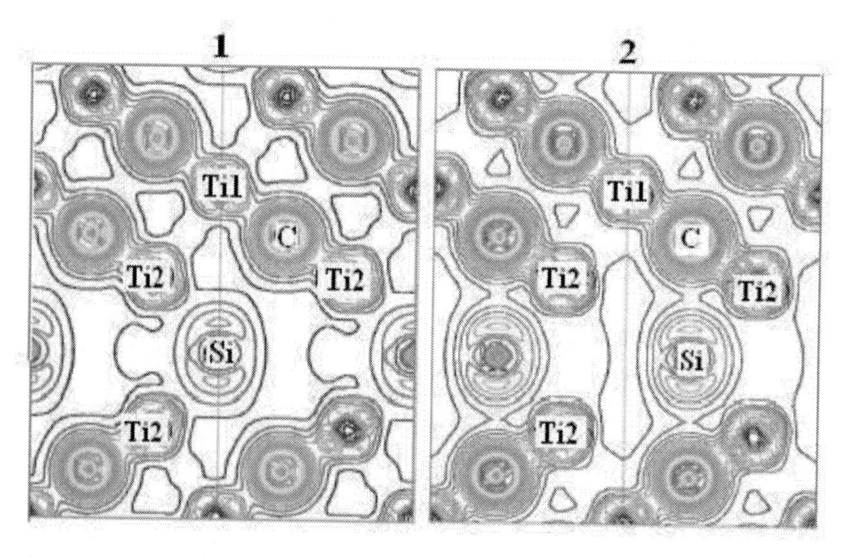

Figure 5. Valence charge density for (1) α- and (2) β- polymorphs of Ti_3SiC_2 [19].

For further understanding of chemical bonding, analysis of the so-called balanced crystal overlap population (BCOOP) is useful, Figure 6. The BCOOP is a function which is positive for bonding states and negative for anti-bonding states. The strength of covalent bonding can be determined by summing up the area under the BCOOP curve [26]. From comparison of the areas under the BCOOP curves, Figure 6, it is visible that the Ti 3d-C 2p bonds are generally much stronger than the Ti 3d- Si3p bonds. The areas under the Ti2-C curves are larger than for the Ti1-C curves; this implies that the Ti2 atoms lose some bond strength to the nearest Si atoms, which to some degree is compensated by a stronger Ti2-C bond.

Let us note that the electronic spectra for Ti_3SiC_2 were also investigated experimentally by x-ray photoelectron spectroscopy (XPS) and x-ray emission spectroscopy (XPS) [7,26,27]; the results obtained are in agreement with above theoretical data.

Recently, attention was focused on the phase stability at high pressure and polymorphism of Ti_3SiC_2 [17-19,26-29]. Figure 7 shows the calculated [29] lattice constants and axial ratio *c/a* varying with external pressure. It is seen that the parameter *a* decreases almost linearly with increasing pressure, in contrast to the *c* parameter, which exhibits a parabolic-like dependence. The *c/a* ratio is smaller than the equilibrium value when the pressure is smaller than about 30 GPa, and then it increases monotonically with pressure. The bond length contraction under pressure is also different for various bond types, Figure 7. These data show that Ti2–Si is the most compressible covalent bond [28]. The electronic structure is also evidently affected by pressure. In particular, the value of $N(E_F)$ decreases rapidly when pressure increases from zero to about 50 GPa, and tends to saturation at a constant value beyond 50 GPa; thus, the electrical conductivity should decrease with pressure – in agreement with experiments, see [23,28].

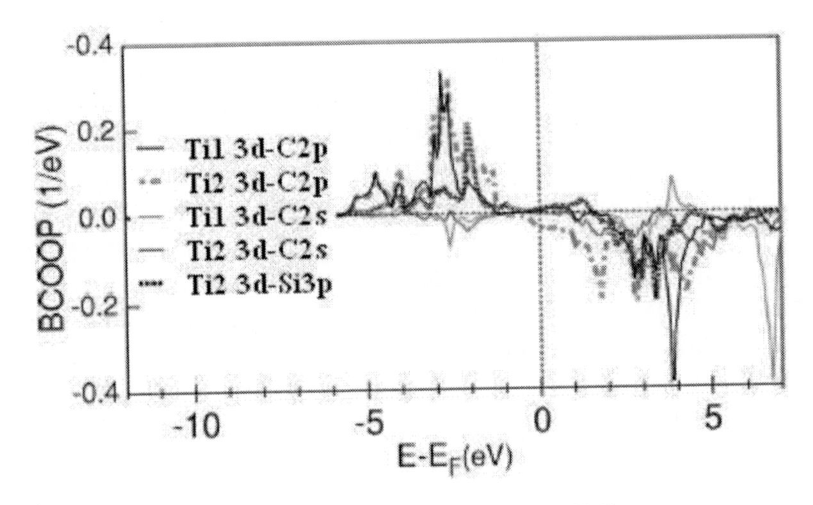

Figure 6. Balanced crystal overlap population (BCOOP) for Ti_3SiC_2 [26].

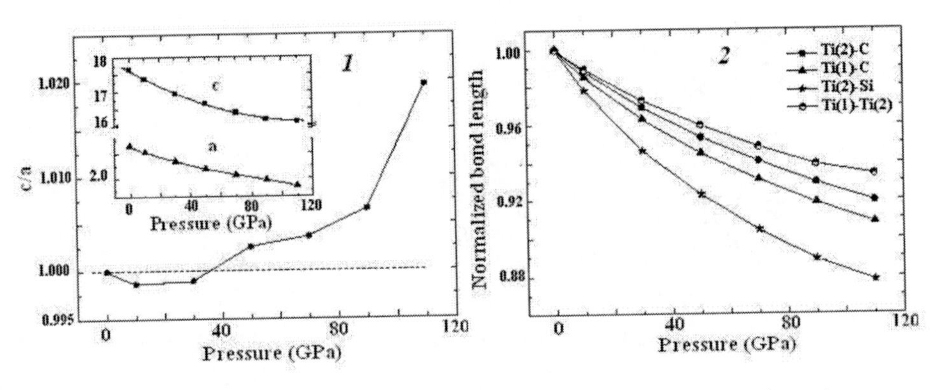

Figure 7. (1) The pressure dependence of the calculated lattice constants (*a, c*, in Å) and the ratio *c/a* for Ti_3SiC_2 and (2) normalized bond lengths in Ti_3SiC_2 as a function of pressure [29].

The phase stability of Ti_3SiC_2 under high hydrostatic pressures and the phase transformation from α- to β-phase (for which the Si atoms are placed in 2*d* Wyckoff position with fractional coordinates (2/3, 1/3, 1/4), Figure 1) have been studied by *ab initio* calculations [17,19,28]. The transition pressure is estimated at about 380 - 400 GPa. The α → β transformation is accompanied by an increase in the cell volume and *c/a* ratio. The total energy of the α-phase is lower than that of the β-phase by 0.663 eV/cell, indicating that α-Ti_3SiC_2 is much more stable than β-Ti_3SiC_2. The distribution of valence charge density for α-Ti_3SiC_2 *versus* β-Ti_3SiC_2 shown in Figure 5 indicates that the strength of the Ti2-Si covalent bond decreases in the β-polymorph. Both polymorphs exhibit a typical metallic behavior. The DOS at the Fermi level corresponds to a local maximum, but in the sequence α-Ti_3SiC_2 → β-Ti_3SiC_2 the value of $N(E_F)$ decreases by about 25-33% [19,28]. As for the mechanical properties, both the bulk and shear moduli of β-Ti_3SiC_2 become smaller than those for α -Ti_3SiC_2, while the Young's modulus is by about 10% larger [19].

The considered Ti_3SiC_2 phase, which contains Ti-C layers alternating with silicon atomic sheets, belongs to a homologous series of the $Ti_{n+1}SiC_n$ phases, where n=1,2,3…and the number of Ti-C layers between the Si sheets increases with n, as shown in Figure 1. The stability and electronic properties of these $Ti_{n+1}SiC_n$ homologues are discussed theoretically [30,31]. The cohesive energy E_{coh} of the homologues depends on the value of n. When n increases, the $Ti_{n+1}SiC_n$ phase becomes more similar to pure TiC as the number of Si sheets per a Ti layer decreases. The extra energy required for insertion of Si sheets is deduced [29] by calculating the difference in E_{coh} between $Ti_{n+1}SiC_n$ and the pure TiC and is plotted in Figure 8 as a function of the number of Si sheets per a Ti layer. This energy difference increases almost linearly with the amount of Si sheets per Ti layers in the structure. This may be interpreted by assuming that the Ti-Si bonds are weaker than the Ti-C bonds and that the Ti-Si and Ti-C bonds have constant strength for all stoichiometries. Next, the stability of different $Ti_{n+1}SiC_n$ homologous, where n=1,2,3 and 4, has been checked up recently [32] by comparing their total energy to that of the appropriate competing phases. The α-Ti_3SiC_2 phase is found to be preferable over all other $Ti_{n+1}SiC_n$ homologous, whereas Ti_5SiC_4 is unstable. The results also suggest that Ti_2SiC can be produced as a metastable compound.

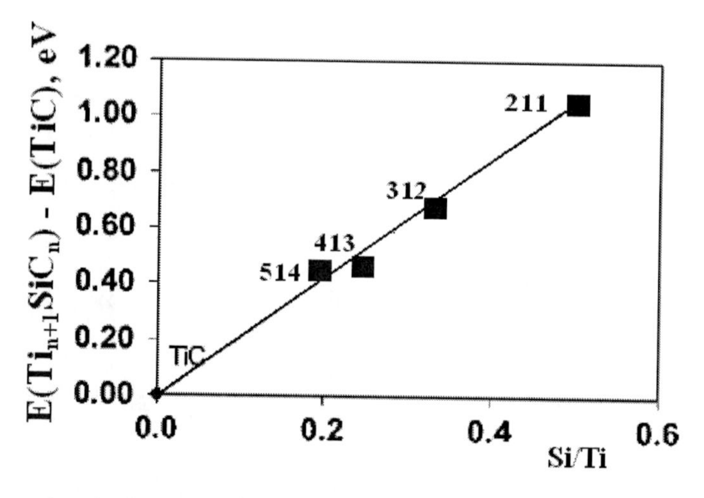

Figure 8. Differences in cohesive energy between the $Ti_{n+1}SiC_n$ homologues and TiC plotted as a function of the number of Si sheets per a Ti layer [30].

Intensive experimental and theoretical studies have been carried out to further improve the properties of Ti_3SiC_2, and the formation of various Ti_3SiC_2 based solid solutions (SSs, such as $Ti_3Si_{1-x}Al_xC_2$, $Ti_3Si_{1-x}Ge_xC_2$, $Ti_3SiC_{2-x}N_x$ *etc.*) is considered to be an effective way, see [1-3,9]. For example, the structural, electronic properties and chemical bonding for hypothetical SSs Ti_3SiCN and Ti_3SiCO, where the carbon atoms are partially replaced by nitrogen or oxygen, are examined [7]. The increase in electron concentration in the sequence $Ti_3SiC_2 \rightarrow Ti_3SiC_{2-x}N_x \rightarrow Ti_3SiC_{2-x}O_x$ is accompanied by an increase in the occupation of the near-Fermi Ti t_{2g}-like band and by an increase in $N(E_F)$. The formation of carbo-nitride SSs $Ti_3SiC_{2-x}N_x$ may cause an increase in their cohesive properties owing to the formation of stronger covalent N-Ti2-C bonds. An opposite situation may take place as a result of oxidation of Ti_3SiC_2, where the partial substitution of O for C may destabilize the hexagonal lattice of Ti_3SiC_2.

The partial replacement of silicon in their planar sheets by Al or Ge was also discussed theoretically [17,18,22,32]. The results [22] show that as the Al concentration increases, all bonds in the solid solution $Ti_3Si_{1-x}Al_xC_2$ weaken, which leads to an unstable structure. Additionally, the increase in the Al content depresses the melting point, Vickers hardness, Young's modulus, compressive strength and tensile strength of $Ti_3Si_{1-x}Al_xC_2$, but improves their workability and self-lubricating properties. The value of $N(E_F)$ indicates that the electronic conductivity of this SS decreases for a small Al content and increases when the Al content increases: $x > 0.67$ [22]. Better oxidation resistance was predicted [18] for $Ti_3Si_{0.75}Al_{0.25}C_2$ than for Ti_3SiC_2. Partial replacement Si $\rightarrow$Ge leads to stabilization of the α-phase, and the $\alpha \rightarrow \beta$ transition for the solid solution $Ti_3Si_{1-x}Ge_xC_2$ corresponds to a larger pressure (~412 GPa) than for the "pure" Ti_3SiC_2 (~397 GPa) or Ti_3GeC_2 (~ 400GPa) [17].

One of the most remarkable features of the cubic mono-carbide TiC is a wide range of homogeneity, which is due to the presence of vacancies in the carbon sublattice up to 50%. The deviations of TiC from stoichiometry over a wide interval of compositions change essentially the physical properties of this carbide. Examination of the carbon non-stoichiometry effect on the electronic and cohesive properties of Ti_3SiC_2 [7] shows that the appearance of carbon vacancies in Ti_3SiC_x, as well as in TiC_x, leads to local perturbations of the electronic structure and affects mainly the DOS distribution of Ti atoms nearest to the vacancy resulting in ''metallization'' of the system: $N(E_F)$ increases from 4.8 states/eV (Ti_3SiC_2) to 6.8 states/eV per unit cell for Ti_3SiC_1. At the same time, the carbon vacancies essentially reduce the cohesive energy of crystal. This may cause destabilization of the Ti_3SiC_2 hexagonal structure and qualitatively explains the absence of a wide homogeneity region for this material [7].

Since Ti_3SiC_2 and related materials are widely used as thin films, hard coatings and composites [2,3], the understanding of their surface properties, such as corrosion, oxidation or epitaxial growth is crucial for applications. Recently, the (1×1) Ti_3SiC_2 (001) surfaces with various terminations (Figure 9) were studied [20] and valuable data about the atomic structure, rumpling, surface stability and electronic properties were obtained. In order to understand the surface properties and the strength of chemical bonds, the cleavage energies (E_{cl}) between different basal planes were calculated (see the next section).

Examination of the bond lengths between the surface and subsurface layers caused by relaxation shows that except for the Si(Ti2) (001) termination (where the bond lengths grow by 3.3 %) all other intra-atomic bonds near the surface become shorter. For example, the variations in the bond lengths of the C(Ti1) and C(Ti2) terminations are larger than 7.3%, while those for other terminations are smaller than 4.5%. It was found that the energies of C(Ti1) and C(Ti2) terminated surfaces are unstable, whereas the Ti2(C), Ti1(C), Ti2(Si), and Si(Ti2) terminations can be stable under some specific conditions [20]. The electronic structure calculations indicate that the density of states and the charge distribution near the surfaces of Ti_3SiC_2 differ essentially from those of the bulk, Figure 10.

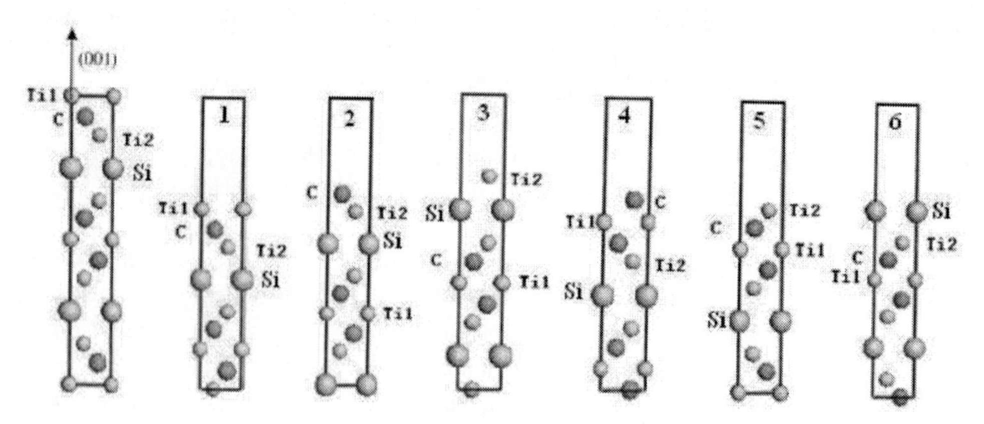

Figure 9. Atomic models of Ti_3SiC_2: bulk (*left*) and (001) surfaces with various terminations: 1- Ti1, 2- C(Ti2), 3- Ti2(Si), 4 - C(Ti1), 5- Ti2(C) and 6- Si(Ti2) [20].

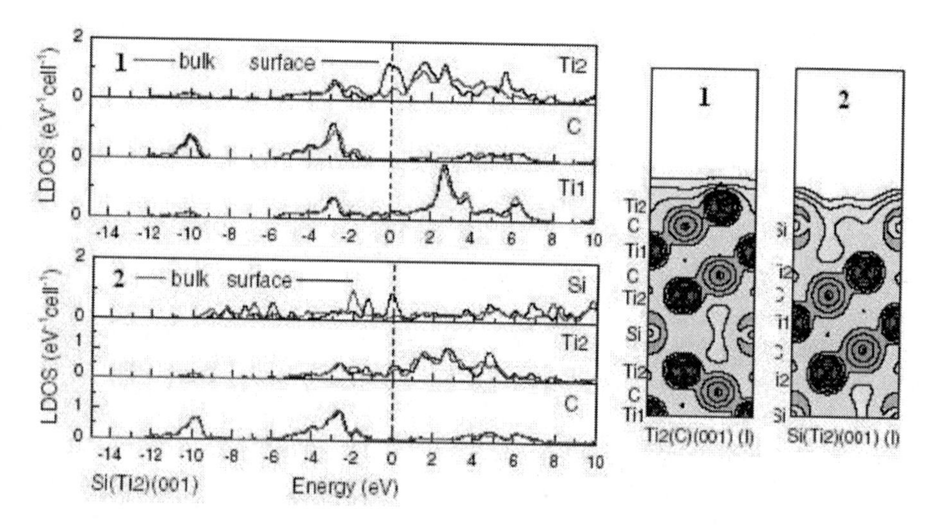

Figure 10. The local density of states (LDOS) of three surface atomic mono-layers of 1 - Ti2(C) (001), and 2- Si(Ti2) (001) terminations and the corresponding charge density maps for Ti_3SiC_2 [20].

Very recently the structural, electronic, and adhesive properties of the SiC/ Ti_3SiC_2 interface were investigated by first-principles calculations [21], where 96 interface geometries were considered. It was established that independent of the SiC terminations, C-terminated Ti_3SiC_2 had the largest adhesion, indicating the fundamental effect of carbon on strong adhesion in this interface system.

Finally, besides the above features, some other physical properties for Ti_3SiC_2 have been investigated theoretically. So, the dielectric functions were determined from first principles calculations and the optical spectra were analyzed by means of the electronic structure [17].

The thermoelectric tensor was calculated and the nature of thermopower (which is negative along the *z* direction and positive in the basal plane) was discussed [25,33].

As Ti_3SiC_2 is mostly interesting as advanced ceramics, we will focus further on the mechanical properties of this material in more details.

2. *Ab Initio* Prediction of Deformation Behavior of Ti_3SiC_2

Ti_3SiC_2 was related to a class of so-called kinking nonlinear elastic solids where plasticity is very anisotropic and kink band formation plays an important role [1,34,35]. It is the formation and annihilation of kink bands that are considered to be responsible for the recently observed fully reversible compressive deformation leading to very high damping properties [36]. Like many ceramics, Ti_3SiC_2 undergoes brittle fracture up to high temperatures (the brittle-ductile transition occurs at about 1100 C), and significant ductility without fracture is observed above 1200 C [1,34,37,38]. Local damage in polycrystals is connected with both transgranular and intergranular cracking where the cleavage and delamination are two main damage modes [1,36-40]. The delamination decreases the fracture toughness and plays an important role in the plastic behavior of Ti_3SiC_2 at high temperature [37,38]. It is clear that multiple deformation and damage mechanisms are related first of all to both peculiarities of the chemical bonding and their response to the stress-strain deformation and can be understood on the atomic level.

Theoretical atomistic simulation of the deformation behavior is very informative and important for understanding the mechanical properties of solids. The deformation models developed in [41-43], in combination with *ab initio* calculations, have been successfully used for studying the brittle properties of metals and ordered intermetallics [44-47]. Using the Rice-Thompson criterion, the brittle-versus-ductile behavior is analyzed by comparison of two competitive processes: crack opening (brittle fracture) and emission of a dislocation near the crack tip (plastic deformation) [42,43]. Two energy characteristics are calculated within the first-principles methods to describe these processes quantitatively: (i) the cleavage energy, which models the crack and (ii) the unstable stacking fault (SF) energy, which represents the maximum energy for the sliding of atomic planes and simulates the lattice resistance to the dislocation emission [41-43]. A comparison of these energy parameters allows one to predict a brittle crack propagation or ductile behavior and to explain the microscopic fracture mechanism. Cleavage fracture in Ti_3SiC_2 was simulated by *ab-initio* methods in [20,48,49]. Simulation of the other deformation mode – sliding – was performed in [50], where the microscopic origin for intrinsic brittleness in Ti_3SiC_2 was considered within the Rice-Thompson fracture criterion in comparison with ductile HCP Ti.

2.1. Cleavage Fracture

The microscopic mechanism of brittle failure in Ti_3SiC_2 has been studied by comparison of ideal cleavage energies *Gc* for the fracture between the different basal layers [20,48,49]. Although the ideal characteristics – cleavage energy and cleavage stress overestimate the observed values, nevertheless they give their upper limits. Additionally, the *ab-initio* approach allows one to connect the cleavage process with the peculiarities of electronic structure and chemical bonding and provides the microscopic reasons for brittle fracture.

First-principles calculations [20,48,49] gave similar results for cleavage energies and chemical bonding in Ti_3SiC_2 despite different methods (SIESTA, VASP and FLAPW were used in [20], [48] and [49], respectively) and calculation schemes. The cleavage energy was calculated in Ref. [48] as $G_C = (E_{broken}-E_{bulk})/2$, where E_{broken} is the total energy of a system with broken bonds. The energy of cleavage for Ti1-C, Ti2-C and Ti2-Si bonds was obtained to be 6.16, 7.16 and 3.16 J/m^2, respectively. These results demonstrate that the Ti-C bonds are twice as strong as the Ti-Si bonds. Within the simple model based on the number of bonds and the comparison of bulk modulus for Ti_3SiC_2 and TiC, the force constants K_1 for Ti-C and K_2 for Ti-Si bonds were estimated to be $K_1= 4\ K_2$ and the dependence of elastic energy E_{el} on strain S was approximately $E_{el} = 9\ K_1S^2$ [48]. This means that the elastic response is stronger than it might be suggested from cleavage energies.

The cleavage energy between X and Y planes was calculated in the theoretical work [20] as $G_C = (E^X_{slab} + E^Y_{slab} - n\ E_{bulk})/2\ S_0$, where E^X_{slab} and E^Y_{slab} are the energies of two symmetric complementary parts of crystal and n is the number of bulk units in supercell. These calculations gave lower cleavage energies (5.07, 6.33 and 2.88 J/m^2 for Ti1-C, Ti2-C and Ti2-Si bonds, respectively) than the previous data. The difference was related to the relaxation effects, which decrease the total energies.

The cleavage process was also simulated by varying the separation distance d between the slabs [49]. The energies of separation of two crystal halves as a functions of separation distance $E_{CL}(d)$ obtained for cleavages between the (0001) Ti1-C, Ti2-C and Ti2-Si hexagonal layers are shown in Figure 11. The calculated points in Figure 11 were fitted with the universal binding energy relation (UBER): $E_{CL}=G_C(1-(1+x)\exp(-x))$, $x=d/\lambda$, where the asymptotic value determines the ideal cleavage energy G_C (defined as the energy required for cleaving an infinite crystal into two semi-infinite parts) and λ is a scale parameter or characteristic length. The maximum derivative of $E_{CL}(d)$ defines the theoretical cleavage stress σ_{max}.

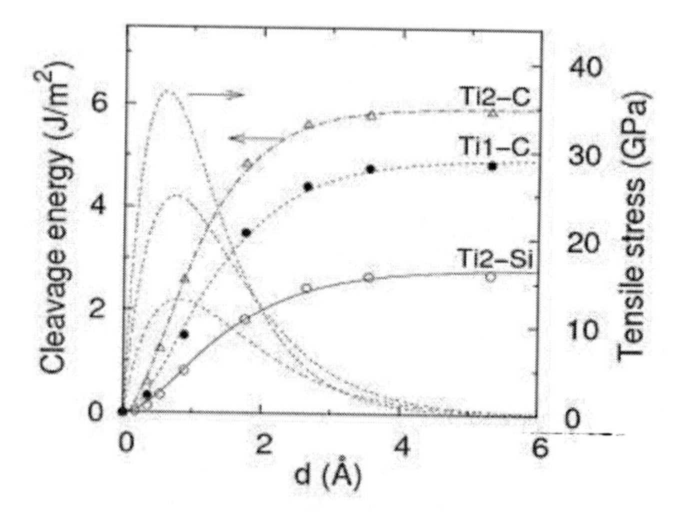

Figure 11. Cleavage energies and tensile stress versus separation distance d ($d = 0$ corresponds to the equilibrium interlayer distances) between the Ti1 and C, Ti2 and C or Ti2 and Si layers in Ti_3SiC_2.

The calculated cleavage characteristics (Table 3) demonstrate the same trend in strength of chemical Ti1-C, Ti2-C and Ti2-Si bonds. All *ab-initio* calculations [20,48,49] predict that crack should be preferentially formed between the Ti2 and Si layers, where the cleavage energy is twice lower than in other cases. The largest value of G_C obtained for the cleavage between the neighboring Ti and C layers corresponds to the strong covalent Ti3d-C2p bonding, and delamination within Ti2-C-Ti1-C-Ti2 layers is unlikely. The calculations [49] predict the ideal tensile strength between the Ti2 and Si layers to be near 12.6 GPa. The maximum stress occurs at an interlayer distance of 4.6 Å, which corresponds to the maximum strain of 21%. Actually, the strain to failure is known from experimental data on tensile stress to be in excess of 15% [1,37]. It should be noted that the ideal parameters always overestimate the real characteristics, which are determined by the least number of broken bonds and also depend on defects and loading type.

To estimate the crystal stretching under tensile stress before failure, the dependences of lattice constant *a* and the interatomic distances were calculated [49] as a function of *c* (Figure 12). When *c* increases from the undistorted value to 20 Å, the Ti2-C and Ti1-C distances vary only slightly, while the Ti2-Si distance increases by 10%. Further strain results in a sharp increase of the Ti2-Si distance, which almost completely corresponds to strain. Therefore, it is the Ti2-Si bond that is responsible for the accommodation of tensile stress, and its large stretching before failure can explain the damage tolerance of Ti_3SiC_2.

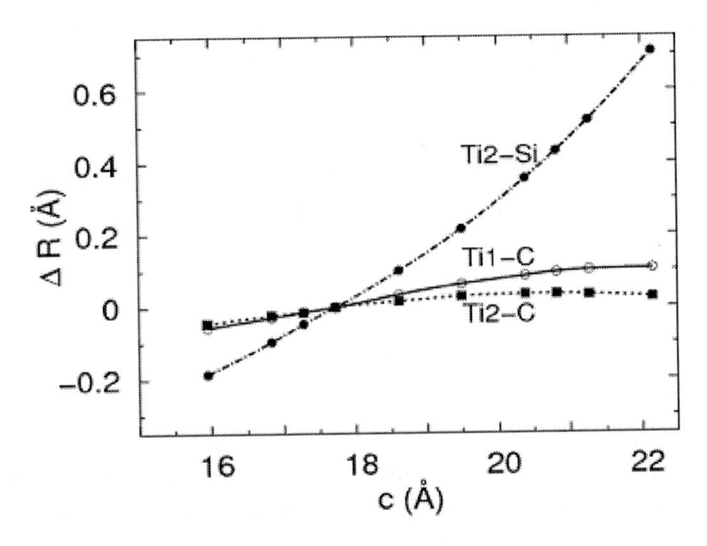

Figure 12. The Ti1-C, Ti2-C and Ti2-Si distances as functions of lattice constant *c* in Ti_3SiC_2.

2.2. Sliding

The generalized stacking fault (GSF) energies correspond to the energy changes with sliding (shearing) of a crystal part along some direction in the slip plane and they are the key parameters, which determine the structure and mobility of dislocations [41]. The relaxed unstable stacking fault energies γ_{US} (the GSF energy maximum) for $[\bar{1}2\bar{1}0]$ direction in the basal Ti2-Si, Ti1-C and Ti2-C planes obtained in [50] are summarized in Table 3.

Table 3. Cleavage energies $G_C = 2\gamma_S$ (J/m^2), unstable stacking fault energies γ_{US} and D=0.3 γ_S/γ_{US} for Ti_3SiC_2

	(0001) plane	G_C	γ_{US} $[\bar{1}2\bar{1}0]$	D
Ti_3SiC_2	Ti1-C	4.680	4.206	0.17
	Ti2-C	5.761	4.849	0.18
	Ti2-Si	2.677	0.944	0.43

It should be noted that other possible dislocations out of the basal plane such as [0001] or ⅓ $[\bar{1}\bar{1}23]$ are unlikely in Ti_3SiC_2 owing to a large c/a ratio. The lowest value of γ_{US} corresponds to sliding along $[\bar{1}2\bar{1}0]$ in the Si plane. The energy barrier for planar shear along the same direction in Ti1-C and Ti2-C basal planes is five times higher, which correlates with the strong covalent Ti3d-C2p bonds and the weaker Ti-Si bond [7,14]. Thus, there is a strong shear anisotropy not only along *a* and *c* axes owing to *c/a*>>1, but also for sliding in the different basal planes, which shows up in the deformation characteristics.

For the hexagonal structure, the basal slip is dominant if the ⅓ $[\bar{1}2\bar{1}0]$ dislocation is split into partials and/or the dislocation core is spread in the basal plane [41,51]. The observed main slip in Ti_3SiC_2 is ⅓ $[\bar{1}2\bar{1}0]$ but no experimental evidence was found for splitting of the basal dislocation into partials [40]. The dislocation dissociation and spreading are controlled by the energy of intrinsic stacking fault γ_{ISF}, which is a local minimum of the GSF energy at $a/\sqrt{6}$ along $[10\bar{1}0]$. The value of γ_{ISF} = 0.45 J/m^2 is close to the lowest GSF energy maximum (0.49 J/m^2) corresponding to the barrier for dislocation nucleation. Furthermore, the dependence of the GSF energy was obtained almost flat near γ_{ISF} and this fact may be a reason for the absence of dislocation dissociation [52]. To explain the observed twist, the dislocation walls were suggested to consist of ⅓ $[2\bar{1}\bar{1}0]$ and ⅓ $[1\bar{2}10]$ making angles with $[11\bar{2}0]$ of 60° and 120°, respectively.

Thus, theoretical modeling predicts that the basal Ti2-Si plane is a weak plane for easy cleavage and preferable for slip. Kink bands are observed typically in metallic systems. However, in this Ti_3SiC_2 ceramics the kink bands can move easily in the basal Si planes owing to a rather low energy barrier and the absence of other plane slips.

Within the Rice-Thompson criteria for the competition between brittle fracture and plastic deformation, the material should be ductile if D=0.3 $\gamma_S/\gamma_{US} > 1$ [42,53]. As follows from Table 3, even though the largest D for the Si basal plane is less than 1, this points out the intrinsic brittle fracture in Ti_3SiC_2. Nevertheless, this carbosilicide should be more ductile than molybdenum silicides, where D=0.04-0.1 [54]. The brittle fracture of Ti_3SiC_2 is attributed to a small value of cleavage energy relative to shear energy.

3. Structural and Electronic Properties of Hypothetical Tubular Nanolaminates

In this part we briefly consider the first theoretical models of hypothetical tubular forms of Ti_3SiC_2 and related titanium silicocarbides [55, 56].

Nano-scale quasi-one-dimensional (1D) structures, such as nanorods, nanowires, and nanotubes have attracted much attention because of their unique electronic, optical, and mechanical properties that differ drastically from their bulk counterparts and belong to the most promising modern nanomaterials with a wide range of potential technical applications [57-59]. The novel properties of these nanostructures could be attributed to combinations of size, dimensionality, and morphology effects. A lot of studies point out that layered compounds are the most suitable systems for formation of the above 1D nano-sized structures. Thus, it is reasonable to assume that the layered Ti_3SiC_2 (as well as the related $Ti_{n+1}SiC_n$ phases: Ti_2SiC and Ti_4SiC_3) under certain conditions can be transformed into nano-tubular forms.

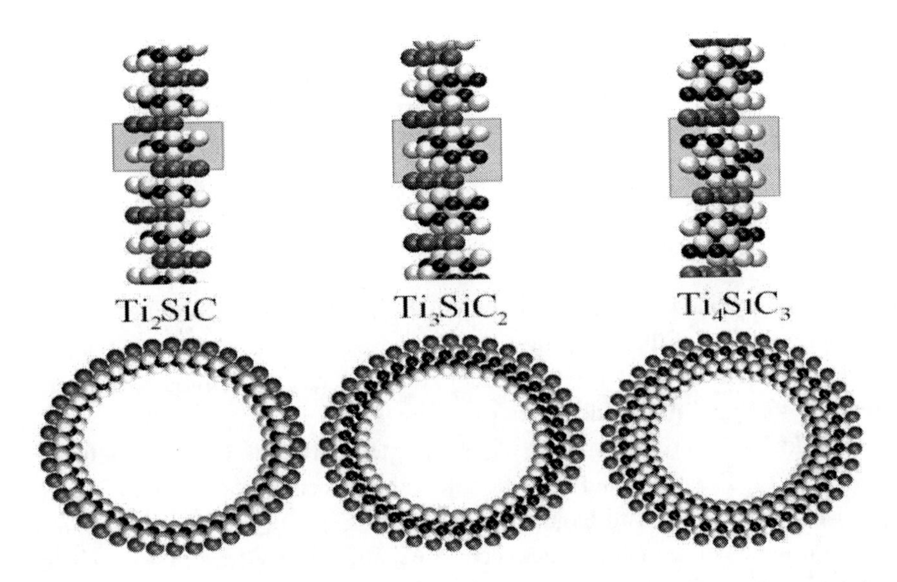

Figure 13. Atomic structures of bulk and nanotubular (20,20) $Ti_{n+1}SiC_n$ laminates. For nanotubes their cross-sections are shown.

Recently, the first atomic models of Ti_3SiC_2-based nanotubes were proposed, and their electronic properties were investigated [55]. The atomic structures of Ti_3SiC_2 infinite-long nanotubes (Ti_3SiC_2 NTs) can be constructed in a conventional way (see, for example [60]) by rolling of a (Si/Ti/C/Ti/C/Ti) layer into cylinders, preserving the stoichiometry of the bulk material. Like a Ti_3SiC_2 flat layer, the walls of these nanotubes can be described as six coaxial atomic cylinders: three Ti, two carbon and one outer Si cylinder, see Figure 13.

In this way, similar to single-walled carbon and other known inorganic nanotubes three groups of Ti_3SiC_2 NTs can be designed depending on the rolling direction: non-chiral *armchair* (n,n)-, *zigzag* (n,0)- and chiral (n,m)-like nanotubes. As an example, Figure 13 illustrates the structures for some non-chiral *zigzag* (n,0) and *armchair* (n,n) single-walled nanotubes optimized using the density-functional tight-binding approach [61], which are representatives of those found for the other Ti_3SiC_2 tubes. These results testify the possibility of the existence of stable tubular nanolaminate structures, which preserve tubular morphology after geometry optimization.

Likewise, using the same "convolution" algorithm, the unit cells of other hexagonal nanolaminates can be designed. For example, Ti_2SiC and Ti_4SiC_3 nanotubes are comprised by four (Si/Ti/C/Ti) and eight (Si/Ti/C/Ti/C/Ti/C/Ti) atomic cylinders, respectively [56].

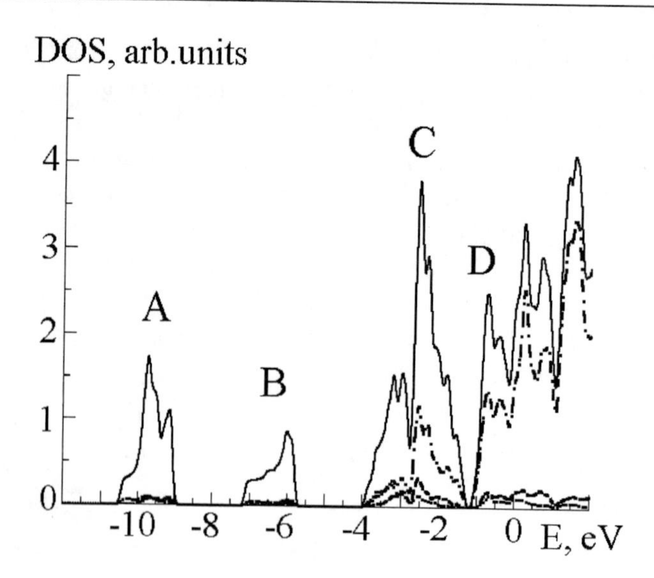

Figure 14. The densities of states (DOS) for (22,0) Ti_3SiC_2 nanotube. Solid line – total DOS,-·-·- Ti*3d*-states,----- Ti*4p*-states, ····· Ti*4s*-states. The Fermi level - 0.0 eV.

According to the band structure calculations [55], all Ti_3SiC_2 nanotubes are metallic like bulk Ti_3SiC_2 regardless of their chirality (*zigzag*- or *armchair*-like) and diameters. However, the electronic structure of tubular Ti_3SiC_2 nanolaminates has some important features. Figure 14 exhibits the total and partial densities of states (DOS) for (22,0) Ti_3SiC_2 nanotube as an example. The lowest-lying states (band A) positioned between -10.5 and -9.0 eV originate from the C*2s* states. The next band B in the interval between -7.2 and 5.5 eV is mainly of the Si*3s* type. The broad near-Fermi band (from -4 eV to E_F) is of a hybridized Ti*3d*–C*2p*–Si*3p* type, where the states in the interval between -4 eV and -1.3 eV (peak C) are responsible for the Ti-C bonding, and the states between -1.3 eV to the Fermi level (peak D) have a non-bonding nature. The main contribution in $N(E_F)$ arises from the states of titanium atomic cylinders; in their turn, the Ti*3d* states are dominating (69.6%) - in comparison with the Ti*4p* (8.7%) and Ti*4s* (4.3%) states. Thus, the DOS of Ti_3SiC_2 nanotubes exhibits a typical metallic character with a finite density of states at the Fermi level $N(E_F)$. However, in the case of a tubular nanolaminate, E_F corresponds to a local minimum of DOS, while crystalline Ti_3SiC_2 has the Fermi level placed in a local maximum of DOS, see also [7, 19].

The electronic spectra of Ti_2SiC and Ti_4SiC_3 nanotubes have been calculated in the same density-functional approach, though without geometry relaxation [56]. It was found that they retain their metal-like character with essential density of states at the Fermi level and also have some important features. In comparison with the respective bulk laminates, the Si3*p* states move into the region of lower bond energies and make an appreciable DOS contribution to the Fermi level, which is practically absent in the crystals. Thus, in contrast to the crystals, for Ti_2SiC and Ti_4SiC_3 nanotubes, conduction will also be realized with the participation of the states of silicon atoms. Moreover, against the profile of the total DOS of the silicocarbide nanotubes, the formation of additional peaks is clearly seen, which are due to analogous energy shifts of the 3*d* states of titanium atoms in the "inner" atomic cylinder of tubes. Further changes in the DOS of these tubes are possible, when the effects of the atomic relaxation of such structures are taken into account, but as shown by calculations for the related TiC and $TiSi_2$ nanotubes [62, 63], their role must be quite insignificant.

Calculated inter-atomic bond populations for all Ti_n+1SiC_n nanotubes indicate the general weakening of the covalent Ti-C bonds, the relative strengthening of the Ti-Si bonds, and the fundamental change in the charge state of the silicon surface atoms in the nanotubes, which adopt negative effective charges, unlike the positively charged Si atoms in the corresponding crystals [56]. The above-mentioned features may give rise to substantial differences in the functional (e.g., mechanical, thermo- and electro-physical) characteristics of the supposed tubular forms of $Ti_{n+1}SiC_n$ as compared with the corresponding crystals.

The estimations [55] of the strain energy E_{str} (*i.e.* the energy difference between a tube and the corresponding flat layer) for Ti_3SiC_2 NTs demonstrate, for example, that the E_{str} for (20,20) tube is close to 0.07 eV/atom, which is between the values of E_{str} for widely known carbon (~ 0.01 eV/atom [57]) and successfully synthesized MoS_2 tubes (~ 0.15 eV/atom [60]) with comparable diameters. This result can be considered as an additional argument for the possibility of existence of the proposed Ti_3SiC_2 tubular materials. Moreover, *zigzag*-like Ti_3SiC_2 NTs are energetically more favorable than the tubes with *armchair* -like chirality [55].

Although the strain energies of Ti_3SiC_2 nanotubes were found to be relatively low, these nanotubes have not been fabricated yet. One may speculate [55,56] that the simulated tubes (as well as the tubes based on other related layered $M_{n+1}A(C,N)_n$ nanolaminates) can most likely be prepared using the Prinz's technology from thin epitaxial films, which can be controllably detached from substrates and rolled into various cylindrical micro- and nanotubes, see [64]. Moreover, such Ti_3SiC_2 thin films have been successfully synthesized [65,66].

Another possible way for a mass production of the nanotubular laminates may be physical or chemical templating, see reviews [58,59]. For example, this method was applied for the synthesis of titanium carbide TiC and vanadium silicide V_5Si_3 nanotubes, i.e. the compounds related with silicocarbides of transition metals. It was observed that the reaction between carbon nanotubes, used as templates and reactants simultaneously, and titanium vapor carried out at 1300°C for 30 h gives at the final stage a mass of TiC nanotubes and nanowires, TiC nanoparticles covering non-reacted carbon nanotubes [67]. Carbon nanotubes served as templates and reactants together with Nb for the production of NbC nanotubes [68]. As another substrate, sapphire can be used: free-standing silicide nanostructures such as hexagonal V_5Si_3 nanowires and nanotubes were grown on a vanadium foil placed on Si powder by chemical vapor deposition method involving VCl_3 [69].

Evidently, fabrication using chemical template-free reactions with certain reagents and under certain conditions may also result in the formation of tubular nanolaminate forms. This route was successful for the fabrication of nanotubular d-metal compounds also related with nanolaminates: tungsten carbide WC and manganese silicide Mn_5Si_3. The low-temperature (900°C) thermal decomposition of $W(CO)_6$ in the presence of Mg powder was carried out to fabricate nanotubes of WC [70]. Mn_5Si_3 nanocages and nanotubes were prepared by the reaction of $MnCl_2$ with Mg_2Si at 650°C [71].

Concluding Remarks

In this review we summarized the results of *ab-initio* simulations of the structural, electronic and mechanical properties of Ti_3SiC_2 - the well-studied representative of new layered ternary $M_{n+1}AX_n$ phases, which demonstrate a unique mechanical behavior, excellent oxidation resistance and non-susceptibility to thermal shock. Here we were concentrated on the peculiarities of electronic structure and chemical bonding of these phases, which determine the microscopic mechanisms governing the observed electronic and mechanical properties.

First of all, we show that *ab-initio* calculations were able to describe correctly the structural properties of these layered phases – the calculations predicted the lattice parameters and atomic positions to be in good agreement with experiment. Two structural phases known for Ti_3SiC_2 were analyzed and the reasons for their different stability were established. The calculations demonstrate that the chemical bonding in $M_{n+1}AX_n$ phases has covalent, ionic and metallic components. In Ti_3SiC_2 the ionic contribution is due to the charge transfer from Ti to Si and C atoms, covalent component is mainly due to the hybridized Ti-C bonds in the hexagonal Ti-C-Ti-C-Ti layers, while the intra-layer Ti-Ti bonds are the strongest metallic bonds. Like many laminates, $M_{n+1}AX_n$ phases demonstrate a strong anisotropy of the intra- and interlayer interactions. Strong bonding exists in the basal Ti and Si layers as well as between the Ti-C-Ti-C-Ti layers, whereas the interlayer Ti-Si bond is rather weak in Ti_3SiC_2. Such chemical bonding anisotropy leads to the anisotropy of elastic and electronic properties.

The calculations have demonstrated how vacancies and substitutional impurities affect the structural and electronic properties of these ternary phases that allow one to predict the solution solubility, stability, hardness and electrical conductivity, and to analyze the possible ways for modification of the known phases. The electronic structure and geometry of surface were simulated by using *ab-initio* methods and important conclusions on the atom relaxation near surface and surface band structure were established.

Theoretical atomistic simulation of the deformation behavior is very informative and important for understanding the mechanical properties of solids. Since *ab-initio* approaches give total energies rather precisely and are applicable for description of rather large supercells, they have used for theoretical modeling of deformation and brittle/ductile characteristics. The atomic mechanism of cleavage failure in Ti_3SiC_2 has been studied by the first principles calculations of cleavage and tensile parameters. The Ti-Si bonds were found to be essentially elongated before failure and accommodate almost all stress that explains the atomic mechanism of damage tolerance. The simulation of sliding showed that the main slip system occurs on the Si plane. As we demonstrated, *ab-initio* methods provide the understanding of microscopic origin for the intrinsic brittleness of Ti_3SiC_2.

Finally, first-principles investigations give a theoretical basis for the search for new nanolaminates with the unusual and valuable properties. Theoretical modeling predicts the possible crystal nanostructures and their stability as well as the ways for their synthesis.

References

[1] Barsoum, M. W. *Prog. Solid State Chem.* 2000, *28*, 201-281.
[2] Zhang, H. B.; Bao, Y. W.; Zhou, Y. C. *J. Mater Sci. Technol.* 2009, *25*, 1-38.

[3] Wang, J. Y.; Zhou, Y. C. *Annual Rev. Mater. Res.*, 2009, *39*, 415-443.
[4] Ivanovsky, A. L.; Novikov, D. L.; Shveikin, G.P. *Mendeleev Commun.* 1995, *3*, 90-91.
[5] Ivanovskii, A. L.; Medvedeva, N. I. *Mendeleev Commun.* 1999, *1*, 36-38.
[6] Ivanovskii, A. L.; Medvedeva, N. I. *Z. Neorgan. Khimii,* 1998, *43*, 462-428.
[7] Medvedeva, N. I.; Novikov, D. L.; Ivanovsky, A. L.; Kuznetsov, M. V.; Freeman, A. J. *Phys. Rev.* 1998, *B58,* 16042-16050.
[8] Ivanovsky, A. L. *Uspekhi Khimii*, 1996, *65*, 499-518.
[9] Ivanovskii, A. L.; Gusev, A. I.; Shveikin, G. P. *Quantum Chemistry in Material Sciences. Ternary carbides and Nitrides of Transition Metals and the Elements of IIIb, IVb Subgroups.* Ural Branch Russ. Academy Sci.: Ekaterinburg, 1996, 214 pp.
[10] Sun, Z. M; Zhou, Y. C. *Phys. Rev.* 1999, *B60*, 1441-1443.
[11] Ahuja, R.; Eriksson, O.; Wills, J. M.; Johansson, B. *Appl. Phys. Lett.*, 2000, *76*, 2226-2228.
[12] Zhou, Y. C.; Sun, Z. M. *J. Phys. - Cond. Matter* 2000, *12*, L457-L462.
[13] Sun, Z.M.; Zhou, Y. C. *J. Mater. Chem.* 2000, *10*, 343-345.
[14] Zhou, Y. C.; Sun, Z. M.; Wang, X. H.; Chen, S. *J. Phys. - Cond. Matter* 2001, *13*, 10001-10010.
[15] Zhou, Y.C.; Wang, X. H.; Sun, Z. M.; Chen, S. Q. *J. Mater. Chem.* 2001, *11*, 2335-2339.
[16] Li, S.; Ahuja, R.; Barsoum, M. W.; Jena, P.; Johansson, B. *Appl. Phys. Lett.,* 2008, *92*, art. 221907.
[17] Ahuja, R.; Sun, Z.; Luo, W. *High Pressure Res.* 2006, *26*, 127-130.
[18] Orellana, W.; Gutierrez, G.; Menendez-Proupin, E.; Rogan, Garcia, G.; Manoun, B.; Saxena, S. *J. Phys. Chem. Solids,* 2006, *67*, 2149-2153.
[19] Wang, J.Y.; Zhou, Y.C. *Phys. Rev.* 2004, *B69*, art. 144108.
[20] Zhang, H. Z.; Wang, S. Q. *Acta Materialia*, 2007, *55*, 4645-4655.
[21] Wang, Z. C.; Tsukimoto, S.; Saito, M.; Ikuhara, Y. *Phys. Rev.,* 2009, *B79*, art. 045318.
[22] Xu, X.; Wu, E.; Du, X.; Tian, Y.; He, J. *J. Phys. Chem. Solids,* 2008, *69*, 1356-1361.
[23] Wang, J, Y.; Zhou, Y. C. *J Phys. - Condens. Matter* 2003, *15*, 1983–1991
[24] Pietzka, M. A.; Schuster, J. C. *J. Phase Equilib.,* 1994, *15*, 392 -396.
[25] Chaput, L.; Hug, G.; Pecheur, P.; H. Scherrer, H. *Phys. Rev.* 2007, *B75*, art. 035107.
[26] Magnuson, M.; Palmquist, J. P.; Mattesini, M.; Li, S.; Ahuja, R.; Eriksson, O.; Emmerlich, J.; Wilhelmsson, O.; Eklund, P.; Högberg, H.; Hultman, L.; Jansson, U. *Phys. Rev.,* 2005, *B72*, art. 245101.
[27] Eklund, P.; Virojanadara, C.; Emmerlich, J.; Johansson, L. J.; Högberg, H.; Hultman, L. *Phys. Rev.,* 2006, *B74*, art. 045417.
[28] Lv, M. Y.; Chen, Z. W.; Liu, R. P. *Mater. Lett.,* 2006, *60*, 538-540.
[29] Wang, J. Y.; Zhou, Y. C. *J. Phys.: Condens. Matter* 2003, *15*, 1983–1991
[30] Palmquist, J. P.; Li, S.; Persson, P. O.; Emmerlich, J.; Wilhelmsson, O.; Högberg, H.; Katsnelson, M. I.; Johansson, B.; Ahuja, R. Eriksson, O.; Hultman, L.; Jansson U. *Phys. Rev.,* 2004, *B70*, art. 165401.
[31] Keast, V. J.; Harris, S.; Smith, D. K.; *Phys. Rev.*, 2009, *B80*, art. 214113.
[32] Wang, J. Y.; Zhou, Y. C. *J Phys. - Condens. Matter* 2003, *15*, 5959–5968.
[33] Chaput, L.; Hug, G.; Pecheur, P.; H. Scherrer, H. *Phys. Rev.* 2005, *B71*, art. 121104(R).
[34] Barsoum, M. W.; Farber, L.; El-Raghy, T. *Metall. Mater. Trans.* A 1999, *34,* 1727-1733.

[35] Barsoum, M. W.; Murugaiah, A; Kalidindi, S.R.; Zhen, T. *Phys. Rev. Lett.* 2004, *92,* art. 255508.
[36] Zhen, T.; Barsoum, M. W.; Kalidindi, S. R.; Radovic, M.; Sun, Z. M.; El-Raghy, T. *Acta Mater.,* 2005, *53,* 4963-4973.
[37] Sun, Z. M.; Zhang, Z. F.; Hashimoto, H.; Abe, T. *Mater. Trans.,* 2002, *43*, 432-435.
[38] Zhang, Z. F.; Sun, Z. M.; Hashimoto H. Mater. Letters, 2003, *57*, 1295-1299.
[39] Zhang, Z. F.; Sun, Z. M.; Zhang, Hashimoto, H. *Advanced Eng. Mater.,* 2004, *6*, 980-983.
[40] Kooi, B. J.; Poppen, R. J.; Carvalho, N. J.; De Hosson, J. T.; Barsoum, M. W. *Acta Mater.*, 2003, *51*, 2859-2872.
[41] Vitek V. *Cryst. Latt. Def.,* 1974, *5*, 1-34.
[42] Rice, J. R.; Thompson R. *Phil. Mag.,* 1973, *29*, 73-84.
[43] Sun, Y.; Rice, J. R.; Trushinovsky, L. *Mater. Res. Soc. Symp. Proc.*, 1991, *231*, 243-248.
[44] Kaxiras, E.; Duesbery, M. S. *Phys. Rev. Lett.,* 1993, *70*, 3752-3755.
[45] Yoo, M. H.; Fu, C. L. *Mater. Sci and Eng.*, 1993, *A153*, 470-478.
[46] Medvedeva, N. I.; Mryasov, O. N.; Gornostyrev, Yu.N.; Novikov, D. L.; Freeman, A. J. *Phys. Rev. B* 1996, *54*, 13506-13514.
[47] Gornostyrev, Yu.N.; Katsnelson, M. I.; Medvedeva, N. I.; Mryasov, O. N.; Freeman, A. J. *Phys. Rev. B,* 2000, *62*, 7802-7808.
[48] Fang, C. M.; Ahuja R.; Eriksson, O.; Li, S.; Jansson U.; Wilhelmsson, O.; Hultman, L. *Phys.Rev B* 2006, *B74*, art. 054106.
[49] Medvedeva, N.I.; Freeman, A.J. *Scripta Materialia*, 2008, *58*, 671-674.
[50] Medvedeva, N.I.; Ivanovskii, A.L. *Deformation and Fracture,* 2008, *1*, 1-4.
[51] Girshick, A.; Pettifor, D.G.; Vitek, V. *Phil. Mag.*, 1998, *77*, 999-1012.
[52] Vitek, V.; Perrin, R.C.; Bowen, D.K. *Phil. Mag.,* 1970, *21*, 1049-1073.
[53] Rice, J.R. *J. Mech. Phys. Solids*, 1992, *40*, 239-255.
[54] Chan, K.S. *Metall. Mater. Trans.,* 2003, *34A*, 2315-2328.
[55] Enyashin, A. N.; Ivanovskii, A. L. *Mater. Lett.* 2008, *62,* 663-665.
[56] Enyashin, A. N.; Ivanovskii, A. L. *Theor. Exp. Chem.* 2009, *45,* 98-102.
[57] Ivanovskii, A. L. *Uspekhi Khimii* 1999, *68,* 119-135.
[58] Patzke G.R.; Krumeich, F.; Nesper, R. *Angew. Chem. Int. Ed.* 2002, *41,* 2446-2461.
[59] Tenne, R.; Seifert, G. *Annu. Rev. Mater. Res.* 2009, *39,* 387-413.
[60] Tenne, R.; Remškar, M.; Enyashin, A. N.; Seifert, G. *Top. Appl. Phys.* 2008, *11,* 631-671.
[61] Seifert, G. *J. Phys. Chem. A* 2007, *111,* 5609-5613.
[62] Enyashin, A. N.; Ivanovskii, A. L. *Physica E* 2005, *30,* 164-168.
[63] Enyashin, A. N.; Gemming, S. *Phys. Status Solidi B* 2007, *244,* 3593-3600.
[64] Prinz, V. Y. *Microelectronic Engineering* 2003, *69,* 466-475.
[65] Emmerlich, J; Högberg, H.; Sasvári, S.; Persson, P. O. A.; Hultman, L.; Palmquist, J. P.; Jansson, U.; Molina-Aldereguia, J. M.; Czigány, Z. *J. Appl. Phys.* 2004, *9,* 4817-4826.
[66] Högberg, H.; Hultman, L.; Emmerlich, J; Joelsson, T.; Eklund, P.; Molina-Aldereguia, J. M.; Palmquist, J. P.; Wilhelmsson,, U.; Jansson, U. *Surf. Sci.* 2008, *62,* 663-665.
[67] Taguchi, T.; Yamomoto, H.; Shamoto, S. *J. Phys. Chem. C* 2007, *111,* 18888-18891.
[68] Shi, L.; Gu, Y.; Chen, L; Yang, Z.; Ma, J.; Qian, Y. *Carbon* 2005, *43,* 195-213.

[69] In, J.; Seo, K.; Lee, S; Yoon, H.; Park, J.; Lee, G.; Kim, B. *J. Phys. Chem. C* 2009, *113,* 12996-13001.

[70] Pol, S. W.; Pol, V. G.; Gedanken, A. *Adv. Mater.* 2006, *18,* 2023-2027.

[71] Yang, Z.; Gu, Y.; Chen, L.; Shi, L.; Ma, J.; Qian, Y. *Solid State Comm.* 2004, *130,* 347-351.

In: MAX Phases: Microstructure, Properties and Applications ISBN 978-1-61324-182-0
Editors: It-Meng (Jim) Low and Yanchun Zhou

Chapter 11

MECHANICAL BEHAVIOR OF Ti_2SC UNDER COMPRESSION

Shrinivas R. Kulkarni
Easy Lab Technologies LTD., Science and Technology Center, University of Reading, Reading RG6 6BZ Berkshire UK

ABSTRACT

In this chapter the high pressure behavior of Ti_2SC is addressed. The synthesis of Ti_2SC for use in performing the experimental studies is also discussed. A brief description of high pressure technique, requirements for conducting high pressure experiments, getting good data and obtaining the bulk modulus is also presented. Finally, it is shown how the high pressure results can help in tailoring the physical properties of Ti_2SC and other similar compounds.

1. INTRODUCTION

Pressure is one of the important physical parameters which can affect the properties of matter in unusual ways. High pressure (produced due to extreme conditions of heat and cold, mechanical stresses, large magnetic and electric fields, and intense radiation environments) are easily discernable in the fields of weapon development, jet engine technology and nuclear sciences. In engineering this field helps in optimizing the design parameters of the components in situations where breakdown in various hostile environments is a significant concern. Progress in understanding of the physics of materials under extreme conditions has great significance for other areas of science as well; notably geophysics, where conditions of high heat and pressure are routine, and astrophysics. However, the use of this parameter was limited due mostly to the lack of suitable materials that could facilitate generation of high pressure and to support the material of study in such systems. But with the advent of Diamond Anvil Cell(s) (DAC) high pressure research is now becoming an important frontier in all the major faculties of science.

The pressure range between ambient pressure (0 GPa) to 50 GPa, is good enough to compress volumes by 5-50% (since the bulk modulus of most solids is in the region 20-200 GPa). This can produce large changes in electronic energies. For instance, the direct band gap of group IV, 111-V and 11-VI tetrahedral semiconductors can be shifted by 0.5-1.5 eV by the application of 10 GPa. Also, phonon energies can be changed typically by - 10-20% by this pressure. Many semiconductors undergo phase transitions to metallic states or direct-gap to indirect-gap crossovers by 30 GPa, so that this often limits the useful pressure range of experiments. Many times nano and macro samples of the same material show different behavior under pressure as well, such as different phase transformation pressures and bulk moduli. Thus, in general it can be said that with the application of a few GPa pressure many novel properties in matter can be uncovered that are of high interest to scientific community.

2. Principle of High Pressure Cells

2.1. Diamond Anvil Cells

The tool used to generate such high pressures is popularly known as the Diamond Anvil Cell (DAC).These cells have changed the way in which pressure above several GPa (now typically reaching 100 GPa) could be generated in a very small volume. This facility coupled with the high energy synchrotron radiation has allowed researchers to extract various information about a variety of samples that could be as small as 50 μm.

The idea of generating very high pressures in materials by squeezing them in between the opposed anvil (made of tungsten carbide) with small area was implemented successfully by Dr. Percy Williams Bridgman [1]. However, the actual impetus to high pressure research was witnessed with the development of the (DAC) and the ruby pressure scale at the National Bureau of Standards. Its usage has become so prevalent nowadays that it is applied in as diverse fields as molecular biology to geophysics. Moreover, this technique can also be used to synthesize new materials, obtain completely new phases with superior properties than the parent material and to fine tune the physical properties of materials. Thus, it can be realized that this technique provides new pathways for the advancement of materials research.

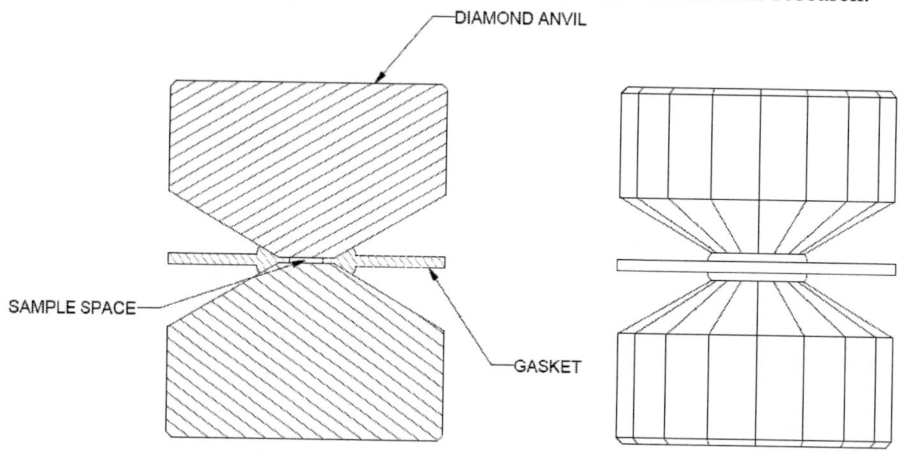

Figure 1. Opposed Anvils and Gasket assembly in a typical DAC.

The main component of DAC is of course diamonds as it fits into the criteria of materials that could withstand high compressive stresses due to it's extremely high hardness and it is also transparent to photons over a wide energy range [21]. In principle the operation of DAC is very simple.

A sample placed between the parallel faces of two opposed diamond anvils is subjected to pressure by the application of force on two opposed anvils (Figure 1). This generates very high stresses on the sample due to very small area of the anvil culet (50 – 500 μm). In order to make this stress hydrostatic the sample space is filled with a pressure transmitting medium which can be solid, liquid or gas. The maximum pressure depends on the size of the diamond culet, type of diamond, gasket material and diamond backing plates or anvil support [21, 23]. Many a times it's the anvil support that limits the maximum pressure and in order to achieve high pressures the anvil support needs be as strong as possible [24]. Tungsten Carbide (WC) bound in a Co matrix is the most commonly used material for anvil support at room temperatures. At higher temperatures different materials such as tungsten, rhenium, molybdenum and their alloys and carbides are used. Boron carbide is the most suitable support material in DACs used for x-ray diffraction due to its x-ray transparency [24, 25]. The rest of the DAC body could be made out of stainless steels, Beryllium Copper, Titanium Copper, inconels and their alloys depending on experimental requirements.

A schematic view of the lever arm body of the DAC is presented in Figure 2 while the piston cylinder assembly along with mounted diamond anvils is shown in the Figure 3.

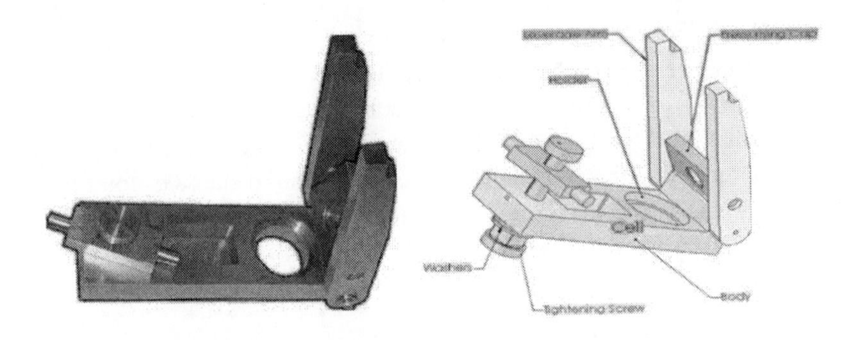

Figure 2. Body of Mao – Bell type Diamond Anvil Cell.

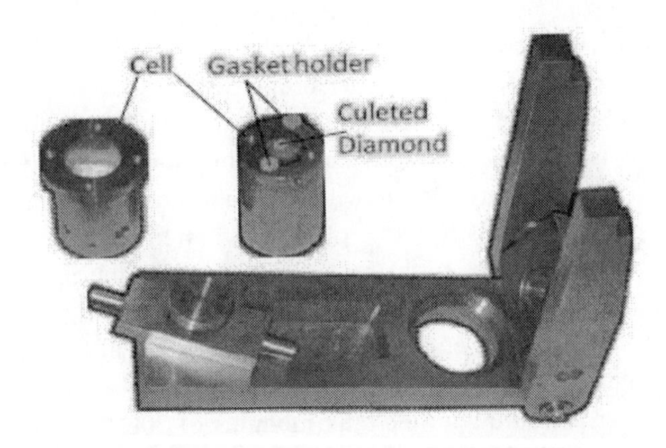

Figure 3. Piston cylinder assembly used to pressurize the sample loaded in the stainless steel gasket.

3. Sample Loading Technique

The preparation for the high pressure experiments starts with choosing a gasket material depending upon the highest pressure to be attained. Various gasket materials such as stainless steel, beryllium, rhenium, and tungsten are commonly used, while more recently boron has been used for radial x-ray diffraction and diamond gasket has been used for high temperature and high pressure experiments. Once the gasket material is picked the next step to be performed is the indentation of this gasket. For this purpose the un-indented gasket of usually 250 μm initial thickness, is pressurized in between the diamond anvils, until the desired indentation thickness is obtained. This thickness depends upon the maximum pressure to be attained in the experiment. Usually an indentation thickness of 50 – 60 μm is optimum for pressures up to 60 GPa. Once the gasket is indented a hole of roughly half the culet size is made at the center of this indentation in which the sample, pressure medium and pressure marker are loaded.

4. Data Collection

There are a variety of techniques by which information about the sample subjected to high pressure and temperature can be obtained. Some of the commonly known techniques are Raman, X-Ray Diffraction (XRD), Mossbauer spectroscopy and XRF etc. In recent days synchrotron XRD is widely used to obtain information about the changes in lattice spacing of the sample with pressure. For such XRD experiments typically a monochromatic (for angle dispersive setup (ADS)) or white (for energy dispersive setup (EDS)) x – rays are focused down to a very small spot size, usually less than 50 μm. The diffraction patterns are then either recorded on an online image plate (ADS) or with a solid state Ge point detector (EDS) and in the case of ADS, the recorded images are integrated using a computer program to obtain intensity vs. 2θ plots such as FIT2D [2].

5. High Pressure Research on MAX Phases

Ever since it was established that the MAX phases, especially Ti_3SiC_2, possessed unique combination of properties in terms of hardness, Young's and shear moduli, and machinability etc. [3] it became perceptible to explore other properties of these materials that would shed more light on their quiescent properties, which in turn would possibly make them more unique. Bulk modulus is one such thermodynamically important property which discloses the materials response to high pressure. By knowing the bulk modulus of a material it is possible to compare it with other materials in terms of its bond strengths and physical responses to extremities of pressure. Moreover, with the x-ray diffraction the amount of anisotropy, if present, can also be determined easily. The first high pressure work on the MAX phases was performed on Ti_3SiC_2 by Onodera et al. [4]. From this study, which was conducted up to 61 GPa in diamond anvil cell using synchrotron x-rays, it was established that this material was structurally stable under pressure with the bulk modulus of 206 GPa. It was also shown that this material exhibited anisotropy in compression wherein the c-axis was more compressible that the a-axis. From other high pressure studies the following points are noteworthy; (a) most of the phases exhibit higher compressibility along the c-axis than that along the a-axis. The phases that display the opposite behavior are Cr_2AlC [5], Nb_2AlC [5] and Nb_2AsC [6]. (b) Only Ta_2AlC[5] and Ta_4AlC_3 [5] exhibit almost similar compressibility along the two axes. (c) The highest bulk modulus belongs to Ta_4AlC_3 (261 GPa) while the lowest is for Zr_2InC (127 GPa) [7] and all other compounds have bulk modulus in between these values. (d) As long as the experimental conditions are hydrostatic or quasi – hydrostatic none of these phases exhibit transitions in crystal structure, however, if the conditions are non – hydrostatic then a shear induced α to β phase transition is observed [8]. A similar structural transition was observed in thinned Ti_3SiC_2 sample [9]. There are no other reports on phase transitions other than these.

Though experimental high pressure studies give information about the structural behavior of the material, much more can be learned about them by conducting theoretical studies, which mostly includes ab – inito calculations. From these calculations information about the evolution of the band structures, changes of free energy and behavior of different bonds etc. with pressure can be obtained. Moreover, once the routine is robustly established, such studies can be performed on the non–existing materials with high enough accuracy. Likewise, when the effect on bulk modulus by changes in the A element was studied it was found that the bulk modulus did not change much implying that the A element does not have much effect on the bulk modulus [10]. In another such study on MAX phases it was observed that bulk moduli of M_2AX $(X=C)$ phases could be classified into two groups. One group comprising only group *IVB* transition metals (*M*), where the bulk modulus was lower than *MC* and the other group that includes transition metals of groups *VB* and *VIB*, where the bulk modulus of the corresponding *MC* was essentially conserved. It was also observed in this study that the *A* element seemed to have no significant effect on the functional dependence between the bulk modulus of M_2AX and *MC* configurations [11]. In another study where the pressure dependence of the valence electron concentration (VEC) in these phases was studied, it was found that the axial compressibilities (compressibility along the a-axis and the c-axis) shows a reversal as the VEC changes from 4 to 6. Moreover, as the pressure is applied the *M* element is shifted away from the C plane to *A* plane due to which the M-C bonds get oriented towards the c-axis which in turn increases its stiffness as the pressure is applied. Additionally it was also found that the A elements show lesser significance in the overall bonding in the MAX phases having higher VEC [12].

Table 1. Summary of the bulk modulus values obtained by both theory and experiments

Compound	Bulk modulus Expt (GPa)	Bulk Modulus Theory (GPa)
Ti_2AlC	186	171
V_2AlC	201	215
Cr_2AlC	165	228
Nb_2AlC	209	179
Ta_2AlC	251	-
Nb_2AsC	224	-
Zr_2InC	127	-
V_2AsC	-	219
Nb_2SnC	-	155

A summary of the bulk modulus values obtained both by theory and experiments is presented in Table 1 from which it can be seen that the bulk modulus values of these phases lie in the range of 120 - 210 GPa.

5.1. Brief History of Ti_2SC

Kudielka et al [13] synthesized Ti_2SC and Zr_2SC in powder form and deciphered their structural parameters. However, no information about their physical properties was presented. These phases were also observed in ultra-high strength steels [14, 15] in which Ti and Zr were used as sulfur getters. Interestingly it was observed that the steels, which had these compounds, showed an improvement in the fracture toughness. This improvement has been attributed to the Ti_2SC particles being more resistant to void nucleation than the usually occurring MnS particles. Another reference to these phases occurs in the field of superconductivity, where they are commonly known as carbosulfides. It was recently shown by [16] that Nb_2SC_x , which falls in the family of Ti_2SC and Zr_2SC, is a type II superconductor.

6. Synthesis of Ti_2SC and High Pressure Experiemtns

Polycrystalline fully dense samples of Ti_2SC were fabricated by hot isostatic pressing - 325 mesh powders (3-ONE-2, Voorhees, NJ) at 1500°C at ~ 45 MPa for 5h. For phase analysis powdered sample of the sintered piece was obtained by drilling at its center. The XRD of this powder showed no peaks other than those associated with Ti_2SC. The same powder was used to perform high pressure experiments in a DAC.

The high pressure experiments were carried out in a Mao-Bell diamond anvil cell with diamonds of 300 μm culet size. A stainless steel gasket of thickness 250 μm was used for loading the sample. The gasket was pre-indented using the diamond anvils to a thickness of 60 μm and a hole of 125 μm was drilled. Finely ground sample was then carefully loaded between the aluminum foils that were 15 μm in initial thickness. Aluminum serves two purposes, first, it acts as a pressure transmitting medium, creating nearly hydrostatic conditions [17, 18] and second, it serves as a pressure marker [19].

The experiment was conducted at room temperature at B2 station at CHESS Cornell University, Ithaca, NY, using angle dispersive x-ray diffraction. The monochromatic x -ray beam of wavelength 0.4959 Å was focused to a spot size of 35 μm. The diffraction patterns recorded on an image plate, were integrated using FIT2D computer program to obtain intensity versus 2θ plot in the range of 10 to 35° 2θ. The pressure was determined using the equation of state of Al [20]. To obtain the cell parameters each peak was refined individually using a least squares method and then these were indexed assuming a $P6_3/mmc$ space group.

The ambient pressure lattice parameters, a_0 and c_0, measured in this experiment were 3.208(1) Å and 11.21(1) Å, respectively. These values compare favorably with those reported earlier. Table 2 lists the summary of the experimental lattice parameters, the relative changes in the lattice parameters and the unit cell volume with pressure respectively.

As the pressure, *P*, was increased, no extra peaks appeared in the pattern, although, the peaks became broader and some of the peaks overlapped. In addition, there was a gradual decrease in the intensities of all the peaks as *P* was increased (Figure 4). From the above observations it was concluded that there were no phase transitions up to the maximum pressure of 47 GPa. Note that the broadening of the Ti_2SC peaks was larger as compared to that of *Al* peaks. The reason for this behavior could be due to the different nature of samples used; Ti_2SC was used as powder where as *Al* was used as foil. The relative variation in lattice parameters with pressure for Ti_2SC is shown in Figure 5, which shows that the compressibility is anisotropic for this sample.

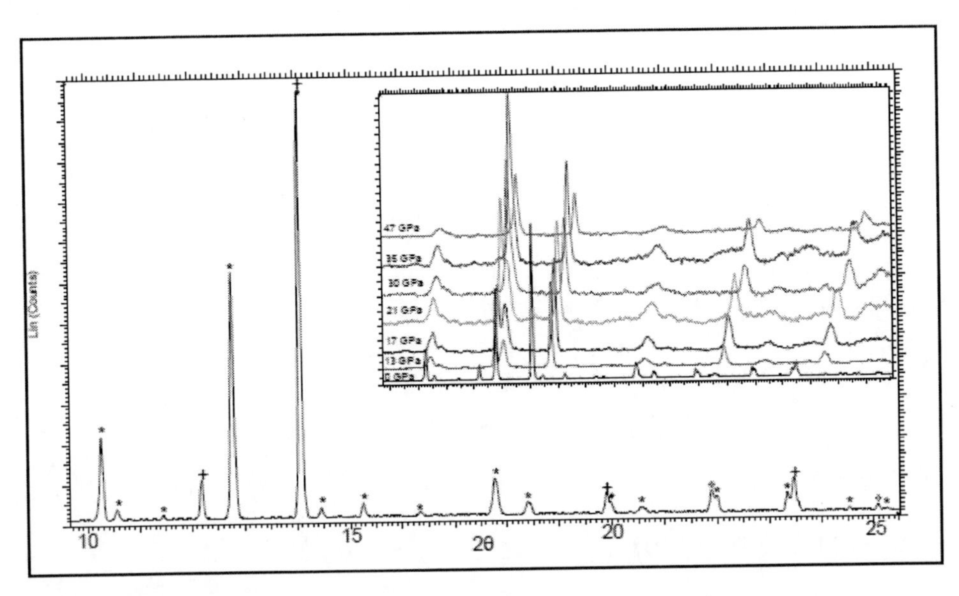

Figure 4. Room pressure XRD pattern of Ti_2SC powder sample placed between aluminum foil in diamond anvil cell. Peaks are marked by (*), (+), (♦) and (❖) for Ti_2SC, Al, Fe gasket and unidentified peaks, respectively. The inset shows XRD patterns at various pressures as indicated. Only the integer values of pressure are shown for clarity, Table 2 shows the actual values.

Table 2. Unit Cell parameters, volume and their ratios respectively of Ti_2SC

P(GPa)	a(Å) ± 0.001	c(Å) ± 0.01	V(Å^3) ± 0.1	a/a_o	c/c_o	V/V_o
0	3.208	11.21	99.9	1	1	1
13.9	3.151	10.92	93.9	0.9822	0.9744	0.9400
17.3	3.132	10.84	92.0	0.9764	0.9668	0.9218
21.8	3.116	10.77	90.5	0.9716	0.9606	0.9067
30.8	3.087	10.63	87.7	0.9623	0.9486	0.8785
35.7	3.074	10.58	86.6	0.9583	0.9444	0.8673
45.6	3.046	10.55	84.8	0.9495	0.9417	0.8490
47.5	3.037	10.49	83.8	0.9469	0.9363	0.8395
44.2	3.044	10.54	84.6	0.9490	0.9403	0.8469
23.5	3.115	10.76	90.4	0.9711	0.9602	0.9055
14.6	3.143	10.88	93.1	0.9800	0.9704	0.9320
4.06	3.191	11.11	98.0	0.9949	0.9911	0.9809
0	3.207	11.21	99.8	0.9999	1	0.9989

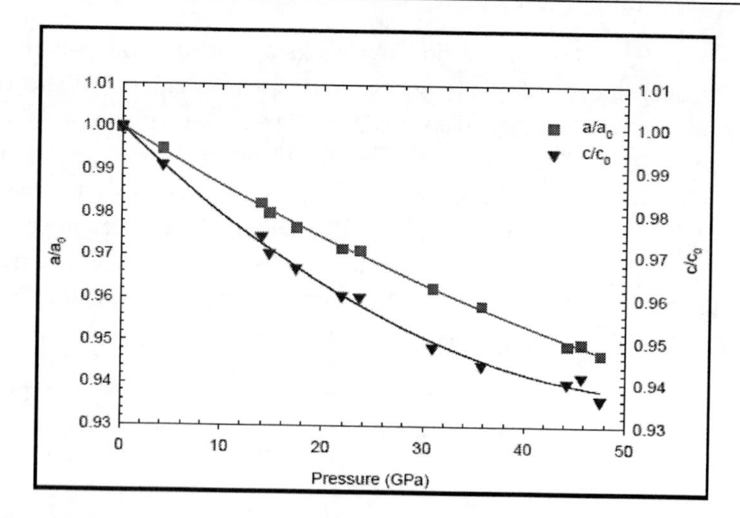

Figure 5. The relative variation in lattice parameters with pressure for Ti_2SC.

The change in a/a_0, where a_0 is the lattice parameter at ambient pressure, shows a linear behavior with pressure while that of c/c_0 shows a quadratic relation to pressure variations. The least squares fitting of the results shown in Figure 5 indicates that:

$$\frac{a}{a_0} = 1 - 0.001419p + 6.4958\times10^{-6}p^2 \qquad R^2 = 0.9993$$

and

$$\frac{c}{c_0} = 1 - 0.002223p + 1.9504\times10^{-5}p^2 \qquad R^2 = 0.9982,$$

where p has the units of GPa.

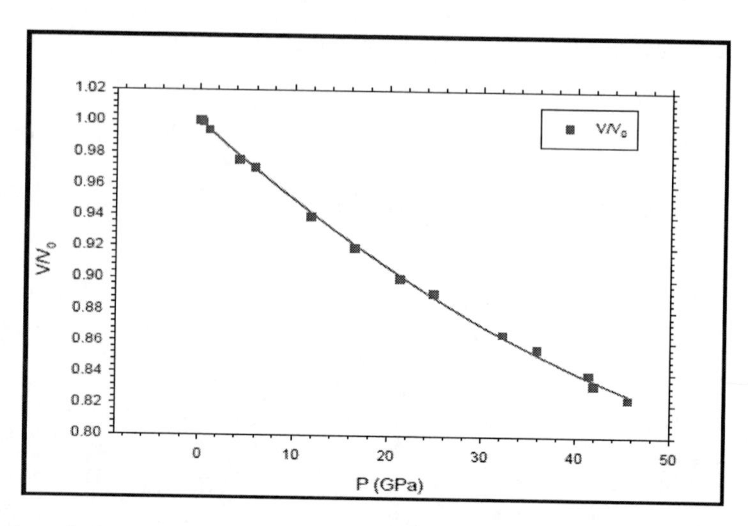

Figure 6. Variation of relative volume with pressure. The solid line indicates Birch – Murnaghan fit to this data.

In order to obtain bulk modulus of this sample, the pressure–volume data was fitted with the *Birch-Murnaghan* equation of state as shown in Figure 6. From this fit, the K_o of Ti_2SC is determined to be 191 ± 4 GPa with a pressure derivative equal to 4.0 ± 0.3 .

Since high pressure XRD gives information about the compressibility of the material in its respective lattice direction, by comparing the compressibilites of two structurally similar materials with similar end members we can get an idea about their bond strengths in respective axial directions. Earlier Manoun *et al.* [5] studied high pressure behavior of Ti_2AlC which has a different '*A*' element when compared to Ti_2SC. By comparing the compressiblities of these two MAX phases an idea about their relative bond strengths can be obtained.

The relative variation in lattice parameters with *P* for these two MAX phases is shown in Figure 7. From this figure it can be seen that for Ti_2SC, the changes in a/a_0 are almost linear with *P*, while that for c/c_0 it shows a second order relation to *P*. This is also the case for Ti_2AlC, however, for the later the compressibility along *c*-axis is larger than that of the former. This suggests that the bonding along the *c*-axis, which is mainly between *M–A* elements, is stronger for Ti_2SC as compared to that of Ti_2AlC. This in turn indicates that by replacing Al with S, bond strength in Ti_2-A-C MAX phases is increased.

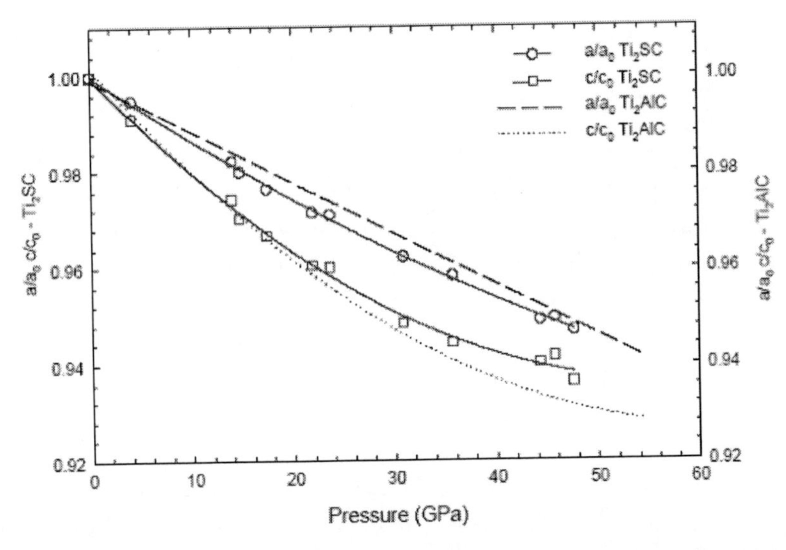

Figure 7. Pressure dependence of relative lattice parameters for Ti_2SC (right axis) and Ti_2AlC (left axis).

A comparison of the variation of V/V_0 with pressure for Ti_2SC and Ti_2AlC is shown in Figure 8. From this figure it can be seen that this response is almost identical, and thus the two K_o's are very similar (K_o for Ti_2AlC is 186±2 GPa). This is a somewhat surprising result given that the unit cell volume of Ti_2SC is ≈ 8.3 % smaller than that of Ti_2AlC. Nevertheless, this result is consistent with recent *ab initio* calculations [21], where K_o was found not to change significantly when the *A* element was changed, provided the latter was from row 3 or 4 of the periodic table. For example, K_o for V_2AlC is 215 GPa; for V_2AsC it is 219 GPa. Note, however, that changing the *M*-element has a larger effect on the bulk modulus [5]. This comparison indicates that by changing the '*A*' element, the bond strength along '*c*-axis' can be changed which in turn changes the bulk modulus.

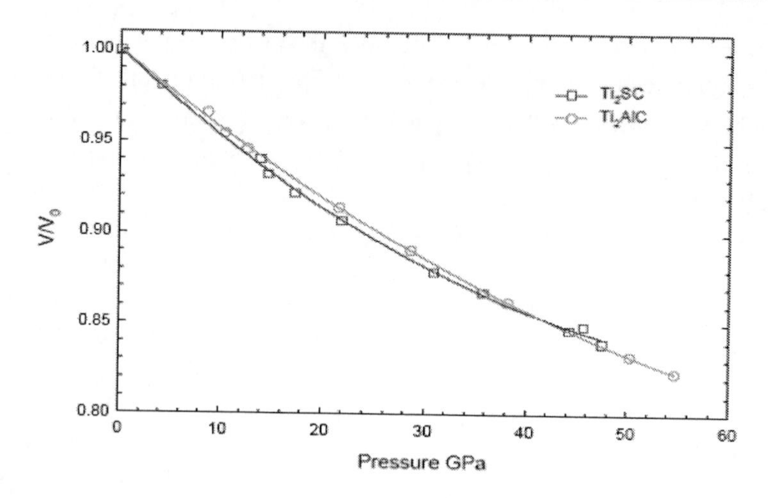

Figure 8. Relative unit cell volume changes as a function of pressure for Ti_2SC (red) and Ti_2AlC (black). The line represents the Birch–Murnaghan fit to the data of Ti_2SC.

Conclusions

In conclusion, the K_o for Ti_2SC determined using DAC and synchrotron XRD is 191±4 GPa, with a $K_0^{'}$ of 4.0 ± 0.3. These values compare well with *ab initio* calculations [22] (K_0 = 183 GPa with $K_0^{'}$ of 4.1). The compressibility along the *c*-axis is greater than that along the *a*-axis. Replacing the *A* element in Ti_2AlC by *S* to form Ti_2SC did not change the bulk modulus significantly, however, an increase in the bond strength along the *c*-axis was observed. This indicates that in the *MAX* phases with same end members, changing the *'A'* element will change the bond strength along *c*-axis.

References

[1] http://nobelprize.org/nobel_prizes/physics/laureates/1946/index.html.
[2] A.P. Hammersely, S.O. Svensson, M. Hanfland, A.N. Fitch, and D. Hausermann, *High Pressure Research*, 14 (1996): p. 235.
[3] M.W. Barsoum, *Prog. Solid State Chem.*, *28* (2000): p. 201.
[4] A. Onodera, H. Hirano, T. Yuasa, N.F. Gao, and Y. Miyamoto, *Appl. Phys. Lett.*, 74 (1999): p. 3782.
[5] B. Manoun, R.P. Gulve, S.K. Saxena, S. Gupta, M.W. Barsoum, and C.S. Zha, *Phys. Rev. B.*, 73 (2006): p. 1.
[6] R.S. Kumar, R. Sandeep, A.L. Cornelius, and M.W. Barsoum, *Appl. Phys. Lett.*, 86 (2005): p. 111904.
[7] B. Manoun, H.P. Lierman, R.P. Gulve, S.K. Saxena, A. Ganguly, M.W. Barsoum, and C.S. Zha, *Appl. Phys. Lett.*, 84 (2004): p. 2799.
[8] Z. Wang, C.S. Zha, and M.W. Barsoum, *Appl. Phys. Lett.*, 85 (2004): p. 3453.

[9] L. Farber, I. Levin, M.W. Barsoum, T. El-raghy, and T. Tzenov, *J. Appl. Phys.*, 86 (1999): p. 2540.
[10] D. Music, Z. Sun, and J.M. Schneider, *Solid State Commun.*, 133 (2005): p. 381.
[11] Z. Sun, D. Music, R. Ahuja, S. Li, and J. Schneider, *Phys. Rev. B*, 70 (2004): p. 092102.
[12] J. Emmerlich, D. Music, A. Houben, R. Dronskowski, and J.M. Schneider, *Phys. Rev. B*, 76 (2007): p. 224111.
[13] H. Kudielka and H. Rohde, *Zeitschrift fuer Kristallographie, kristallgeometrie, Kristallphysik, Kristallchemie*, 114 (1960): p. 447.
[14] J.W. Bray, J.L. Maloney, K.S. Raghanav, and W.M. Garrison, *Metall. Mater. Trans. A*, 22 (1991): p. 2277.
[15] J.L. Maloney and W.M. Garrison, *Scripta Materialia*, 23 (1989): p. 2097.
[16] K. Sakamaki, H. Wada, H. Nozaki, Y. Onuki, and M. Kawai, *Mol. Cryst. and Liq. Cryst.*, 341 (2000): p. 99.
[17] W.A. Passett, M.S. Weathers, T.C. Wu, and T.J. Holmquist, *J. Appl. Phys.*, 74 (1993): p. 3824.
[18] A.K. Singh and G.C. Kennedy, *J. Appl. Phys.*, 48 (1977): p. 3362.
[19] H.P. Lierman, A.K. Singh, B. Manoun, S.K. Saxena, V.B. Prakapenka, and G. Shen, *Int. J. Refract. Met. Hard Mater*, 22 (2004): p. 129.
[20] R.G. Greene, H. Luo, and A.L. Ruoff, *Phys. Rev. Lett.*, 75 (1994): p. 2075.
[21] S.E. Lofland, J.D. Hettinger, K. Harrell, P. Finkel, S. Gupta, M.W. Barsoum, and G. Hug, *Appl. Phys. Lett.*, 84 (2004): p. 508.
[22] S.R. Kulkarni, R.S. Vennila, N.A. Phatak, S.K. Saxena, C.S. Zha, T. El-Raghy, M.W. Barsoum, W. Luo, and R. Ahuja, *J. Alloys Compd.*, 448 (2008): p. L1.

In: MAX Phases: Microstructure, Properties and Applications ISBN 978-1-61324-182-0
Editors: It-Meng (Jim) Low and Yanchun Zhou

Chapter 12

MECHANICAL PROPERTIES OF CR_2ALC CERAMICS

Peiling Wang[1] and Wubian Tian[2]

[1]State Key Lab of High Performance Ceramics and Superfine Microstructure, Shanghai Institute of Ceramics, Chinese Academy of Sciences, Shanghai 200050, China

[2]National Institute of Advanced Industrial Science and Technology (AIST), Nagoya 4638560, Japan

ABSTRACT

In this chapter, the latest comprehensive studies on the mechanical behaviors of Cr_2AlC, such as fracture strength and deformation behaviour under compressive and tensile stress conditions, in a broad temperature range from room temperature to 1000°C and under various strain rates are reviewed. The fundamental mechanical properties, such as hardness, damage tolerance, thermal shock resistance and solid solution hardening are also presented. It was found that the room-temperature mechanical properties of Cr_2AlC studied are usually comparable to other Al-containing MAX phases, such as Ti_2AlC and Ti_3AlC_2, and strongly dependent on the microstructure developed. The good machinability of Cr_2AlC is attributed to low hardness, layered microstructure and good thermal conductivity. However, compared with Ti_3AlC_2 and Nb_2AlC, the thermal shock resistance property of Cr_2AlC is inferior due to the relatively smaller thermal conductivity and higher coefficient of thermal expansion.

The compressive strength of Cr_2AlC decreases continuously from room temperature to 900 °C when tested at a constant strain rate. The ductile-to-brittle transition temperature is measured to be in the range of 700~800°C. When tested at different strain rates, Cr_2AlC fails catastrophically at room temperature and the deformation mode changes with strain rate at 800°C. In addition, the compressive strength increases slightly with increasing strain rate at room temperature and it is less dependent on strain rate when tested at 800°C. The variation of flexural strength of Cr_2AlC with testing temperature exhibits similar decreasing tendency to that of compressive strength and the plastic deformation occurs at temperatures higher than 800°C. The plastic deformation mechanism of Cr_2AlC at elevated temperatures is discussed on basis of microstructural observations.

1. INTRODUCTION

Ternary compounds MAX phases ($M_{n+1}AX_n$, n=1-3, M is an early transition metal, A is an IIIA or IVA element, and X is C and/or N) have been studied since 1960s and significant progresses were made in recent years. These compounds are provided with good electrical and thermal conductivity [1], good oxidation and/or corrosion resistance [2, 3] as well as excellent mechanical performance at room temperature and high temperatures [4, 5].

Cr_2AlC, as one of the M_2AX phases (or called as 211 phase), has aroused more and more attention because of the following characters: high electrical ($1.4\text{-}2.3\times10^6\ \Omega^{-1}m^{-1}$) and thermal (17.5-22.5 W/mK) conductivity [6-8], excellent oxidation resistance [7, 9, 10] and corrosion resistance [7, 11]. These properties point to a fact that Cr_2AlC is likely a kind of useful structure materials that can be used at room temperature or elevated temperature under severe service conditions.

In this chapter, the comprehensive studies till to now on the mechanical behaviors of Cr_2AlC, such as fracture strength and deformation behaviour under compressive and tensile stress conditions, in a broad temperature range from room temperature to 1000 °C and under various strain rates were reviewed. The fundamental mechanical properties, such as hardness, damage tolerance, thermal shock resistance and solid solution hardening were also presented.

2. ROOM-TEMPERATURE MECHANICAL PROPERTIES OF CR_2ALC

2.1. Hardness

The measured hardness of Cr_2AlC is changed with the density, microstructure and resulted phase constituent of synthesized sample that are related to the starting material, sintering processing and synthesis technique applied. The reported hardness values of Cr_2AlC locate in the range of 3.2-5.6 GPa, with the lowest in the sample hot-pressed (HPed) from commercially available Cr-Al-C materials [12] and the highest in the sample spark-plasma-sintered (SPSed) from finer elemental powders [13], if assuming that the hardness is asymptotically stable as indentation load is higher than 10 N. The hardness of Cr_2AlC is close to that of Ti_2AlC [14] and Ti_3AlC_2 [5] when they share similar microstructure after sintering.

Generally speaking, if Cr_2AlC samples were sintered by SPS or Hot-pressing at the same temperature using the same starting materials, the SPSed sample possesses higher hardness than the HPed one. By using the same sintering technique, the hardness of the sample synthesized from finer starting materials is higher than that from coarser one [12], [13]. The responsible mechanism for the differences in hardness between Cr_2AlC samples is likely the Hall-Petch effect (i.e. the hardness has inverse relation with the grain size).

In addition, similar to other MAX phases, it was found that no indentation cracks were observed and it was therefore hard to estimate the microcrack length of the indentation, so as to calculate the toughness of the material. The typical damage region around the indentation is shown in Figure 1. Instead of sinking into the surface, the indentations accumulate as a result of a combination of grain pullout, break up and buckling, extending to a limited distance and absorbing most of damage energy.

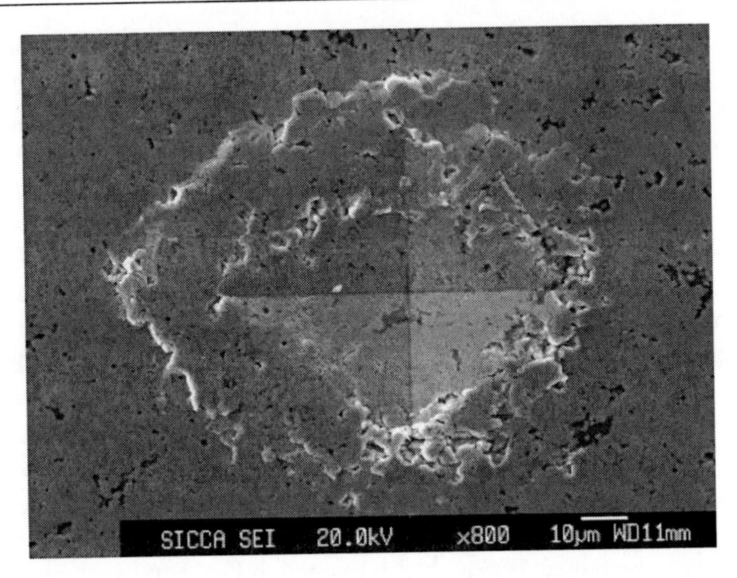

Figure 1. SEM image of the damage region of Cr2AlC around the indentation.

2.2. Solid Solution Hardening

The formation of solid solutions among the ternary phases of Cr_2AlC, V_2AlC and Ti_2AlC had been investigated experimentally [15] and theoretically [16]. And the calculation of the elastic stiffness and electronic band structure of the same system was also carried out [17], in which the authors predicted that solid solution hardening might be operative in $(Cr_{1-x}V_x)_2AlC$ system. Experimental substantiation of this strengthening effect was recently conducted by synthesizing single-phase $(Cr_{1-x}V_x)_2AlC$ solid solutions using a SPS technique. Their properties, such as hardness and thermal conductivity, were also examined [18].

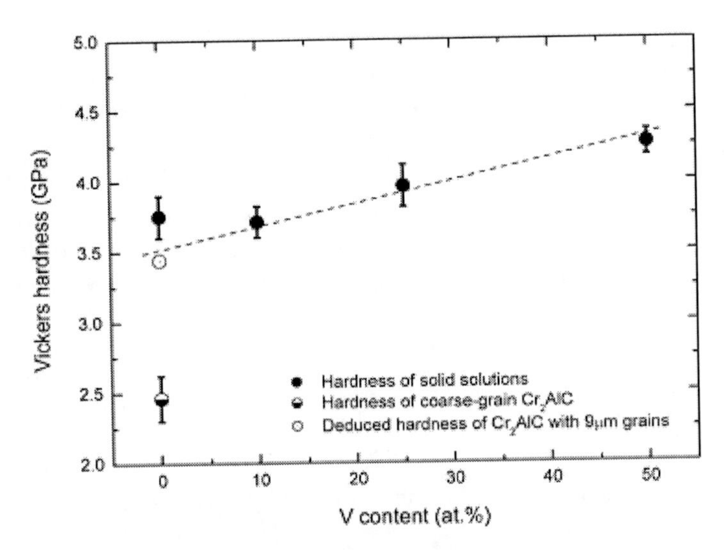

Figure 2. Vickers hardness of $(Cr_{1-x}V_x)_2AlC$ solid solutions as a function of V content. The dash line represents the linear fit of the deduced hardness of 9 μm Cr_2AlC and those of V-containing solid solutions [18].

The average grain size of synthesized Cr_2AlC is around 7 μm, which is slightly smaller than those of V-containing solid solutions (with the close grain size of 9 μm). The Vickers hardness of the synthesized solid solutions measured at a load of 300 N as a function of V content is shown in Figure 2. It is found that the hardness of V-containing samples with the similar grain sizes increases almost linearly with the V content in the solid solutions. It is well established that smaller grain size (d) of a material normally leads to higher strength and hardness (H_v) according to the Hall-Petch relation (i.e. linear dependence of H_v on $d^{-1/2}$) [19]. However, the harness value of Cr_2AlC deviated from the linear relationship, to a higher level, which is most likely caused by the small grain size in this material. If the grain size of sample Cr_2AlC is increased, its hardness is naturally expected to decrease. To confirm the relationship between hardness and the V content in $(Cr_{1-x}V_x)_2AlC$ solid solutions, with comparable grain size, one monolithic Cr_2AlC sample was synthesized aiming at a microstructure with grain size comparable to that of the V-containing solid solutions. However, it was difficult to get the grain size of Cr_2AlC sample close to that of $(Cr_{1-x}V_x)_2AlC$, instead, a microstructure with grain size of ~27 μm was obtained. The Vickers hardness of this coarse-grained Cr_2AlC was measured to be 2.5 GPa, as plotted in Figure 2 (the half-solid circle). With the Vickers hardness values at grain sizes of 7.0 μm and 27 μm, a hardness value of Cr_2AlC at a grain size of 9 μm is deduced, by the assumption that the hardness values obey the Hall-Petch relation with grain size, and this value is also plotted in Figure 2 (open circle). In this way, it is found that the Vickers hardness of the $(Cr_{1-x}V_x)_2AlC$ solid solutions demonstrate good linear relationship with the molar fraction of V in the compounds, in the range of 0 - 50 mol.% V replacement for Cr. By excluding the possible effect of grain size on the hardness, it is fair to conclude that the strengthening effect is attributed to solid-solution hardening in the $(Cr_{1-x}V_x)_2AlC$ compounds [18].

2.3. Flexural Strength

The room-temperature flexural strength of Cr_2AlC is 378 MPa when the sample was hot-pressed from relatively coarse Cr-Al-C powder mixture at 1400 °C for 60 min under 20 MPa [6]. This value is fairly close to that of Ti_3AlC_2 (375±15 MPa) [5]. By using finer raw powders [20] or replacing chromium by commercially available chromium carbide [7] as starting materials, the density and microstructure of fabricated Cr_2AlC sample are improved simultaneously, and flexural strength increases to 483-494 MPa correspondingly. If optimizing the processing parameters (temperature, pressure, etc.) [21], the maximum flexural strength of HPed bulk Cr_2AlC was reported to exceed 600 MPa. By applying SPS techniques, the flexural strength of Cr_2AlC reaches 555±11 MPa even sintered at as low as 1250 °C [22].

2.4. Damage Tolerance

The damage tolerance of Cr_2AlC was investigated by indentation of the tensile surface of bending specimen at varying loads (20, 50 and 100 N), followed by three-point bending testing [20]. The dependence of post-indentation strength on indentation load is shown in Figure 3, in which the post-indentation strengths of Cr_2AlC samples are constant to indentation loads of 20N and then decrease linearly by 31% (from 483.3 MPa to 332.0 MPa) when the indentation load increases to 100 N. Similar damage tolerance has been observed in other MAX systems. For example, an indentation load of 100 N reduces the retained strength of Ti_3AlC_2 by ~7% [5] or of Nb_2AlC by ~26% [23].

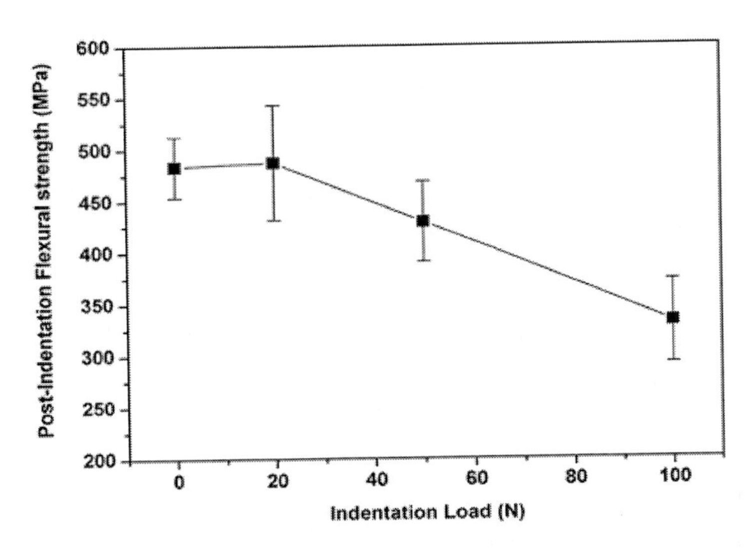

Figure 3. Effect of indentation load on post-indentation strength of HPed Cr_2AlC sample [20].

2.5. Thermal Shock Resistance

Susceptibility to thermal shock was determined by quenching Cr_2AlC specimens from high temperatures (300, 500, 700, 900 and 1100 °C) into water held at ambient temperature [20]. The effect of quench temperature on the post-quench strength of Cr_2AlC is plotted in Figure 4, in which the thermal shock properties of Ti_3AlC_2 and Nb_2AlC are also included for comparison. It can be found that the flexural strengths of Cr_2AlC samples quenched from 300 °C to room temperature decrease from 483 MPa to 455 MPa, while the retained strengths decrease quickly to 199 MPa when the quench temperature increases to 500 °C. A further increase in quench temperature up to 1100 °C results in a small reduction of strength (154 MPa, 138 MPa and 159 MPa for 700 °C, 900 °C and 1100 °C, respectively). These results indicate that Cr_2AlC is not susceptible to thermal shock at a quench temperature below 300 °C, while its retained strength reduces almost 60% as the quench temperature increased to 500 °C, revealing that the critical thermal shock temperature of Cr_2AlC is between 300 °C and 500 °C. In comparison to Ti_3AlC_2 and Nb_2AlC, the thermal shock resistance (TSR) of Cr_2AlC is inferior to that of Ti_3AlC_2 and Nb_2AlC because the former contains lower post-quench strengths at temperatures above 400 °C, as shown in Figure 4 [20].

It has well-known that the thermal shock resistance of the material is related to the coefficient of thermal expansion (CTE) and Young's modulus as well as thermal conductivity according to the equations that define thermal shock resistance [24]. Accordingly, both the higher CTE of Cr_2AlC (13.3×10^{-6} K^{-1}) [6] as comparison to Ti_3AlC_2 (9.0×10^{-6} K^{-1}) [5] and Nb_2AlC (8.7×10^{-6} K^{-1}) [23] and the smaller thermal conductivity of Cr_2AlC (18~22 W/mk) [6], [8] as compared to Ti_3AlC_2 (40 W/mk) [5] and Nb_2AlC (23 W/mk) [25] would result in the poor TSR of Cr_2AlC since all three materials have similar Young's moduli.

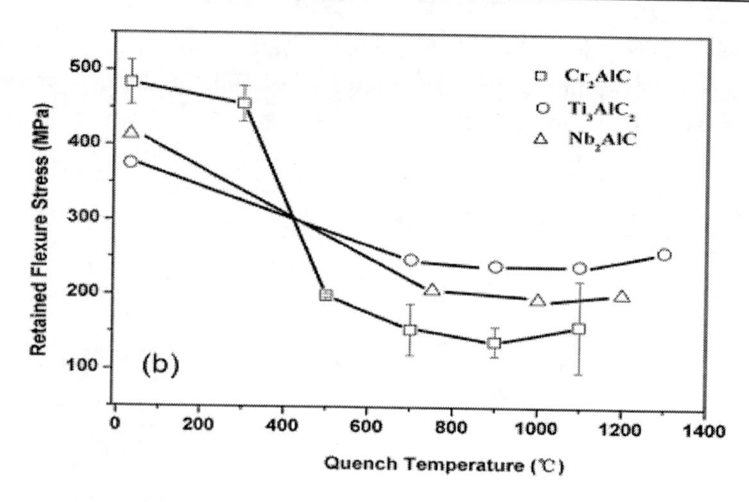

Figure 4. Effect of quench temperature on post-quench strength of Cr_2AlC sample, in which the thermal shock properties of Nb_2AlC and Ti_3AlC_2 are included for comparison [20].

2.6. Compressive Strength

Lin *et al.* [7] measured the compressive strength of Cr_2AlC to be 625±44 MPa, which is higher than that of Ti_2AlC (540±21 MPa) [14] andTi_3AlC_2 (560±20 MPa) [5], whereas a significantly higher compressive strength value of 1159±23 MPa [20] and of 997±29 MPa [20, 26] was reported by Tian et al.. A typical compressive stress-strain curve at room temperature is illustrated in Figure 5. It is noted that the compressive stress did not drop abruptly to a minimum value after the ultimate stress but decreased slightly followed by a small linear increase in stress and then sudden decrease (see inset in Figure 5). This compressive stress-strain behavior seems to be similar to that of Ti_3SiC_2 measured at 900 °C, in which two modes of failures are observed: a shear mode at the lower temperatures and a plastic mode at temperatures > 1000°C [5].

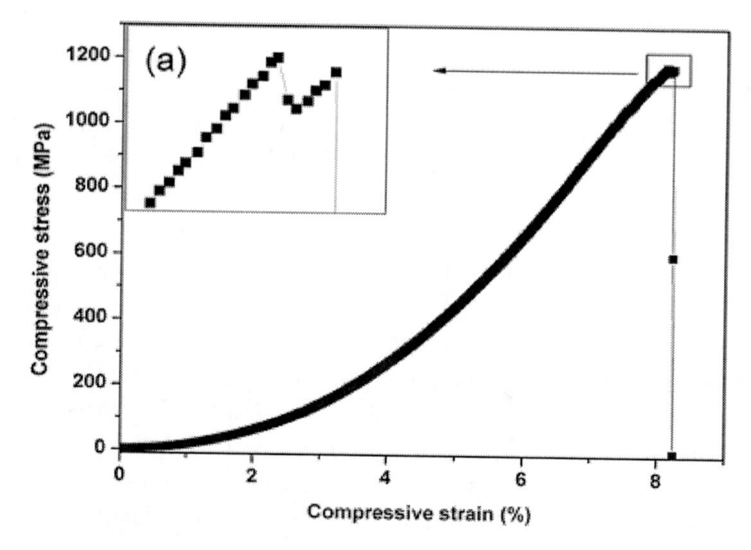

Figure 5. Typical compressive stress-strain curve of Cr_2AlC ceramics at room temperature [20].

2.7. Machinability

In addition, it was found that Cr_2AlC was readily machinable by using conventional high-speed tool steel without lubrication. This machinable property is similar to that of Ti_3SiC_2 [5] and most other MAX phases [5, 23]. These properties are likely related to their intrinsic properties, such as low hardness and high thermal conductivity, and the special machining mechanism [27], i.e. by the breaking off of tiny microscopic flakes, which is similar to graphite.

Summary of the room-temperature mechanical properties of Cr_2AlC are presented in Table 1.

3. Mechanical Properties of Cr_2ALC at Elevated Temperatures

Besides the abovementioned good mechanical properties at room temperature, Cr_2AlC is also reported to exhibit excellent oxidation resistance at 800-1300 °C [7, 9, 10, 28] and corrosion resistance at 900-1000 °C [7, 11]. As a potential candidate of high temperature materials, the investigation of the mechanical properties at elevated temperature of this compound is a must before its practical application. Therefore, the high-temperature mechanical behaviors of Cr_2AlC were shown in following sections.

3.1. Compressive Properties

The compressive behaviors of Cr_2AlC at different temperatures and/or different strain rates were examined on spark-plasma synthesized samples synthesized from elemental Cr-Al-C powder mixture [26]. XRD pattern of the sintered Cr_2AlC is shown in Figure 6(a), in which all the reflections can be indexed as Cr_2AlC according to JCPDS29-0017 and the calculated index in reference [28]. Measured density of the sample is 5.14 g/cm^3, which is 98% of the theoretical density. A typical backscattered SEM image of the etched surface of sintered Cr_2AlC sample is shown in Figure 6(b), in which the gray matrix phase corresponds to Cr_2AlC with a small amount of dark pores inside grains and some bright phases correspond to Cr_7C_3 that were usually found in the synthesized Cr_2AlC. Some dark gray grains that mainly distributed at tri-grain boundaries are identified to be Al_2O_3 according to EDS results (as discussed in next section). This might be a result of the reaction between Al and the absorbed oxygen in starting powder.

3.2. Compression of Synthesized Cr_2AlC at Different Temperatures and Strain Rates

The relationship between compressive stress and engineering strain in the temperature range of room temperature to 900 °C at a strain rate of 5.6×10^{-4} s^{-1} is shown in Figure 7(a) (for clarity, curves are shifted). It is evident that the Cr_2AlC specimens fail abruptly in a brittle mode after the maximum compressive stress is reached when tested below 700 °C.

However, an obviously plastic deformation region is observed on the stress-strain curves when tested at temperatures higher than 800 °C, and this region becomes broadened at 900 °C. In fact, the specimen tested at 900 °C does not break but is deformed to a spindle-like shape. Therefore, the ductile-to-brittle transition temperature (DBTT) of Cr_2AlC locates in the range of 700~800°C. Accordingly, two temperatures of room temperature and 800°C were selected to study the effect of strain rate on the compressive behavior of Cr_2AlC.

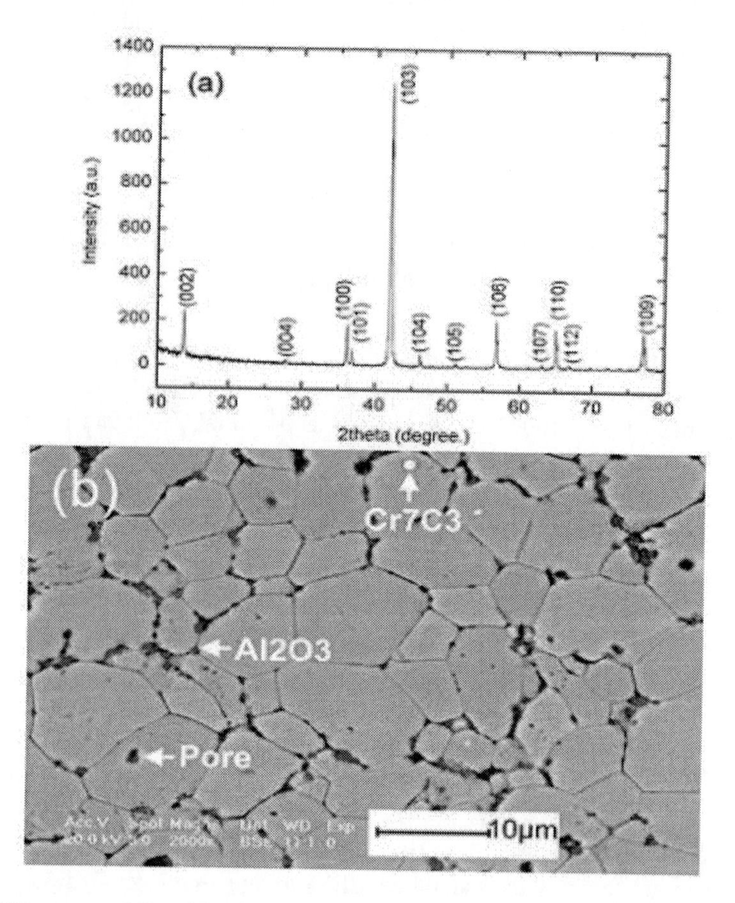

Figure 6. (a). XRD pattern of Cr_2AlC sample SPSed at 1250°C for 30min under a pressure of 50MPa [26]. (b) Backscattered SEM image on the surface of polished and etched sample [26].

The dependence of compressive strength on testing temperature is presented in Figure 7(b). It can be found that the compressive strength of Cr_2AlC specimen decreases continuously from 997±29 MPa at room temperature to 749±21 MPa at 800 °C and then drops significantly to 523±7 MPa at 900 °C, indicating that the compressive strength in the ductile mode is much more sensitive to temperature than that in the brittle regime. For comparison, the compressive strengths of Ti_3SiC_2 [29], Ti_3AlC_2 [30] and Ti_2AlC [31] at different temperatures are also plotted in Figure 7(b). It can be found that the compressive strengths of Cr_2AlC are not only higher than those of Ti_3AlC_2 and Ti_2AlC but also slightly higher than those of Ti_3SiC_2 at temperatures below 800 °C. However, it becomes lower than that of Ti_3SiC_2 at the testing temperature of 900 °C, although it is still higher than that of Ti_3AlC_2 and Ti_2AlC. The high compressive strength of Cr_2AlC can be associated with the refined microstructure [26].

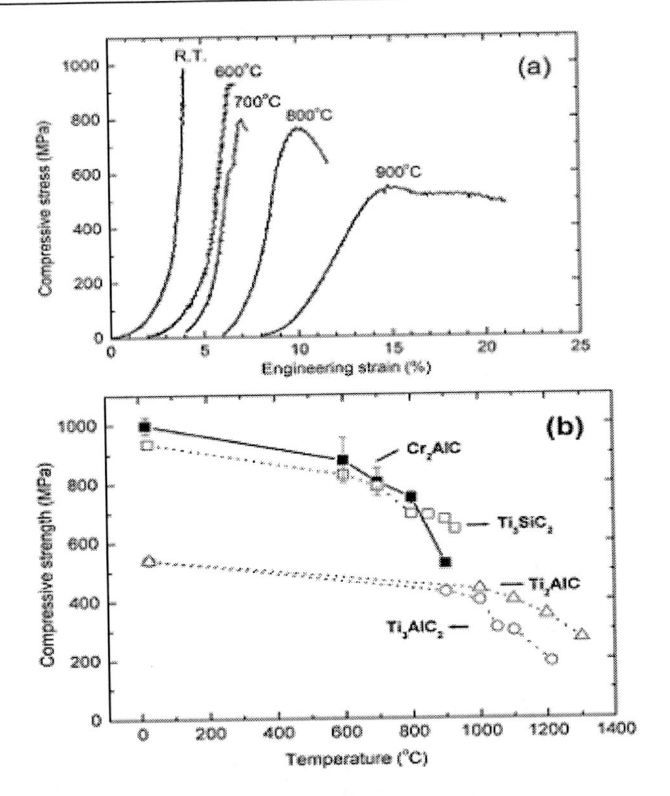

Figure 7. (a) The relationship between compressive stress and engineering strain in the temperature range of room temperature to 900°C at a strain rate of $5.6\times10^{-4}s^{-1}$ [26]. (b) The dependence of compressive strength of Cr_2AlC specimens on the testing temperature. For comparison, the compressive strengths of Ti_3SiC_2, Ti_3AlC_2 and Ti_2AlC at different temperature are also illustrated [26].

Figure 8(a) shows the stress-strain curves of Cr_2AlC specimens tested at room temperature at different strain rates. Despite the scatter in maximum compressive stresses recorded at different strain rates, these stress-strain curves of specimens tested at either higher strain rates or lower strain rates are similar to the specimens tested at a strain rate of 5.6×10^{-4} s^{-1}, which is used as a standard strain rate for testing at different temperatures. This result indicates that the brittle deformation mode does not change in the strain rate range of 5.6×10^{-5}~5.6×10^{-3} s^{-1} when the sample is tested at room temperature [26].

The stress-strain curves of Cr_2AlC specimens tested at 800°C at different strain rates are shown in Figure 8(b). It is evident that the stress-strain curve varies from an elastic way to a plastic one as the strain rate decreases, and the plastic deformation region becomes more broadened as strain rate is below 5.6×10^{-4} s^{-1}, illustrating that the deformation mode changes from a ductile mode to a brittle mode when the strain rate is above 5.6×10^{-4} s^{-1}. Note that the specimens tested at 800 °C at low strain rates, such as 1.4×10^{-4} s^{-1} and 5.6×10^{-5} s^{-1}, do not break but display a spindle-like shape, similar to the specimen tested at 900 °C at a strain rate of 5.6×10^{-4} s^{-1}, revealing that high temperature and low strain rate have similar effect on the plastic deformation of Cr_2AlC [26].

The dependences of compressive strength of Cr_2AlC specimens on the strain rate at room temperature and at 800°C are shown in Figure 8(c), in which the compressive strength increases gradually with the increase of strain rate at room temperature and is less dependent on strain rate when tested at 800°C.

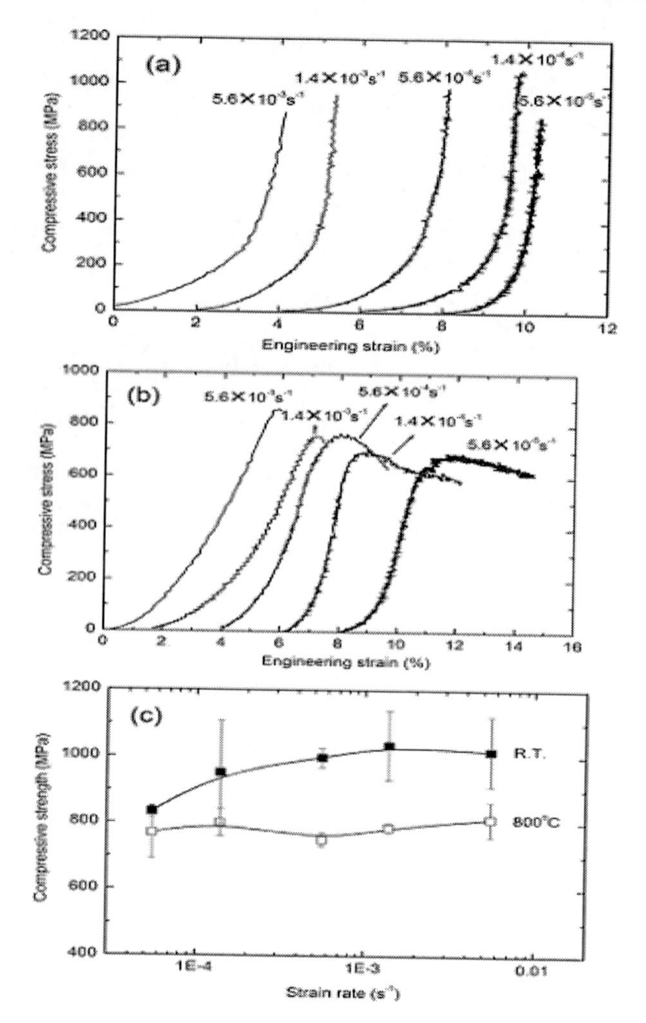

Figure 8. (a) The compressive stress-strain curves of Cr_2AlC specimens tested at a different strain rate (b) at room temperature and (b) at 800 °C [26]. (c) The dependences of compressive strength on strain rate of Cr_2AlC specimens at room temperature and 800 °C [26].

3.3. Microstructural Examination on the Deformed and Fracture Surface

At a certain strain rate ($5.6\times10^{-4}s^{-1}$), the deformed surface and the fracture surface of the selected but typical samples tested at room temperature and 800°C are shown in Figure 9(a)-(d). When tested at room temperature, one main crack propagates through the specimen and makes about a 45° angle with the load direction (indicated as white arrow) (Figure 9(a)). In contrast, when tested at 800°C, a large number of cracks appear and most of them are aligned on the load direction (Figure 9(b)). In addition, the fracture surfaces of the samples tested at all the testing temperatures are characterized predominantly by intergranular cracking, as represented by Figure 9(c) (room temperature) and Figure 9(d) (800°C). Grain boundary decohesion is observed after compression at high temperature (Figure 9(e)). Furthermore, delamination within grains on the fracture surface (Figure 9(f)) can be found frequently in the specimens tested at 800°C [26].

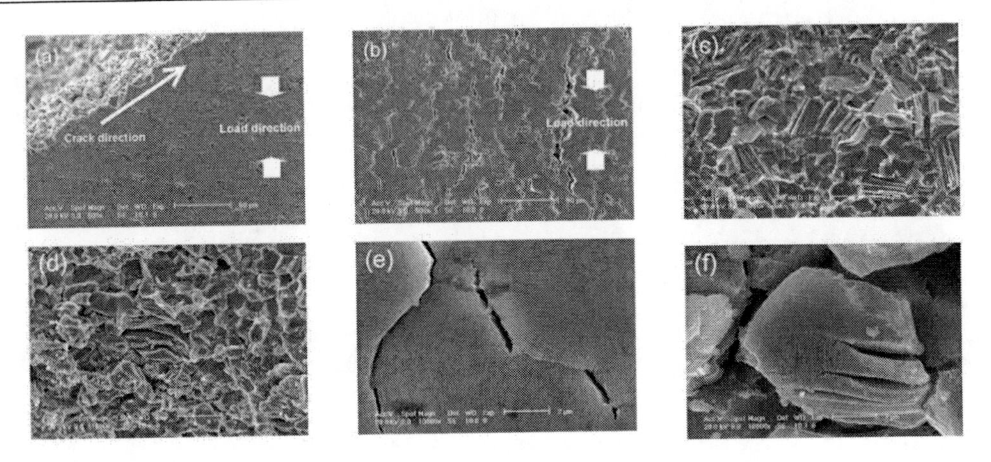

Figure 9. The deformed surfaces of Cr_2AlC specimens tested (a) at room temperature and (b) 800°C. The fracture surfaces of the sample tested (c) at room temperature and (d) 800°C. (e) Typical grain-boundary decohension and (f) delamination within grain in the Cr_2AlC specimens tested at 800°C. All tests at a strain rate of $5.6\times10^{-4}s^{-1}$ [26].

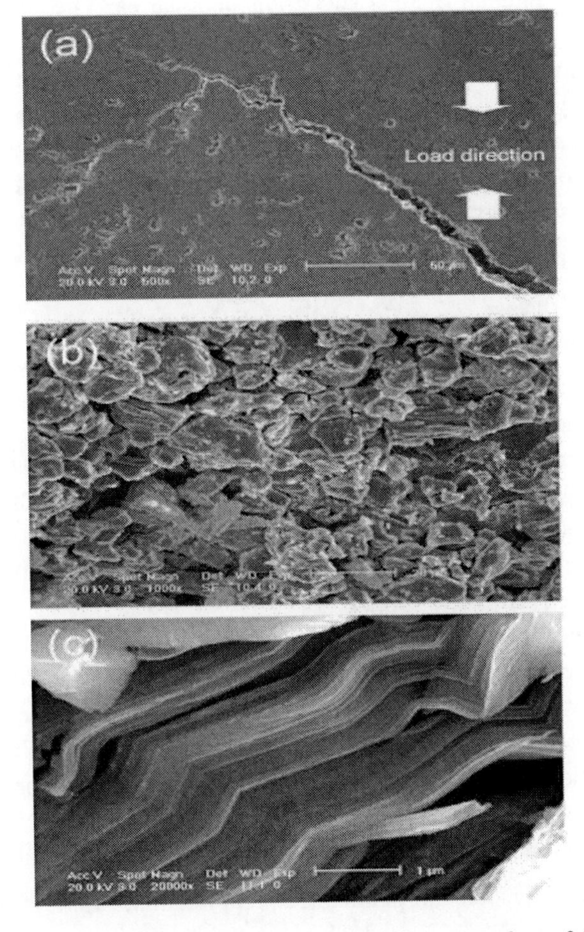

Figure 10. (a) The deformed surface morphology and (b) the fracture surface of the Cr_2AlC specimen tested at 800°C at a strain rate of $5.6\times10^{-3}s^{-1}$. (c) Typical kink band within grain on the fracture surface [26].

The microstructural images of specimens tested at room temperature at other strain rates are similar to those tested at $5.6\times10^{-4}s^{-1}$ (Figure 9(a) and (c)), and hence they are not shown. Figure 10 shows the deformed surface and the fracture surface of the sample tested at 800°C at a high strain rate of $5.6\times10^{-3}s^{-1}$, in which the deformed region is confined in the vicinity of cracking area (Figure 10(a)) and intergranular cracking characteristic, similar to the specimen tested at the same temperature at a strain rate of $5.6\times10^{-4}s^{-1}$ (Figure 9(d)), is also shown (Figure 10(b)). Moreover, the kink band within grain is usually observed on the fracture surface, as shown in Figure 10(c) [26].

3.4. Effect of Temperature and Strain Rate on Compressive Properties of Cr_2AlC

The difficult motion of dislocation at low temperatures and the lack of five independent slip systems are responsible for the absence of macro-plasticity in Cr_2AlC at a temperature below 700°C, i.e. fracture in a brittle mode. With increasing the testing temperature, the motion of dislocations that is believed to be operative only at the basal plane [29-31] becomes readier and they will pile up against the grain boundaries. The pile-ups would produce microcracks in various ways [32], such as grain buckling, kind band formation, etc. On the other hand, the cohesion strength of grain boundary decreases and the propagation of microcracks along grain boundaries becomes effective with increasing temperature. Furthermore, the microcracks will propagate with increasing applied stress during loading, leading to the grain boundary decohesion (Figure 9(e)). It should be pointed out that the small amount of Al_2O_3 phases in grain boundary could also affect the mechanical strengths in two aspects: introducing tensile thermal stress in the Cr_2AlC matrix and weakening the grain boundary [26].

Note that a creep deformation, dislocation activity or diffusion-controlled creep, should be considered when the compressive tests were carried out at high temperature. It is known that the diffusion-controlled creep needs to maintain the cohesion between deformed grains for the transport of material either along the grain boundary or throng the lattices of grains. However, based on the microstructural examination, it can be found that there is no obvious plastically deformed grains appeared but mostly consisting of intergranular cracking (Figure 9(b) and (e)) and delamination (Figure 9(f)) within grains. Therefore it is reasonable to believe that the dominant high temperature deformation mechanism in Cr_2AlC is related to the dislocation activities [26].

At room temperature, the deformation mode is not sensitive to strain rate because of the difficult motion of dislocation. On the other hand, when tested at 800°C the deformation behavior depends on strain rate. This is contributed to more dislocations in the system activated at this elevated temperature. The brittle failure at a strain rate higher than $1.4\times10^{-3}s^{-1}$ is attributed to insufficient time for the motion of activated dislocations to form microcracks and hence macro-structural plastic deformation [26].

Generally, the motion of dislocation is not only a thermally activated process but also a stress-assisted process. Johnston and Gilman [33] reported that strain rate ( ) is related to the dislocation velocity (v), which is in turn related to the shear stress for dislocation motion (τ) [34] according to the following equations:

$$\dot{\varepsilon} = bnv \quad (1)$$

$$v = \left(\frac{\tau}{\tau_0}\right)^P \quad (2)$$

where b is the Burgers vector, n is the number of dislocation per unit area, τ_0 and p are experimentally determined materials constants. Accordingly, the shear stress required for dislocation motion increases with the increasing strain rate. Therefore, the compressive strength of Cr_2AlC increases with the increase of strain rate at room temperature, although the tendency is not so obvious at 800 °C (Figure 8(c)). The mechanism for the less strain rate sensitivity at 800 °C than at room temperature needs further exploration, but it is likely related to the microstructural deformations, such as microcracks, delamination and kink band [26].

4. HIGH-TEMPERATURE FLEXURAL STRENGTH

The deformation and fracture behavior of SPSed Cr_2AlC under tensile stresses by four-point bending tests, as well as its dependence on temperature up to 1000 °C was investigated in literature [22]. XRD result reveals that the obtained sample consists of almost pure Cr_2AlC phase. The measured density of the sample is 5.12 g/cm^3, which is 98% of its theoretical density.

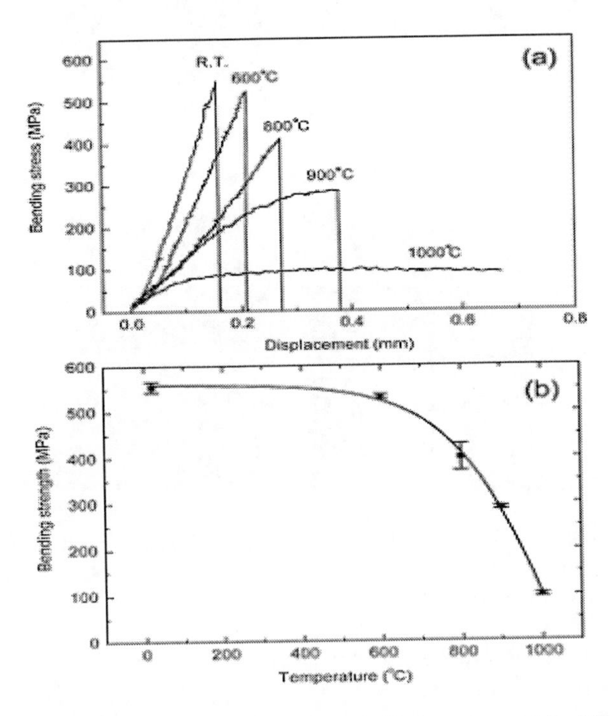

Figure 11. (a)The relationship between flexural stress and displacement of Cr_2AlC specimens tested at different temperatures [22]. (b) Dependence of flexural strength on testing temperature [22].

The relationship between flexural stress and displacement of the tested Cr_2AlC specimens in the temperature range of 20°C to 1000°C is shown in Figure 11(a). It is found that the specimens tested below 800°C failed in a brittle mode after the maximum flexural stress is reached. However, obvious plastic deformation can be recognized on the stress-displacement curves when tested above 900°C. In other words, the ductile-to-brittle transition temperature (DBTT) of Cr_2AlC locates in the range of 800~900°C. The flexural strengths of Cr_2AlC at different testing temperatures are summarized in Figure 11(b), in which the flexural strength decreases with the increasing testing temperature and this strength decreasing tendency is enhanced when tested at temperatures higher than 900 °C. These results demonstrate that Cr_2AlC shows good mechanical strength at testing temperatures up to nearly 900 °C [22].

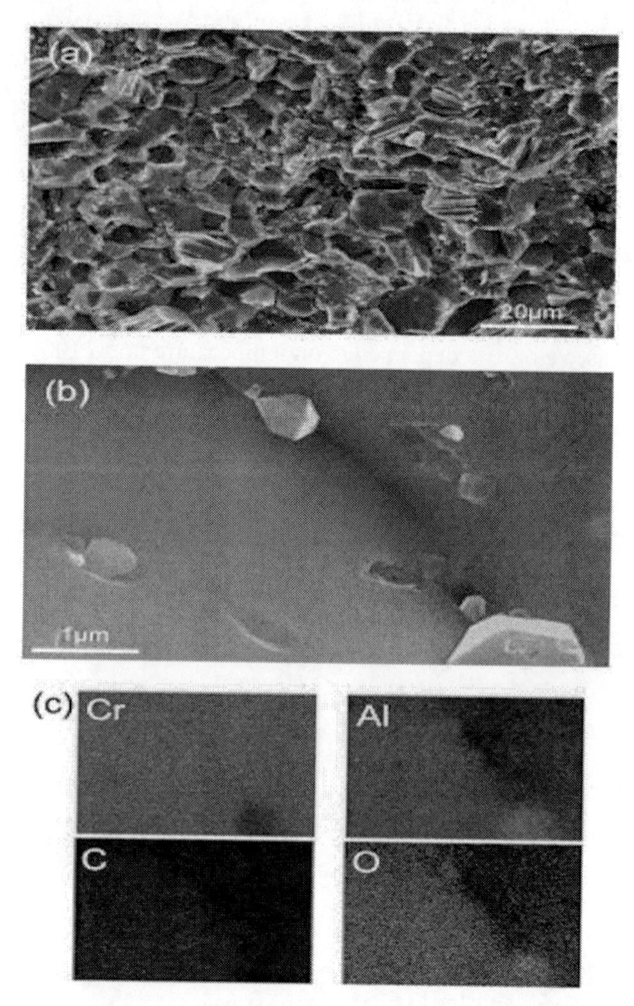

Figure 12. (a) The fracture surface and (b) the detailed examination of Cr_2AlC specimen bended at room temperature. (c) Elements mapping of the same area in Fig. 12(b). Note that the brighter the color in Figure 12(c) corresponds to the richer of elements [22].

Figure 12(a) shows the fracture surface of a Cr_2AlC specimen tested at room temperature, revealing predominantly intergranular fracture. This characteristic is similar to that of Cr_2AlC after compression [26] and such microstructural characteristic was also observed in other ternary compounds, such as Ti_3AlC_2 [5] and Ti_3SiC_2 [4]. Careful examination of the fracture surface of tested Cr_2AlC specimens revealed some small particles on the grain surfaces, as shown in Figure 12(b). Elemental mapping results (Figure 12(c)) exhibit that these particles are deficient in Cr and C while rich in Al and O. Therefore, it is not unreasonable to conclude that they are aluminum oxides, most probably in the form of Al_2O_3, which might be a result of the reaction between Al and the absorbed oxygen in starting powder. The presence of Al_2O_3 particles on grain boundary and the subsequently introduced tensile thermal stress in the Cr_2AlC matrix lead to the weak boundary structure, which can be partially responsible for the fracture characteristic (Figure 12(a)) [22].

When tested at temperatures higher than 900°C, Cr_2AlC shows significant plastic deformation ability (Figure 11). A typical SEM image of the tensile surface of a specimen tested at 1000°C is shown in Figure 13(a), in which large amount of microcracks initiated and propagated in the direction roughly perpendicular to the direction of tensile stress (horizontal direction). This microstructural characteristic is similar to the response of Ti_3SiC_2 in high temperature tensile test [35] and tensile creep test [36]. Under high magnification, it can be seen that the microcracks mainly propagated along grain boundaries (Figure 13(b)). This can be attributed to the dislocation activities and the following pile-ups against grain boundary [4, 5, 26]. The initiation and propagation of large amount of microcracks account for the macro-plastic deformation of Cr_2AlC [22].

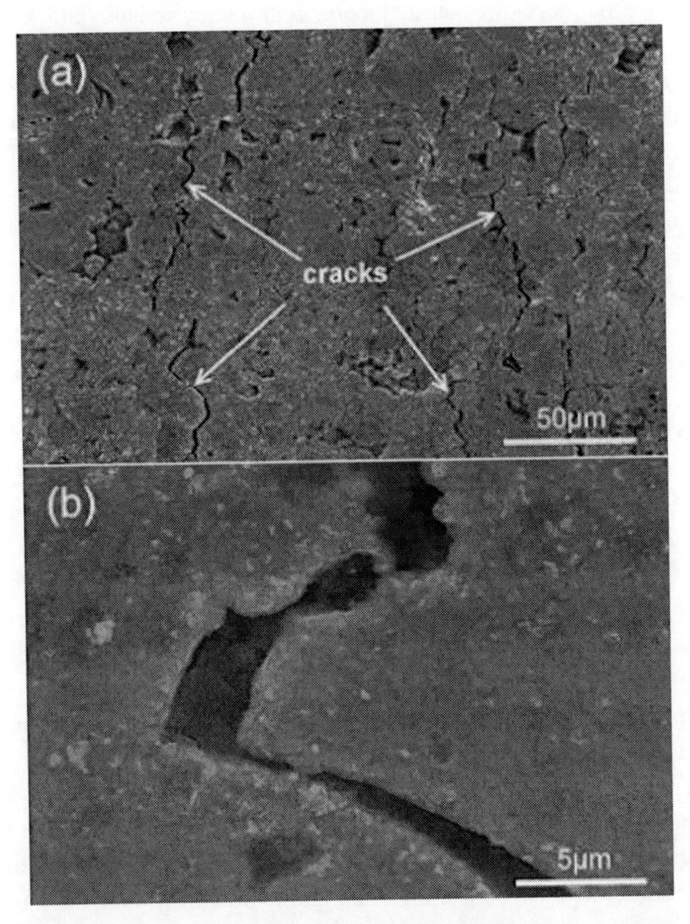

Figure 13. (a) A typical SEM image of the tensile surface of Cr_2AlC specimen tested at 1000°C [22]. (b) SEM observation on grain boundary decohesion under higher magnification [22].

It was found in previous compressive test [26] that a large number of cracks appear and most of them are aligned on the load direction when tested at 800°C. In fact, the crack propagation in compression and in tension is substantially consistent and can be explained as follows. There are essentially three basic fracture modes, namely tensile opening mode (Mode I fracture), shear type of fracture (Mode II fracture) and anti-plane strain fracture (Mode III fracture). Generally, cracks propagate more easily according to Mode I fracture than Mode II fracture in ceramics. As a crack is inclined at an angle $\neq 90°$ in tension or at angle $\neq 0°$ in compression to the stress axis, both Mode I and Mode II fracture contribute to crack advance [34]. However, the dominant direction of Mode I fracture in tension is nearly normal to stress axis and that in compression is nearly parallel to stress axis. Therefore, the crack propagation direction of Cr_2AlC in tension is perpendicular to the tensile stress direction, while that in compression is parallel to compressive stress direction [22].

CONCLUSIONS

This chapter sums up the mechanical properties of Cr_2AlC, including the hardness, room-temperature flexural and compressive strength, thermal shock resistance, damage tolerance, and the elevated-temperature mechanical behaviors under tensile and compressive stress conditions.

The studied mechanical properties of Cr_2AlC at room temperature, with the hardness of 3.2~5.6 GPa, the flexural strength of 378~600 MPa and the compressive strength of 625~1156 MPa, are usually comparable to those of Ti_2AlC and Ti_3AlC_2 ceramics and strongly dependent on the microstructure developed. The low hardness and the layered microstructure, combined with the good thermal conductivity, endue Cr_2AlC with good machinability. In addition, Cr_2AlC exhibits good damage tolerance with retaining ~70 % of the flexural strength under a 100 N indentation load introduced. However, the relatively smaller thermal conductivity and higher coefficient of thermal expansion lead to the inferior thermal shock resistance properties of Cr_2AlC as comparing with those of Ti_3AlC_2 and Nb_2AlC.

When tested at a strain rate of $5.6\times10^{-4}s^{-1}$, the compressive strength of Cr_2AlC decreases continuously from 997±29MPa at room temperature to 523±7 MPa at 900°C. The ductile-to-brittle transition temperature is measured to be in the range of 700~800°C. When tested in the strain rate range of $5.6\times10^{-5}s^{-1}$ to $5.6\times10^{-3}s^{-1}$, Cr_2AlC fails in a brittle mode at room temperature, whereas the deformation mode changes from a brittle to a ductile as the strain rate is lower than $5.6\times10^{-4}s^{-1}$ when compressed at 800°C. The compressive strength increases slightly with increasing strain rate at room temperature and it is less dependent on strain rate when tested at 800°C.

The variation of flexural strength of Cr_2AlC with testing temperature exhibits similar decreasing tendency to that of compressive strength, except for the higher brittle-to-ductile transition temperature locating in the range of 800~900°C. The plastic deformations of Cr_2AlC under both tensile and compressive stress conditions share the same mechanism intrinsically that is attributed to a dislocation related activities, such as initiation and propagation of microcracks, decohesion along grain boundary and delamination and kink band within grains.

Based on many efforts made on the investigation of the mechanical properties of Cr_2AlC, as reviewed in this chapter, the potential application fields of this material is expected in the following aspects: 1) substitute for machinable ceramics due to its machinability; 2) electrodes or kiln furniture related to its electrical conductivity, high-temperature mechanical properties, oxidation resistance; 3) exhaust gas filters for automobiles correlated with its thermal conductivities, damage tolerance and thermal shock resistance; 4) reinforcement or matrix in composite material for various applications combined with its mechanical properties and conductivity. In addition, industrially applicable large area depositions of Cr_2AlC [37] and Ti_2AlC [38] on steel have been conducted and positive results were reported. Moreover, the application of Ti_2AlC as heating element suggests more application areas of Cr_2AlC is needed to be explored [39, 40].

REFERENCES

[1] Barsoum, M.W. *Prog. Solid State Chem.* 2000, *28,* 201-281.

[2] Barsoum, M.W.; Tzenov, N.; Procopio, A.; El-Raghy, T.; Ali, M. *J. Electrochem. Soc.* 2001, *148,* C551-C562.

[3] Sun, Z.M.; Zhou, Y.C.; Li, M.S. *Corros. Sci.* 2001, *43,* 1095-1109.

[4] Barsoum, M.W.; ElRaghy, T. *J. Am. Ceram. Soc.* 1996, *79,* 1953-1956.

[5] Tzenov, N.V.; Barsoum, M.W. *J. Am. Ceram. Soc.* 2000, *83,* 825-832.

[6] Tian, W.B.; Wang, P.L.; Zhang, G.J.; Kan, Y.M.; Li, Y.X.; Yan, D.S. *Scripta Mater.* 2006, *54,* 841-846.

[7] Lin, Z.J.; Li, M.S.; Wang, J.Y.; Zhou, Y.C. *Acta Mater.* 2007, *55,* 6182-6191.

[8] Hettinger, J.D.; Lofland, S.E.; Finkel, P.; Meehan, T.; Palma, J.; Harrell, K.; Gupta, S.; Ganguly, A.; El-Raghy, T.; Barsoum, M.W. *Phys. Rev. B* 2005, *72,* 115120.

[9] Lee, D.B.; Nguyen, T.D. *J. Alloys Compd.* 2008, *464,* 434-439.

[10] Tian, W.B.; Wang, P.L.; Kan, Y.M.; Zhang, G.J. *J. Mater. Sci.* 2008, *43,* 2785-2791.

[11] Jovic, V.D.; Jovic, B.M.; Gupta, S.; El-Raghy, T.; Barsoum, M.W. *Corros. Sci.* 2006, *48,* 4274-4282.

[12] Tian, W.B.; Wang, P.L.; Zhang, G.J.; Kan, Y.M.; Li, Y.X.; Yan, D.S. *Mater. Sci. Eng. A* 2007, *454,* 132-138.

[13] Tian, W.B.; Vanmeensel, K.; Wang, P.L.; Zhang, G.J.; Li, Y.X.; Vleugels, J.; Van der Biest, O. *Mater. Lett.* 2007, *61,* 4442-4445.

[14] Barsoum, M.W.; Ali, M.; El-Raghy, T. *Met. Mat. Trans. A* 2000, *31,* 1857-1865.

[15] Schuster, J.C.; Nowotny, H.; Vaccaro, C. *J. Solid State Chem.* 1980, *32,* 213-219.

[16] Sun, Z.M.; Ahuja, R.; Schneider, J.M. *Phys. Rev. B* 2003, *68,* 224112.

[17] Wang, J.Y.; Zhou, Y.C. *J. Phys.-Cond. Matter* 2004, *16,* 2819-2827.

[18] Tian, W.B.; Sun, Z.M.; Hashimoto, H.; Du, Y.L. *J. Alloys Compd.* 2009, *484,* 130-133.

[19] Rice, R.W.; Wu, C.C.; Boichelt, F. *J. Am. Ceram. Soc.* 1994, *77,* 2539-2553.

[20] Tian, W.B.; Wang, P.L.; Zhang, G.J.; Kan, Y.M.; Li, Y.X. *J. Am. Ceram. Soc.* 2007, *90,* 1663-1666.

[21] Han, J.H.; Park, S.W.; Kim, Y.D. *Synthesis and mechanical properties of nano laminating Cr2AlC using CrCx/Al powder mixtures.* in *Power Metallurgy World*

Congress and Exhibition 2006. 2006. Busan, SOUTH KOREA: Trans Tech Publications Ltd.

[22] Tian, W.B.; Sun, Z.M.; Du, Y.L.; Hashimoto, H. *Mater. Lett.* 2009, *63,* 670-672.

[23] Salama, I.; El-Raghy, T.; Barsoum, M.W. *J. Alloys Compd.* 2002, *347,* 271-278.

[24] Kingery, W.D. *J. Am. Ceram. Soc.* 1955, *38,* 13.

[25] Barsoum, M.W.; Salama, I.; El-Raghy, T.; Golczewski, J.; Porter, W.D.; Wang, H.; Seifert, H.J.; Aldinger, F. *Met. Mat. Trans. A* 2002, *33,* 2775-2779.

[26] Tian, W.B.; Sun, Z.M.; Hashimoto, H.; Du, Y.L. *J. Mater. Sci.* 2009, *44,* 102-107.

[27] Barsoum, M.W.; El-Raghy, T.; Radovic, M. *Interceram* 2000, *49,* 226-233.

[28] Lin, Z.J.; Zhou, Y.C.; Li, M.S.; Wang, J.Y. *Z. Metallkd.* 2005, *96,* 291-296.

[29] Radovic, M.; Barsoum, M.W.; El-Raghy, T.; Wiederhom, S.M.; Luecke, W.E. *Acta Mater.* 2002, *50,* 1297-1306.

[30] Barsoum, M.W.; Zhen, T.; Kalidindi, S.R.; Radovic, M.; Murugaiah, A. *Nature Mat.* 2003, *2,* 107-111.

[31] Zhen, T.; Barsoum, M.W.; Kalidindi, S.R. *Acta Mater.* 2005, *53,* 4163-4171.

[32] Sun, Z.M.; Zhang, Z.F.; Hashimoto, H.; Abe, T. *Mater. Trans.* 2002, *43,* 432-435.

[33] Johnston, W.G.; Gilman, J.J. *J. Appl. Phys.* 1959, *30,* 129-144.

[34] Courtney, T.H. *Mechanical Behavior of Materials*; McGraw-Hill publishing company, New York; State, 1990; pp

[35] Radovic, M.; Barsoum, M.W.; El-Raghy, T.; Seidensticker, J.; Wiederhorn, S. *Acta Mater.* 2000, *48,* 453-459.

[36] Radovic, M.; Barsoum, M.W.; El-Raghy, T.; Wiederhorn, S. *Acta Mater.* 2001, *49,* 4103-4112.

[37] Walter, C.; Sigumonrong, D.P.; El-Raghy, T.; Schneider, J.M. *Thin Solid Films* 2006, *515,* 389-393.

[38] Walter, C.; Martinez, C.; El-Raghy, T.; Schneider, J.M. *Steel Res. Int.* 2005, *76,* 225-228.

[39] Sundberg, M.; Malmqvist, G.; Magnusson, A.; El-Raghy, T. *Ceram. Int.* 2004, *30,* 1899-1904.

[40] Frodelius, J.; Sonestedt, M.; Björklund, S.; Palmquist, J.-P.; Stiller, K.; Högberg, H.; Hultman, L. *Surf. Coat. Technol.* 2008, *202,* 5976-5981.

In: MAX Phases: Microstructure, Properties and Applications ISBN 978-1-61324-182-0
Editors: It-Meng (Jim) Low and Yanchun Zhou

Chapter 13

TRIBOLOGICAL CHARACTERISTICS AND WEAR MECHANISMS IN TI_3SIC_2 AND TI_3ALC_2

Zhenying Huang and Hongxiang Zhai
Center of Materials Engineering, School of Mechanical, Electronic and Control Engineering Beijing Jiaotong University, Beijing, China

ABSTRACT

The tribological behaviors and the relevant mechanisms of highly pure bulk Ti_3SiC_2, Ti_3AlC_2 and the influence of TiC impurities were experimentally investigated, using a low carbon steel disk as the counterpart at room temperature. The highly pure Ti_3SiC_2 exhibits a decreasing friction coefficient (0.53-0.09) and an increasing wear rate (0.6-2.5×10^{-6} mm^3/Nm) with the sliding speed increasing from 5 to 60 m/s, and the normal pressure, in the range of 0.1~0.8 MPa, also has a complex but relatively weak influence on them. The changes can be attributed to the presence and the coverage of a frictional oxide film consisting of an amorphous mixture of Ti, Si and Fe oxides on the Ti_3SiC_2 friction surface. TiC impurities cause both of the friction coefficient and the wear rate to increase significantly in comparison with the highly pure Ti_3SiC_2 in the case of higher sliding speed or larger normal pressure due to an interlocking action and the pulling-out of TiC grains.

The friction coefficient is as low as 0.1~0.4 and the wear rate of Ti_3AlC_2 is only (2.3~2.5) $\times10^{-6}$ mm^3/Nm in the sliding speed range of 20~60 m/s when Ti_3AlC_2 sliding against low carbon steel disk. Such unusual friction and wear properties were confirmed to be dependant predominantly upon the presence of a frictional oxide film consisting of amorphous Ti, Al, and Fe oxides on the friction surfaces. The oxide film is in a fused state during the sliding friction at a fused temperature of 238-324°C, so it takes a significant self-lubricating effect.

1. INTRODUCTION

There were many literatures showing that the layered ternary carbide Ti_3SiC_2 had an unusual combination of properties in electrical and thermal conductivity, thermal shock resistance, damage tolerance, machinability and other useful properties [1-10]. In particular, the basal plans of Ti_3SiC_2 have an ultra-low friction coefficient [11]. Such an unusual combination suggests that the bulk Ti_3SiC_2 could be appropriately applicable material for some special tribological applications.

However, in the previous literature on the friction and wear of Ti_3SiC_2 [12-13], the polycrystalline bulk Ti_3SiC_2 did not exhibit the expected tribological properties. El-Raghy *et al.* [12] investigated the effects of grain size on the friction and wear for a high-purity Ti_3SiC_2 sample sliding against a 440C stainless steel pin in pin-on-disk type tests. The results showed that, irrespective of the grain size, the steady friction coefficient is as high as about 0.83, and the wear rates are as large as 4.25×10^{-3} mm^3/Nm and 1.34×10^{-3} mm^3/Nm for fine- and coarse-grained Ti_3SiC_2, respectively. Using a AISI 52100 steel pin as the counterpart in pin-on-disk type tests, Sun and Zhou [13] measured the friction coefficient and wear rate of a Ti_3SiC_2 sample contained 7.0 wt% TiC impurities. The results showed that the steady friction coefficient is 0.4-0.5 and the wear rate is 9.9×10^{-5} mm^3/Nm. Is it true that Ti_3SiC_2 has such a large friction coefficient and the wear rate and the Ti_3SiC_2 contained TiC impurities has smaller friction coefficient and wear rate than the high-purity Ti_3SiC_2? For the complexity of the tribological problem, a further study is necessary to understand some tribological behaviors and the relevant mechanisms of Ti_3SiC_2.

In general, for any a friction-pair making up of real materials, the friction and wear behavior would be influenced strongly by the lubricating state of the friction surfaces [14-16]. The results of the previous work by Lim *et al.* [14] have shown that, the sliding friction and wear behaviors between dry metal surfaces are determined, at low sliding speeds ($v<$ 1m/s for steel) by surface roughness and by the plastic (and perhaps elastic) properties of the surfaces. At higher speeds($v>$1 m/s for steel), the surface condition is modified by local heating(which can cause oxidation or even melting); then the coefficient of friction depends in a reproducible way on sliding velocity and bearing pressure. It is conceivable that, if either of the friction surfaces is able to generate a film which has antifriction effect during friction, the coefficient of friction and the wear rate will be reduced. Otherwise, a larger coefficient of friction and wear rate may appear due to severe interferences of asperities between friction surfaces. This may also be a dominant factor to determine whether or not the Ti_3SiC_2 exhibiting a good tribological performance.

The material transfer between friction surfaces is worth notice. This is a quite complex tribological problem depending upon not only material's nature but also testing conditions such as normal load, sliding speed, environmental temperature and relative humidity. In ductile materials, plastic deformation and changes in structure during early stages of sliding may be precursors to processes such as transfer and mechanical mixing in which the chemical composition of near-surface material is modified [17-18]. Such processes may also occur in Ti_3SiC_2 when a ductile metal material is used as friction counterpart. Thus, the oxidation of Ti_3SiC_2 may not be independent during sliding friction, the material transfer and the mechanical alloying between the transferred materials and the Ti_3SiC_2 surface may accelerate Ti_3SiC_2 to be oxidized and further form a frictional oxide film. The hard TiC particles contained in Ti_3SiC_2 matrix as impurities may aggravate the transfer.

In this chapter, we describe some tribological properties and the relevant self-antifriction mechanism of the highly pure, polycrystalline bulk Ti_3SiC_2 and Ti_3AlC_2 and report the influence of TiC impurities in order to show the potential of Ti_3SiC_2 for tribological applications.

2. EXPERIMENTAL PROCEDURE

2.1. Sample Preparation

The Ti_3SiC_2 sample used was prepared by reactively hot-pressing a mixture of elemental titanium, silicon, carbon (graphite) and aluminum powders which served as reactive accelerators. The previous investigation [19] showed that the proper addition of aluminum could increase the purity of Ti_3SiC_2 in the reactive product. The powders were mixed with Ti:Si:C:Al = 3:1:2:0.2 in mole ratio, as this ratio resulted in the highest phase purity of Ti_3SiC_2 [19]. The prepared powders were pre-compressed at 8 MPa in a graphite die and subsequently hot-pressed in the same graphite die under 1450 °C and 25 MPa pressure for 2 h with flowing argon gas.

The Ti_3SiC_2 sample containing the TiC phase was prepared using an earlier recipe which contained 3.0 vol% B_2O_3 additives while not contained Al. The original purpose for adding B_2O_3 was to produce a high-purity Ti_3SiC_2 product, but it was obtained with difficulty. The processing procedures were roughly identical for the preparation of the highly pure Ti_3SiC_2. For brevity, the highly pure Ti_3SiC_2 samples denoted as Ti_3SiC_2 and the Ti_3SiC_2 containing TiC are henceforth denoted as TiC-Ti_3SiC_2.

Figure 1a is a XRD pattern showing the phase composition of the prepared Ti_3SiC_2 sample. According to the standard XRD data, the diffraction peaks primarily belong to Ti_3SiC_2 phase. Although there were very weak diffraction peaks belonging to TiC phase which is as the impurity generated during the reactive synthesis of the Ti_3SiC_2 sample, the purity of the Ti_3SiC_2 sample is indeed quite high in the present stage. Based on the calibrated relationship between the relative peak intensity and the weight percentage of the TiC phase for the two-phase mixture of Ti_3SiC_2 and TiC [20], the relative content of the TiC impurity contained in the prepared Ti_3SiC_2 sample was estimated to be less than 2.0 vol%. The effect of thus a small amount of TiC impurity may be inappreciable on the tribological behavior.

Figure 1b is a XRD pattern showing the phase compositions of the prepared TiC-Ti_3SiC_2 sample. The diffraction peaks essentially belong to Ti_3SiC_2 phase and TiC phase. The relative content of the TiC phase in the TiC-Ti_3SiC_2 sample was estimated to be about 20.0 vol% according to the calibrated relationship mentioned above [20]. The very small amounts of other impurities including the residual carbon and the titanium silicides may be inappreciable for the tribological behavior.

Figure 2a is a typical SEM micrograph exhibiting the microstructures of the Ti_3SiC_2 sample. The Ti_3SiC_2 sample primarily consists of Ti_3SiC_2 grains with a plate-like shape and layered microstructures, few TiC grains or other impurities were observed, which is in agreement with the XRD analysis for the Ti_3SiC_2 sample. The Ti_3SiC_2 grains in dimension were relatively uniform. The average grain size was estimated to be about 20 μm in the elongated direction. The specific density of the measured Ti_3SiC_2 sample was of 4.31 g/cm^3.

Figure 2b is a typical SEM micrograph exhibiting the microstructures of the TiC- Ti_3SiC_2 sample. Being in agreement with the XRD analysis, the TiC- Ti_3SiC_2 sample consists of dominant Ti_3SiC_2 grains and TiC grains. The Ti_3SiC_2 grains were also plate-like, while the TiC grains (the light particles as marked in Figure 2b) exhibited an equiaxed shape. The average size was estimated to be about 25 μm for the Ti_3SiC_2 grains, and about 4 μm for the TiC grains. The specific density of the measured TiC- Ti_3SiC_2 sample was of 4.21 g/cm^3.

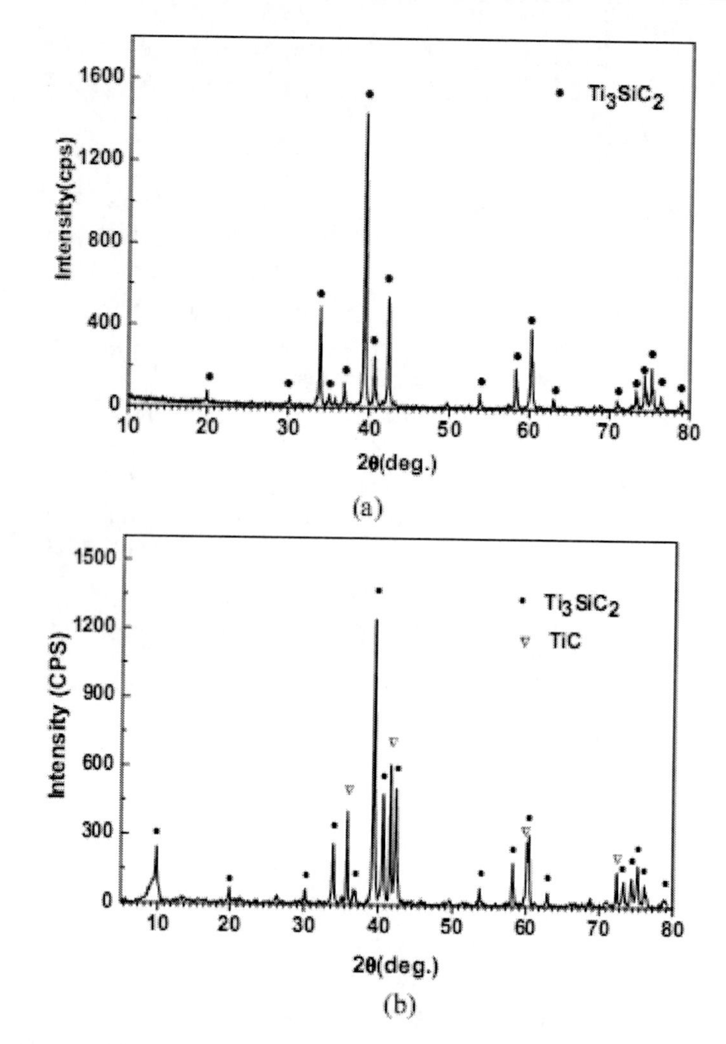

Figure 1. XRD patterns of the hot pressing synthesized (a) Ti_3SiC_2 and (b) TiC-Ti_3SiC_2 samples.

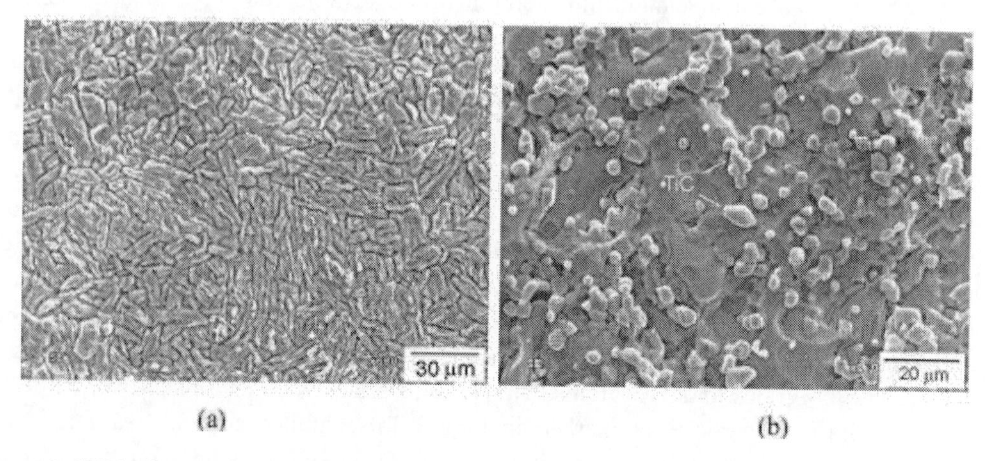

Figure 2. SEM micrographs showing the microstructures of (a) Ti_3SiC_2 and (b) TiC-Ti_3SiC_2; the observed surfaces have been etched by the HF and HNO_3 mixed acid.

A bulk polycrystalline Ti_3AlC_2 sample was prepared by hot-pressing a mixture of the elemental titanium, aluminum, and graphite powders. Commercial titanium, aluminum, and graphite powders were used as the starting materials. The powders were mixed with a mole ratio of Ti:Al:C = 3:1.1:2, and ball milled for 4 h in an ethanol solution. The dried powders were pre-compressed at a pressure of 8 MPa in a graphite die, and then hot-pressed at 1425 °C and 20 MPa for 30 min with flowing argon gas. The heating rate was 40 °C/min, and the cooling rate was about 10 °C/min. Thus, a bulk polycrystalline Ti_3AlC_2 sample was prepared.

Figure 3a shows an XRD pattern of the powders taken from the bulk Ti_3AlC_2 sample. Although, as well known, most diffraction peaks of Ti_3AlC_2, Ti_2AlC and TiC phase are very close to 2θ values in terms of their standard XRD patterns, based on their characteristic peaks, where $2\theta = 9.5°$ for Ti_3AlC_2{002}, $2\theta = 13°$ for Ti_2AlC {002}, and $2\theta = 35.9°$ for TiC {111}[21], it was identified that the diffraction peaks primarily belonged to the Ti_3AlC_2 phase, there were only very weak peaks belonging to the Ti_2AlC_2 phase, and no TiC and other impurity phases were determinable.

Figure 3b shows an SEM micrograph showing microstructures of the bulk Ti_3AlC_2 sample. The observed surface has been etched using a mixed acid of HF and HNO_3 in a volume ration of HF: HNO_3:H_2O = 1:1:3. This typical micrograph shows that most of the Ti_3AlC_2 grains have a banding or plate-like shape and layered structure. The average size of the grains was estimated to be ~20 μm in the elongated direction. The density was measured to be 4.19 g/cm^3, being about 98.6% of the theoretical density of 4.25 g/cm^3 [22].

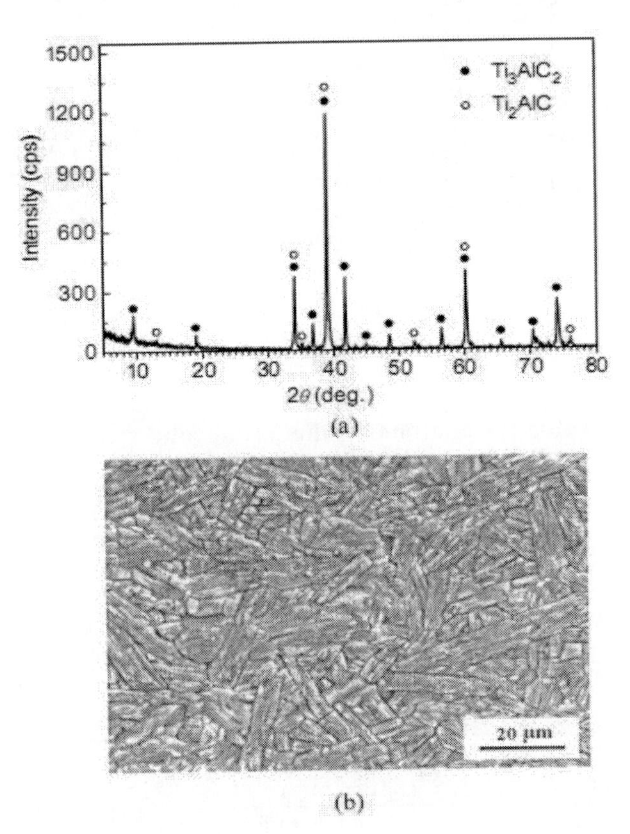

Figure 3. XRD pattern (a) and SEM micrograph (b) of the hot pressing synthesized bulk Ti_3AlC_2 sample [23].

2.2. Friction and Wear Tests

The Ti_3SiC_2, TiC-Ti_3SiC_2 and Ti_3AlC_2 samples were cut into blocks, and used to slide dryly against a steel disk, performed on a block-on-disk type friction tester, shown in Figure 4. The dimensions, chemical composition of the steel disk and the detailed test conditions are listed in Table 1. The coefficient of friction was automatically measured and recorded in real time by the computer system of the friction tester. The wear quantity of the Ti_3SiC_2, TiC-Ti_3SiC_2 or Ti_3AlC_2 block was measured by a weighing method, which measured the mass loss of the Ti_3SiC_2, TiC-Ti_3SiC_2 or Ti_3AlC_2 block for every friction process. The longer sliding distance for one continuous friction process is necessary to accurately measure the mass loss, as the samples prepared were very wear resistant. The tests for every given condition were repeated three times to obtain reliable data, and the average value was used as the evaluating data. The pre-abrasion is necessary whenever a test condition is changed in order to remove possible influences of load history on the friction surface state.

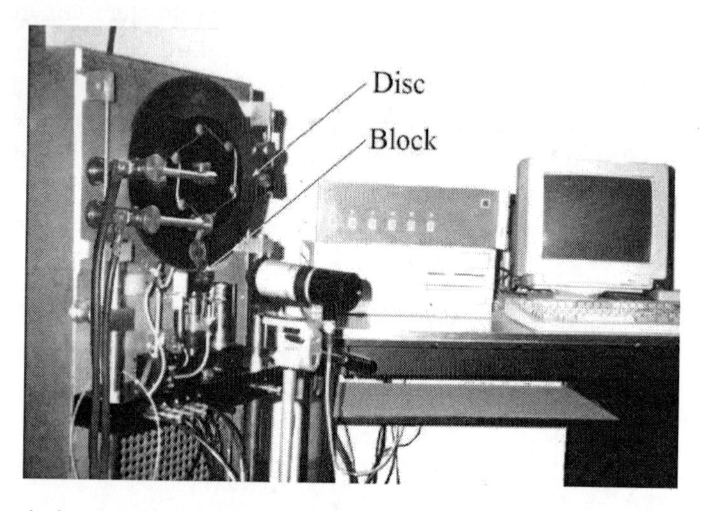

Figure 4. Photograph showing the appearance of the block-on-disc type, current-carrying, high-speed friction tester.

Table 1. Conditions of the friction and wear test

Test type	Block-on-disk type high-speed tester [24]
The sample (block)	Ti_3SiC_2, TiC-Ti_3SiC_2 or Ti_3AlC_2
Dimensions of the block $10 \times 10 \times 12$ mm^3 in length × width × height	
The counterpart (disk)	Low carbon steel (containing 0.2 % carbon)
Dimension of the disk	Φ300 mm × 10 mm in diameter × thickness
Sliding speed	5 ~ 60 m/s
Normal pressure	0.1 ~ 0.8 MPa
Environmental temperature	20 ~ 25 °C
Relative humidity	23~25 %
Sliding distance	24,000 m for one continuous friction process
Pre-abrasion	24,000 m whenever a test condition is changed

3. TRIBOLOGICAL PROPERTIES OF TI_3SIC_2

3.1. Friction Behaviors

Figure 5 shows the friction coefficients as functions of normal pressure for different sliding speeds. Notably, the friction coefficients are speed and pressure dependent. For the low sliding speed of 5 m/s, the coefficient of friction monotonously increases from 0.42 to 0.53 with increasing the normal pressure from 0.1 MPa to 0.8 MPa. However, for the medium sliding speed of 20 m/s, only when the normal pressure increases from 0.1 MPa to 0.2 MPa the coefficient of friction increases from 0.25 to 0.36, then it gradually reduces to 0.27 with increasing the normal pressure to 0.8MPa. A similar pressure-dependent change is exhibited at the sliding speed of 40 m/s, but the maximum value and the corresponding pressure are changed. The coefficient of friction increases from a quite small value of 0.08 to 0.21 with increasing the normal pressure from 0.1 MPa to 0.4 MPa, and then gradually reduces to 0.18 with increasing the normal pressure to 0.8 MPa. An inverse change is found at the higher sliding speed of 60 m/s. The coefficient of friction slowly increases from 0.12 to 0.2 with increasing the normal pressure from 0.1 MPa to 0.6 MPa, and then shows a faster increasing tendency, but only increases to 0.29 when the normal pressure increases to 0.8 MPa. Clearly, the sliding speed is an assignable factor. The coefficient of friction notably reduces with increase in the sliding speed, unless a larger normal pressure is applied when the sliding speed is high. These evidences suggest that the friction behavior could be governed by a changing antifriction mechanism, which is dependent upon both of the sliding speed and the normal pressure.

3.2. Wear Behaviors

The wear rates of the Ti_3SiC_2 are shown in Figure 6 as functions of the normal pressure for different sliding speeds. Clearly, the wear rates increase with increase in the sliding speed unless the normal pressure applied is small at the sliding speed of 60 m/s. This speed-dependent change is reversed to that of the coefficient of friction in magnitude, i.e., a smaller wear rate of Ti_3SiC_2 corresponds to a larger coefficient of friction, and vice versa. For the low speed of 5 m/s, the wear rate increases slightly from 0.64×10^{-6} mm^3/Nm to 0.90×10^{-6} mm^3/Nm with an increase in the normal pressure from 0.1 MPa to 0.8 MPa. However, for the medium speeds, the wear rates decrease slightly from 1.8×10^{-6} mm^3/Nm to 1.37×10^{-6} mm^3/Nm and from 2.5×10^{-6} mm^3/Nm to 2.0×10^{-6} mm^3/Nm, with the normal pressure increasing from 0.1 MPa to 0.8 MPa, for sliding speeds of 20 m/s and 40 m/s, respectively. An inverse change similar to the behavior in the coefficient of friction is found at high speed of 60 m/s. The wear rate increases from a relatively low value of 1.0×10^{-6} mm^3/Nm to 2.11×10^{-6} mm^3/Nm with increasing the normal pressure from 0.1 MPa to 0.5 MPa, then increased at a larger rate to 3.62×10^{-6} mm^3/Nm with further increasing the normal pressure to 0.8MPa. These behaviors suggest that the wear rate of Ti_3SiC_2 could be governed by a complex mechanism relating to the state of friction surfaces.

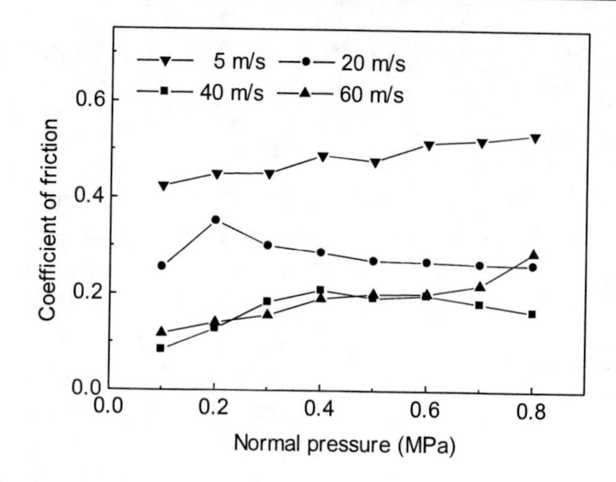

Figure 5. Friction coefficients of Ti_3SiC_2 sliding against low carbon steel in the different sliding speed and normal pressure conditions [25].

speed and normal pressure conditions [25].

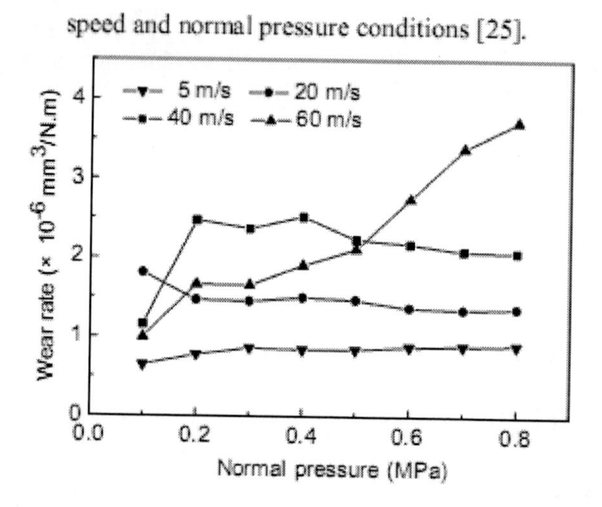

Figure 6. Wear rates of Ti_3SiC_2 sliding against low carbon steel in the different sliding speed and normal pressure conditions [25].

3.3. Friction Surface Behaviors

The friction surfaces wear observed by SEM associating with several typical behaviors of the coefficient of friction and the wear rate. Figure 7 is the observed micrographs for the Ti_3SiC_2 friction surfaces. It can be seen that the friction surfaces are different in image with the sliding speed and normal pressure changing. Clearly, there is a frictional film covering the Ti_3SiC_2 friction surface for the medium speed of 20 m/s, and the percentage of coverage is increased with the increase of the normal pressure. As shown in Figure 7(c and d), the film only partially covers the friction surface in the case of 0.2 MPa, while it completely covers the friction surface in the case of 0.8 MPa and the film surface is very smooth accordingly. In fact, the similar statuses are also presented at the sliding speed of 40 m/s. However, when the sliding speed increases to 60 m/s, the statuses are changed. There is yet a completely covered and smoother film existing in the friction surface for the low normal pressure of 0.2 MPa, but the film becomes poor and non-uniform when the normal pressure increases to 0.8 MPa, as shown in Figure 7(e and f). It is worth to note that there are only few prophyritic films existing in the Ti_3SiC_2 friction surface when the sliding speed is 5 m/s, irrespective of the normal pressure of 0.2 MPa or 0.8 MPa.

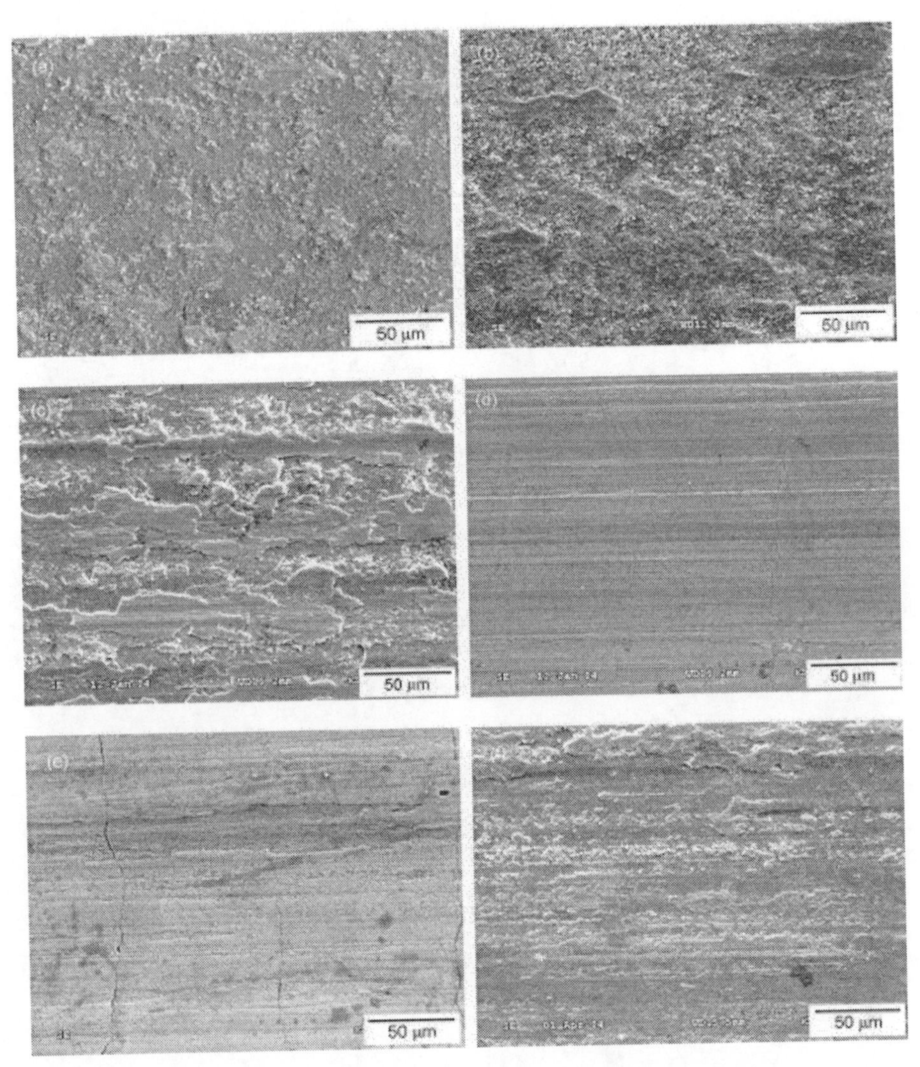

Figure 7. SEM micrographs showing the Ti_3SiC_2 worn surfaces after 72,000 m sliding distance in the different sliding speed and normal pressure conditions. (a) 5 m/s and 0.2 MPa, (b) 5 m/s and 0.8 MPa, (c) 20 m/s and 0.2 MPa, (d) 20 m/s and 0.8 MPa, (e) 60 m/s and 0.2 MPa and (f) 60 m/s and 0.8 MPa [25].

The thickness of the film was observed from the cross-section of the film. Figure 8 is a micrograph showing the cross-section state of the film formed under the sliding speed of 20 m/s and the normal pressure of 0.8 MPa. The thickness of the film is estimated to be about 1.0 µm. This is a typical thickness for the film, when it completely covered the Ti_3SiC_2 friction surface.

Figure 9 are typical SEM micrographs showing the friction surfaces of the steel disk for the lower sliding speed of 5m/s and the medium sliding speed of 20 m/s, corresponding to the film-lacking in Ti_3SiC_2 friction surface(Figure 7(b)) and the film-full in Ti_3SiC_2 friction surface (Figure 7(d)), respectively. The friction surface for the sliding speed of 5 m/s is quite rough and there are obvious scratches on the surface, while the friction surface for the speed of 20 m/s is smoother, and obviously there is a film covering the friction surface.

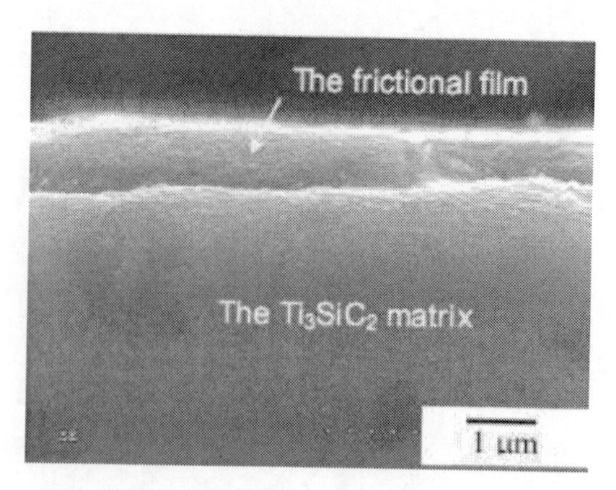

Figure 8. A typical SEM micrograph showing the cross section of the frictional film on the worn surface of Ti_3SiC_2 [25].

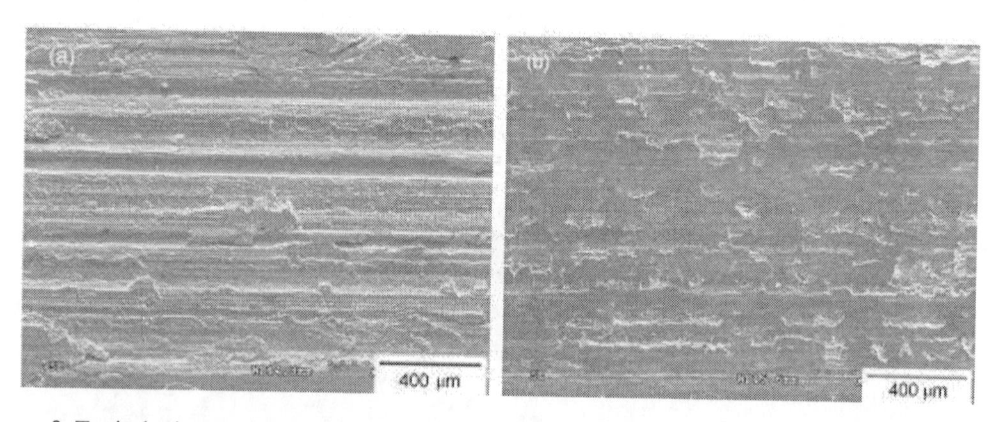

Figure 9. Typical micrographs exhibiting the worn surfaces of the low carbon steel counterpart; (a) 5 m/s, 0.8 MPa and (b) 20 m/s, 0.8 MPa [26].

3.4. Compositions of Films

The compositions of films covering the friction surfaces of the Ti_3SiC_2 and the steel disk were identified by the EDS. Figure 10 (a and b) are the analyzed results for the Ti_3SiC_2 friction surface and the steel disk friction surface, which have been shown in Figures 7(d) and 9(b), respectively. There are only elements of oxygen, titanium, silicon and a little amount of iron, but no carbon existing in the Ti_3SiC_2 friction surface, indicating that the film consists of the oxides of titanium, silicon and iron. There are the same elements existing in the friction surfaces of the steel disk, indicating that the film covering the friction surface of the steel disk consists of the same oxides as the Ti_3SiC_2 friction surface. The only difference between the two surfaces is the content of the elements. It is conceivable that the oxides of titanium and silicon existing in the steel disk surface are transferred from the Ti_3SiC_2 friction surface, and the ferric oxide in the Ti_3SiC_2 friction surface is transferred from the steel disk surface.

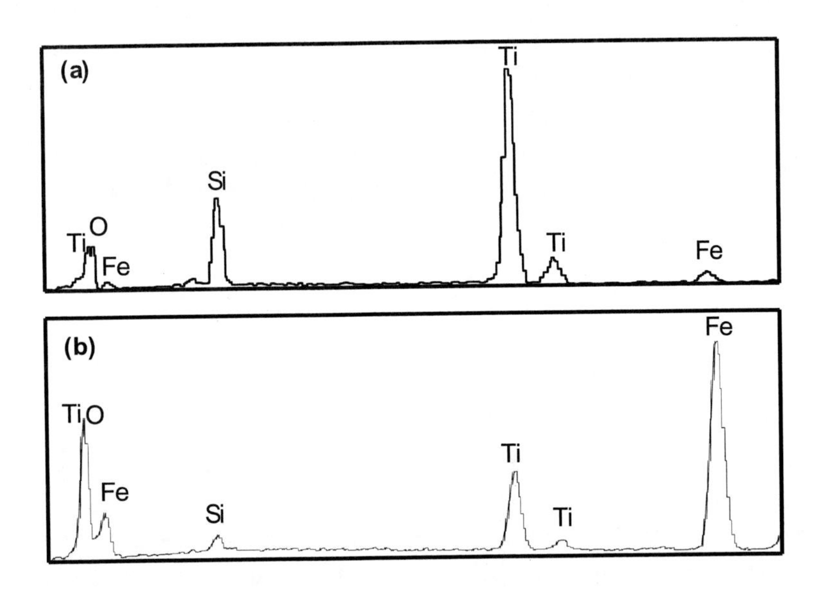

Figure 10. EDS patterns of wear surfaces of Ti_3SiC_2 and the low carbon steel counterpart; (a) Ti_3SiC_2 and (b) the low carbon steel counterpart.

3.5. Discussion

It has been shown that the oxide film is soft and flowable during the sliding friction [19, 27-28]. Such a film would play an antifriction part in the friction surfaces, while the speed and pressure-dependent behaviors of the friction coefficient could be related to the change in covering percentage of the oxide film in the friction surface. It is conceivable that a good oxide film with a higher percentage of coverage results in a smaller coefficient of friction, and a poor oxide film with a lower percentage of coverage results in a larger friction coefficient. Indeed, there was such a corresponding relation between the coefficient of friction and the percentage of coverage. In the cases of the medium speeds of 20 m/s and 40 m/s, the increased percentages of coverage resulted in the reducing of the coefficients of friction with increasing the normal pressure. In the case of the higher speed of 60 m/s, the reducing percentage of coverage resulted in the increasing of the coefficient of friction with increasing the normal pressure. As the same causal relation, in the case of the lower sliding speed of 5m/s, the larger coefficient of friction was due to the lack of the oxide film. It is worth to indicate that the coefficient of friction as high as 0.83 measured by El Raghy *et al.* [12] could be attributed to the poor friction surfaces of the Ti_3SiC_2 as well as the friction counterpart. In fact, there were almost no observable films existing in the friction surfaces.

The changing behaviors of the wear rates of Ti_3SiC_2 are also related closely with the existing of the oxide film. For the lack of the oxide film, the Ti_3SiC_2 friction surface for the low sliding speed of 5 m/s would be harder than the steel disk, and consequently the friction surface of the steel disk was scratched remarkably as shown in Figure 9(a). Clearly, in this case, the lower wear rate of Ti_3SiC_2 was at the price of the severe wear of the steel disk. As the soft oxide film is formed and completed with increasing the normal pressure, the hardness relation between the Ti_3SiC_2 and the steel disk was changed for the medium speeds of 20 m/s and 40 m/s. Since the softer oxide film would be readily worn, the wear rate of Ti_3SiC_2 was increased accordingly. There would be more oxides generated at the higher sliding speed of 60 m/s due to the higher frictional temperature, but they were difficult to be maintained in the friction surface when the normal pressure was larger, due to the crowding out effect of the friction surfaces during the sliding friction process. Consequently, the wear rate of Ti_3SiC_2 increased fast. In addition, the change in the viscosity of the oxide film could also be an influencing factor. The viscosity would be reduced with increasing the sliding speed and/or the normal pressure, since the frictional temperature in the friction surface rises. Then the oxide film is difficult to be maintained for the larger sliding speed and normal pressure could be related with the change of the viscosity as well.

Some literatures [29-30] have shown that Ti_3SiC_2 can be oxidized at a certain temperature in air, and forms a dense, adhesive and layered scale consisting of SiO_2 and TiO_2 on the surface of Ti_3SiC_2. Essentially, when two surfaces slide together, most of the work done against friction is turned into heat [14-16]. The resulting rise in temperature may modify the mechanical and metallurgical properties of the sliding surfaces, and it may make them oxidize or even melt; all these things influence the friction of coefficient and the rate of wear. Lim *et al.* [14] have developed wear mechanism and temperature maps for different materials in dry-sliding contact, where bulk surface(T_b) and flash (T_f) temperature are plotted as isothermal contours on normalized sliding velocity and contact pressure axes. The oxidation occurring on the Ti_3SiC_2 friction surface may also due to the generating of frictional heat, which is proportional to the friction work μLV, where L is the normal load, V the sliding speed, and μ is the coefficient of friction [31]. Thus, the rate of generating oxides is an increasing function of the L, V and μ, and hence the larger L and/or V as well as μ the larger generating rate of the oxides in unit time. On the other hand, simultaneously with the generating, the oxides would be consumed continuously from the friction surface as wear, and the consuming rate in unit time would be also an increasing function of the L and V. Hence, the larger L and/or V the larger consuming rate. The oxide film can be formed on the friction surface, only when the generating rate is larger than the consuming rate. However, the thickness of the oxide film is limited by the viscosity of the oxides during friction as well as the roughness of friction surface, cannot exceed a characteristic thickness, and the redundant oxides would be removed from the friction surface as the wear debris. Therefore, the oxide film can be maintained at the friction surface, if only the generating rate is equal to the consuming rate when a steady oxide film has been formed. This means that a larger generating rate could result in a larger wear rate. The oxide film cannot be formed, or a formed oxide film will be damaged, if the generating rate is smaller than the consuming rate. These states can be characterized as:

$$\begin{cases} \omega_g > \omega_c, \text{ the oxide film can be formed;} \\ \omega_g = \omega_c, \text{ the oxide film can be maintained;} \\ \omega_g < \omega_c, \text{ the oxide film cannot be formed or will be damaged.} \end{cases} \tag{1}$$

where ω_g and ω_c is the generating rate and the consuming rate, respectively.

The state could be changed with the sliding speed and/or the normal pressure, since the ω_g and ω_c are different function of the L and V. The states for the low sliding speed of 5 m/s and the normal pressures applied from 0.1 MPa to 0.8 MPa can be classified to $\omega_g < \omega_c$. However, the states for the medium speeds of 20 m/s and 40 m/s could be changed from $\omega_g < \omega_c$ to $\omega_g = \omega_c$ with increasing the normal pressure. Particularly, the states for the higher sliding speed of 60 m/s could be changed from $\omega_g > \omega_c$ to $\omega_g = \omega_c$ and further to $\omega_g < \omega_c$ with the increase in normal pressure. These changes should be further characterized.

3.6. Influences of Tic Impurities

The friction coefficient of the TiC-Ti_3SiC_2 exhibited similar speed- and pressure-dependent changes as the Ti_3SiC_2, but the changing magnitude was apparently larger than that of Ti_3SiC_2, as shown in Figure 11. Overall, the friction coefficient in magnitude is apparently larger than that of Ti_3SiC_2 for every sliding speed. In addition, the friction coefficient exhibited a critical transition at the normal pressure of 0.2 MPa for the sliding speed of 60 m/s, and the critical transition behavior was abrupt and out of control when the critical pressure was reached.

The TiC-Ti_3SiC_2 wear rate exhibited a slight difference from the Ti_3SiC_2 wear rate in speed- and pressure-dependent behavior which increased with increases in both the sliding speed and the normal pressure, as shown in Figure 12. In magnitude, the TiC-Ti_3SiC_2 wear rate was smaller than or equal to the Ti_3SiC_2 wear rate when the normal pressure was smaller, but larger than the Ti_3SiC_2 wear rate with an increase in the normal pressure. It is worth to note that, corresponding to the critical transition of friction coefficient, the TiC-Ti_3SiC_2 wear rate suddenly increased at the normal pressure of 0.2 MPa for the sliding speed of 60 m/s.

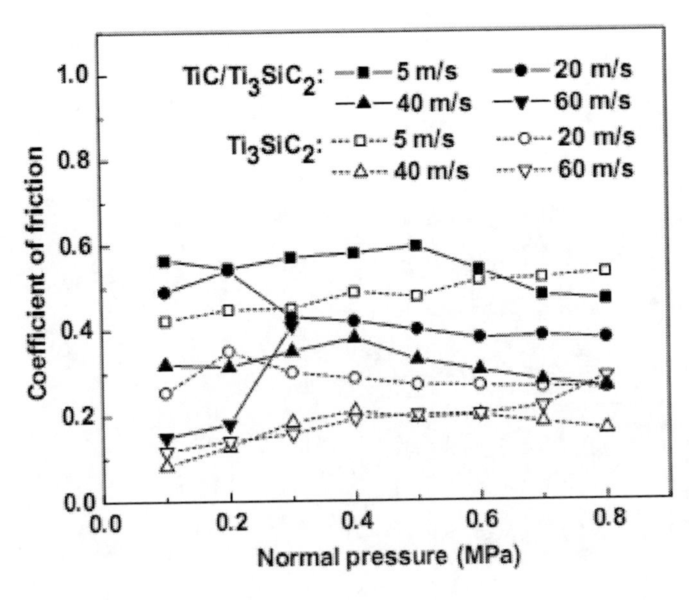

Figure 11. Friction coefficients of TiC/Ti_3SiC_2 sliding against the low carbon steel counterpart, as well as a comparison with Ti_3SiC_2.

There were also frictional films to be found on the TiC-Ti_3SiC_2 friction surfaces, and their speed- and pressure-dependent changes were essentially identical to the Ti_3SiC_2 friction surfaces. However, these films were considerably poor, in comparison with the films on the Ti_3SiC_2 friction surfaces. Figure 13a and b exhibit typical SEM micrograph images of the TiC-Ti_3SiC_2 friction surfaces which have undergone a sliding distance of 72,000 m under the normal pressures of 0.2 and 0.8 MPa for a medium sliding speed of 20 m/s, respectively. Apparently the films were thinner in comparison with the Ti_3SiC_2 friction surfaces shown in Figure 7c and d, indicating that the frictional film was formed with difficulty on the TiC-Ti_3SiC_2 friction surface than on the Ti_3SiC_2 friction surface. In this case, a certain material transfer from the steel disk to the TiC-Ti_3SiC_2 friction surface could occur, because the scrapping action of the harder TiC impurities. Thus, the frictional films in chemical composition and nature may have a difference between the Ti_3SiC_2 and TiC-Ti_3SiC_2 friction surfaces.

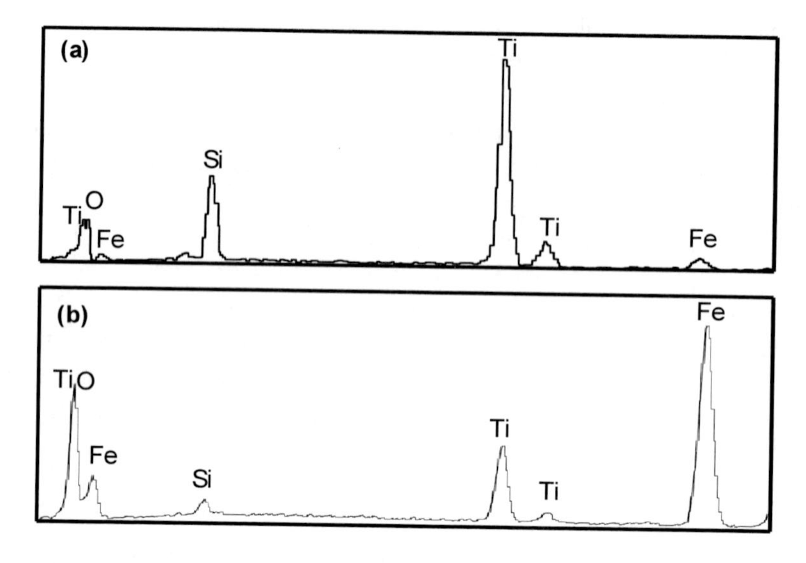

Figure 12. Wear rates of TiC/Ti_3SiC_2 sliding friction against the low carbon steel, as well as a comparison with Ti_3SiC_2.

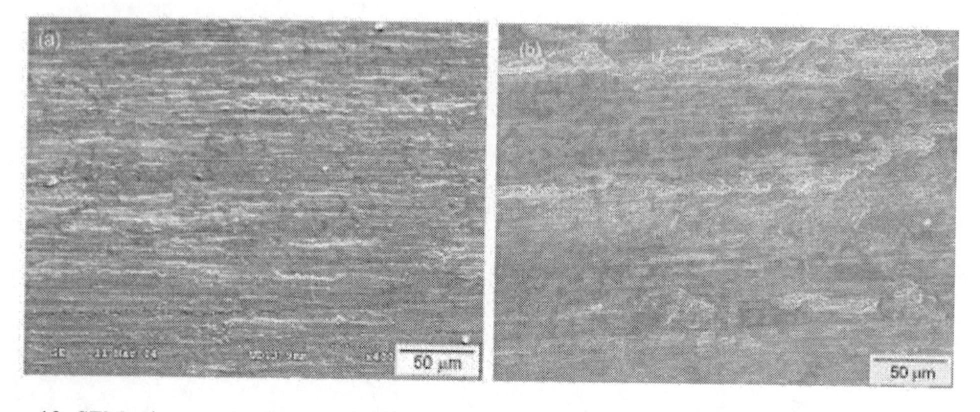

Figure 13. SEM micrographs showing TiC/Ti_3SiC_2 worn surfaces. (a) 20 m/s, 0.2 MPa and (b) 20 m/s, 0.8 MPa [26].

XRD analyses were also used to identify the phase behavior of the frictional films. In fact, it was quite difficult to obtain an XRD pattern of the frictional film without the contribution of the matrix underneath the film because the X-ray could readily penetrate the film and reach into the surface of the matrix. However, a reference method is applicable by comparing the changes in the friction surface before and after a friction test. Figure 14a and b show the results of the XRD analyses for the Ti_3SiC_2 and TiC-Ti_3SiC_2 friction surfaces shown in Figs. 7d and 13b, respectively. In comparison with the original state shown in Figure 1a, the XRD pattern of the Ti_3SiC_2 friction surface exhibited only a very small change, i.e., there was a slight raise in the area of 2θ=27-29° with no evidence showing the presence of Ti, Si and Fe oxides, indicating that the frictional film is amorphous. Otherwise, some diffraction peaks of the oxides should appear. By contrast, the XRD pattern of the TiC-Ti_3SiC_2 friction surface showed the relative intensities of the peaks belonging to the TiC phase significantly increased comparing with the original state shown in Figure 1b. At the same time, some new diffraction peaks appeared, and detailed analyses showed that the new peaks belong to iron titanate ($Fe_{2.25}Ti_{0.75}O_4$) and iron silicate ($FeSiO_4$), respectively, indicating that the frictional film covering the TiC-Ti_3SiC_2 friction surface consisted of co-crystalline or partially co-crystalline iron titanate and iron silicate. The forming of $Fe_{2.25}Ti_{0.75}O_4$ and $FeSiO_4$ could be related to the presence of TiC impurities and a mechanical alloying reaction during the impact of asperities between the friction surfaces of TiC-Ti_3SiC_2 and the steel disk. The increased TiC on the TiC-Ti_3SiC_2 friction surface could be attributed to the accumulation of TiC particulates on the friction surface rather than the products of Ti_3SiC_2 decomposition. Otherwise, the TiC particulates could also appear in the Ti_3SiC_2 friction surface, if the increase of TiC was generated from the decomposition of Ti_3SiC_2.

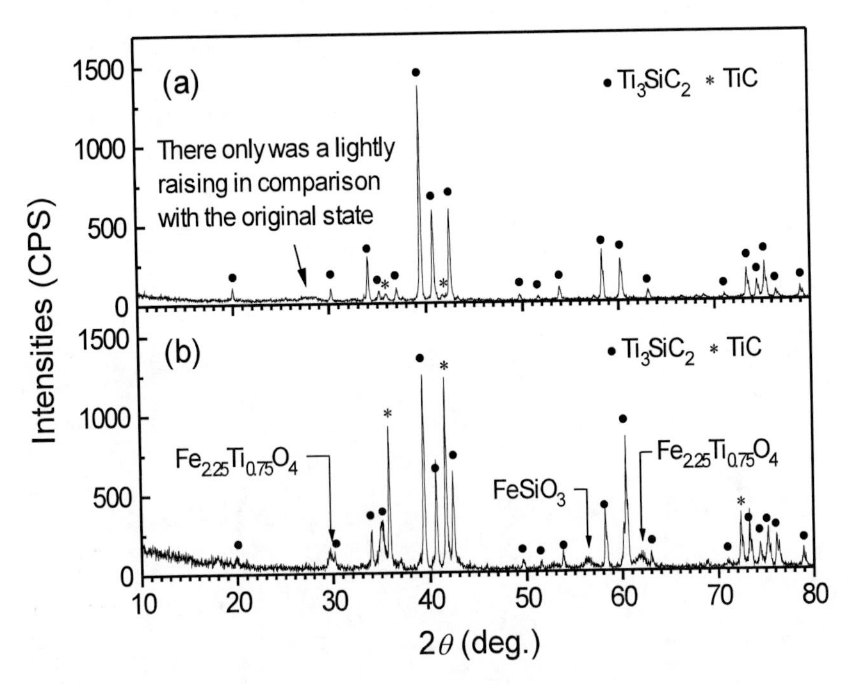

Figure 14. XRD patterns of the worn surfaces; (a) Ti_3SiC_2 and (b) TiC/Ti_3SiC_2 [26].

Although oxide film was also present, the TiC-Ti_3SiC_2 friction surface appeared to be harder, as the rubbing tracks were considerably shallower in comparison with the rubbing tracks on the Ti_3SiC_2 friction surface for the same sliding speed and normal pressures (shown in Figures 7d and 13b). Furthermore, the presence of additional Fe transferred from the low carbon steel disk also reflected that the TiC-Ti_3SiC_2 friction surface was harder. Obviously, the higher hardness in the TiC-Ti_3SiC_2 friction surface can be attributed to the presence of hard TiC impurities; perhaps, the crystallizing behavior of the oxide film is a factor as well.

Figure 15a and b exhibits typical SEM micrographs of the matrix surfaces underneath the frictional oxide film for Ti_3SiC_2 and TiC-Ti_3SiC_2, respectively. In order to reveal these surfaces, the oxide films originally covering the friction surfaces have been removed using HF and HNO_3 mixed acid with a volume ratio of HF:HNO_3:H_2O=1:1:3. The images of the matrix surfaces were apparently different between Ti_3SiC_2 and TiC-Ti_3SiC_2. The matrix surface of Ti_3SiC_2 appeared to be quite smooth although there were some slight ridges, but the matrix surface of TiC-Ti_3SiC_2 was very rough with a large number of TiC grains protruding from the surface. The hard protruding TiC grains would increase the sliding resistance and thus increase the friction coefficient due to their interlocking action on the low carbon steel disk friction surface. In addition, the presence of these protruding TiC grains could impede the flow of oxides making up the frictional oxide film, arresting the oxides around the TiC grains for a longer time, and thus allowing for a relatively static space and sufficient time to crystallize and harden, which also increases the friction resistance due to the loss in the fluidity of the oxides.

Hard TiC particles could have a load-bearing action, which might increase wear resistance on the friction surface, but in fact the wear resistance of the TiC-Ti_3SiC_2 was weakened when the normal pressure was larger than about 0.5MPa, particularly the critical wear pressure which was considerably reduced in comparison with Ti_3SiC_2 (shown in Figure 7). These appearances could be attributed to the pullout of TiC particles, together with the flaking of the frictional oxide film form the friction surface. Figure 16 shows a typical SEM micrograph of a local breakage induced by the pullout of TiC particles and film flaking. A similar status can usually be observed from the friction surface of TiC-Ti_3SiC_2 when the normal pressure is larger. It is conceivable that the hard TiC particles wear with difficulty but are easy to entirely pull out from the sliding friction surface when they bear a stronger shear force parallel to the sliding surface.

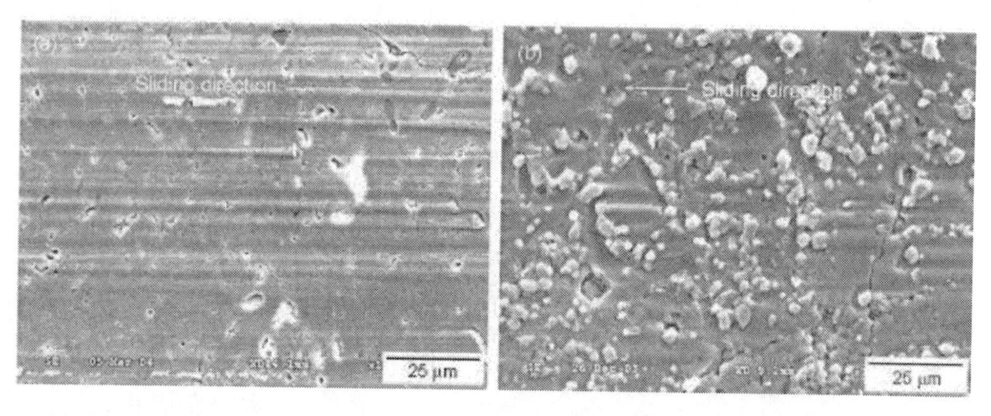

Figure 15. SEM micrographs showing the matrix status underneath the frictional oxide film; (a) Ti_3SiC_2 and (b) TiC/Ti_3SiC_2 [26].

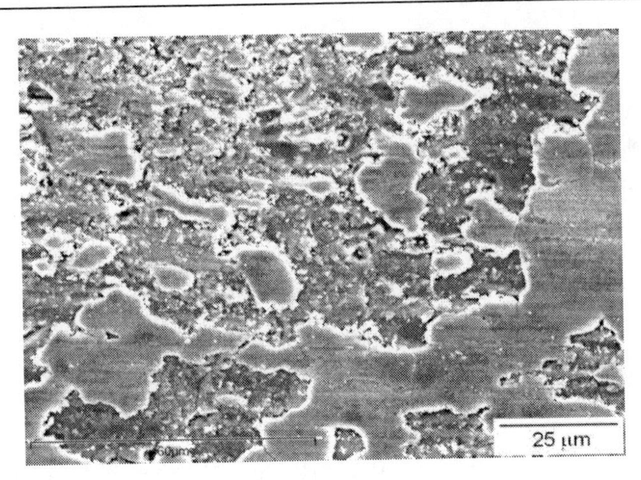

Figure 16. The partial disrepair of the frictional oxides film on TiC/Ti_3SiC_2 worn surface [26].

3.7. Summary and Conclusions

(1) High-purity bulk Ti_3SiC_2 exhibits a changed friction coefficient of 0.09~0.53 and a wear rate of 0.6~2.5×10^{-6} mm^3/Nm for different sliding speeds of 5~60 m/s and normal pressures of 0.1~0.8MPa when it slides dryly against low carbon steel.

(2) The friction coefficient is significantly reduced while the wear rate increases with an increase in the sliding speed for the Ti_3SiC_2; the normal pressure also has a complex but relatively weak influence on them. These changing behaviors can be attributed to the presence and the coverage of a frictional oxide film on the Ti_3SiC_2 friction surface.

(3) The frictional oxide film of Ti_3SiC_2 consists of an amorphous mixture of Ti, Si and Fe oxides which is adhesive during the sliding friction, thus providing a self-antifriction mechanism resulting in the reduction of the friction coefficient and an increase in the wear rate due to its wearing consumption.

(4) TiC-Ti_3SiC_2 exhibits a larger friction coefficient than the high-purity Ti_3SiC_2 in the same test condition. The wear rate is also larger than that of the high-purity Ti_3SiC_2 when the normal pressure is larger than 0.5 MPa. A critical transition occurs at the normal pressure of about 0.2 MPa for a sliding speed of 60 m/s; at this point, both the friction coefficient and the wear rate increase rapidly.

(5) The larger friction coefficient of Tic-Ti_3SiC_2 can be attributed to the interlocking action of the hard TiC impurities, while the larger TiC-Ti_3SiC_2 wear rate is due to the pullout of the TiC particles.

4. TRIBOLOGICAL PROPERTIES OF TI_3ALC_2

4.1. Tribological Behaviors

The friction coefficients as a function of the normal pressure, for different sliding speeds, are shown in Figure 17. It can be clearly seen that the friction coefficient was pressure and/or speed dependent. In the case of 20 m/s, the friction coefficient showed a monotonous decrease from 0.21 to 0.14 with increasing pressure from 0.1 to 0.8 MPa. However, in the case of 40 m/s, the friction coefficient showed a slight rise from about 0.1 with a small increase of the pressure from 0.1 MPa, and then essentially remained constant at about 0.13. When the speed was up to 60 m/s, the friction coefficient showed a further low value of only about 0.07 for the pressure of 0.1 MPa, then rose to about 0.1 with increasing pressure to 0.4 MPa, and then essentially remained 0.1 until 0.8 MPa. It is unusual, in ceramics, that the bulk polycrystalline Ti_3AlC_2 has such a low friction coefficient.

Figure 18 shows the wear rate of Ti_3AlC_2 as a function of the normal pressure, for different sliding speeds. The data were calculated from the weight loss per sliding distance and per normal load (normal pressure × area of friction surface) dividing by the real density of Ti_3AlC_2. In the case of 20 m/s, the wear rate was a sensitive function of the pressure. It exhibited a very low value of only about 0.48×10^{-6} mm^3/Nm when the pressure was 0.1 MPa. However, it was increased to 2.35×10^{-6} mm^3/Nm on increasing the pressure to 0.8 MPa. Being different from such a sensitive increase, when the speed was up to 40 m/s and 60 m/s, the wear rate exhibited only a smaller change with the increasing pressure, and essentially remained at a level of about 2.5×10^{-6} mm^3/Nm. The wear rates were roughly identical for the different speeds when the pressure was larger than 0.7MPa, although they increase with speed for smaller pressures. A wear rate of $\sim2.5\times10^{-6}$ mm^3/Nm seems to be a saturation value for stable friction and wear of Ti_3AlC_2 in terms of the present test conditions.

It is conceivable that such a low friction coefficient and wear rate could have arisen from an unusual state of the friction surface. Figure 19(a) shows a typical SEM micrograph of the friction surface of Ti_3AlC_2 after it underwent a sliding distance of 72000 m (three continuously sliding processes) at a speed of 60 m/s and a pressure of 0.8MPa. It was indeed excellent, as could be remarked, that the surface was covered by a compact, self-generating film, which exhibited a peculiar feature similar to a wet-skidding sod. The film is quite thin and uniform. The inset in Figure 19(a) shows a micrograph exhibiting the cross-section of the film. The thickness of the film was estimated to be about 0.5μm. In fact, the friction surface of the low carbon steel disk was also covered by a frictional film. The typical image is shown in Figure 19(b).

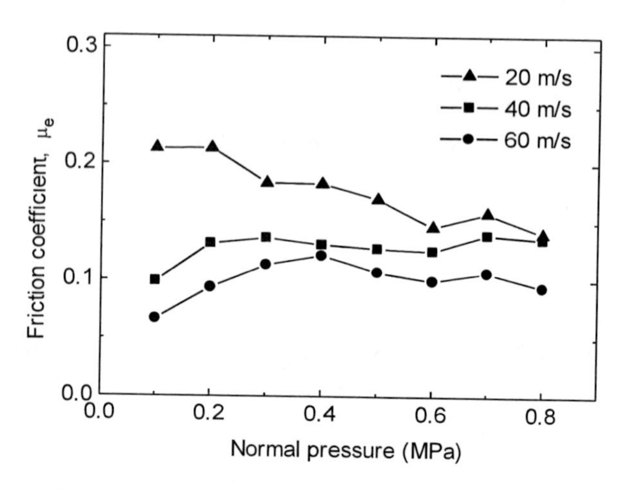

Figure 17. Friction coefficients of Ti_3AlC_2 sliding against low carbon steel in the different sliding speed and normal pressure conditions [32].

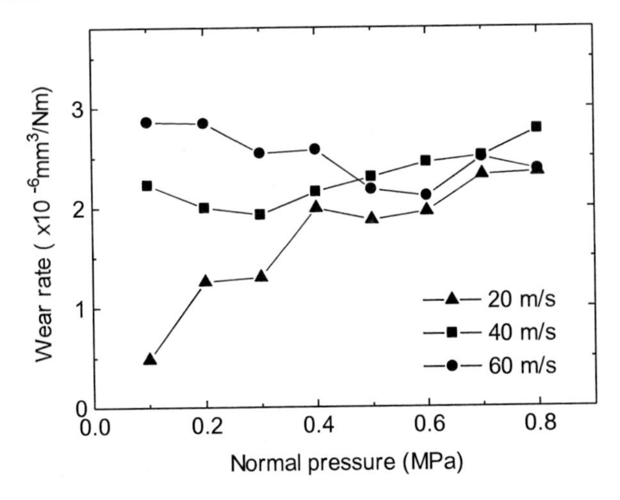

Figure 18. Wear rates of Ti_3AlC_2 sliding against low carbon steel in the different sliding speed and normal pressures [32].

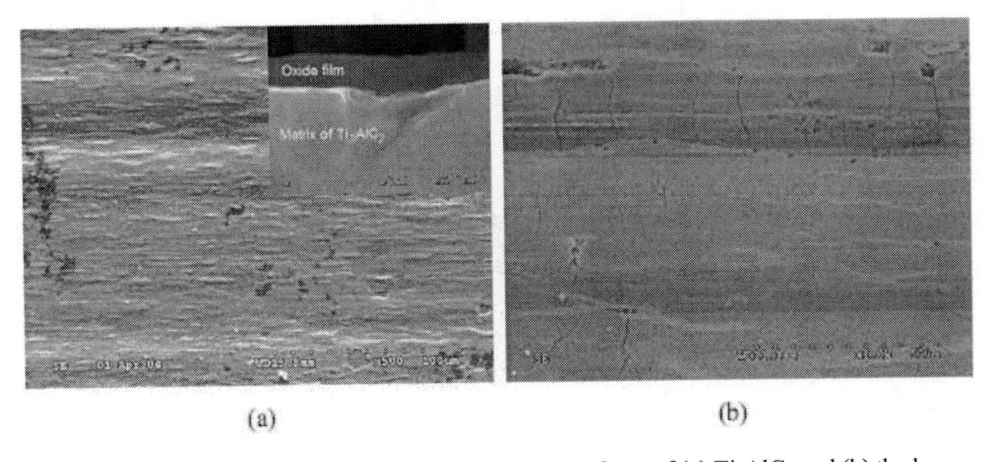

Figure 19. Typical SEM micrographs showing the friction surfaces of (a) Ti_3AlC_2 and (b) the low carbon steel disk, after underwent the sliding distance of 72,000 meters (three continuously sliding process) under the sliding speed of 60 m/s and the normal pressure of 0.8 MPa.

EDS analyses were carried out for identifying compositions of the films. The results are shown in Figures 20(a) and (b) for Ti_3AlC_2 and the disk, respectively. As can be noted, there were only four elements of O, Ti, Al and a small amount of Fe in the Ti_3AlC_2 film, and no carbon constituting Ti_3AlC_2 existed. This demonstrates that the film consists of oxides of titanium, aluminum, and iron. Such a frictional oxide film was first found in bulk polycrystalline Ti_3AlC_2. The film of the disk contained the same four elements of O, Ti, Al and Fe, demonstrating that the film consists of the same oxides, although the contents of the oxides were different between the films of Ti_3AlC_2 and the disk. The peculiar friction and wear properties of Ti_3AlC_2 could be attributed to the presence of unusual films.

The measured microhardness of the surface of Ti_3AlC_2 before and after the friction and wear tests is shown in Figure 21. As can be seen, the hardness is reduced with increasing sliding speed. In the case of 0.2 MPa pressure, the hardness reduced from about 1.48 GPa to 0.93 GPa, corresponding to a speed from 20 to 60 m/s. Similarly, in the case of 0.8MPa pressure, the hardness reduced from about 1.38 to 0.96 GPa. This suggests that the thickness of the oxide film could be increased with increasing sliding speed, because a softer oxide film could contribute more to the measured hardness when the film is thicker, but the harder matrix could make more contributions to the measured hardness when the film is thinner. The typical indentation images are shown in Figure 22. Clearly, for the same indentation load of 0.49 N, the depth of indentation on the surface with a thicker film is larger than that on the surface with a thinner film. It is conceivable that the self-generating oxide film actually played the role of an antifriction material on the friction surfaces, and the antifriction effect was enhanced with increasing sliding speed due to the thickening of the film.

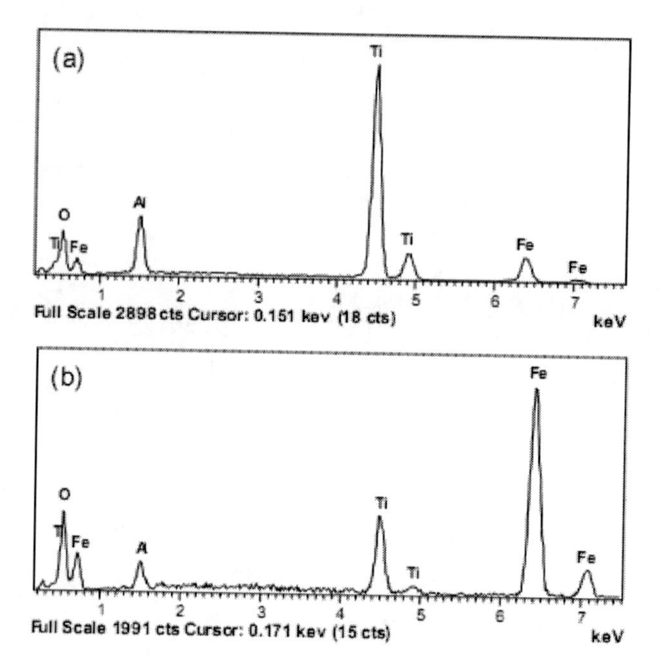

Figure 20. EDS patterns of wear surfaces of Ti_3AlC_2 and the low carbon steel counterpart; (a) Ti_3AlC_2 and (b) the low carbon steel counterpart [33].

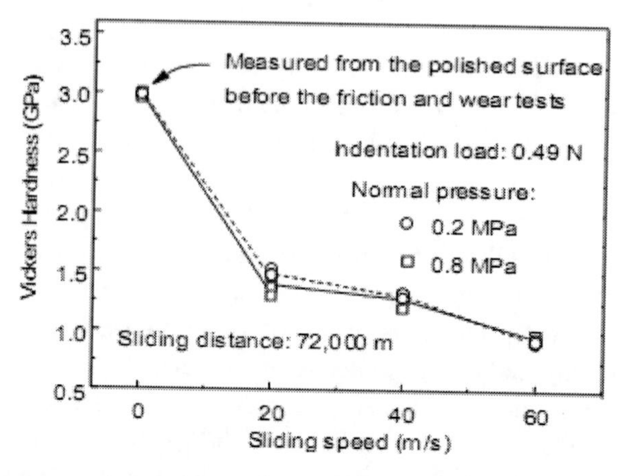

Figure 21. Changes in Vickers hardness of the Ti_3AlC_2 worn surface (including the frictional film) before and after the sliding friction test [32].

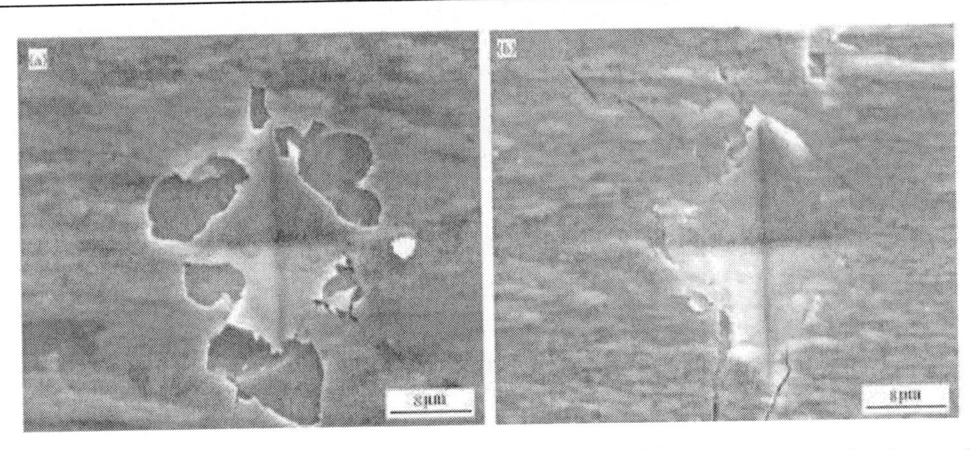

Figure 22. Indentation images of the Ti_3AlC_2 worn surface for the different sliding speed and normal pressure; (a) 20 m/s, 0.8 MPa and (b) 60 m/s, 0.8 MPa [32].

In order to study effects of the oxide film on the friction behavior, the frictional oxide film on the Ti_3AlC_2 friction surface was removed by dissolving $HF:HNO_3 = 1:3$ acid mixture, and the oxide film on the steel disk surface was removed by a 1000 grit SiC abrasive paper. The image of the Ti_3AlC_2 matrix surface underneath the oxide film removed is shown in Figure 23. After removing the oxide film, the friction test was again conducted to examine the change in friction behavior with a new frictional oxide film reforming from the newly exposed Ti_3AlC_2 matrix surface. Figure 24 shows the kinetic coefficient of friction measured at the sliding speed of 20 m/s, after removing the oxide film. The three-recorded data curves exhibit the same trend, that is, the coefficient of friction is reduced from an initial value of about 0.45 to a steady value of about 0.14 with the associated increase of the sliding distance. Such a changing tendency means that the oxide film is reformed gradually on the friction surfaces, and thus leads the coefficient of friction to reduce gradually due to the lubricating effect of the oxide film. This inference is confirmed through the observations for the friction surfaces after each test.

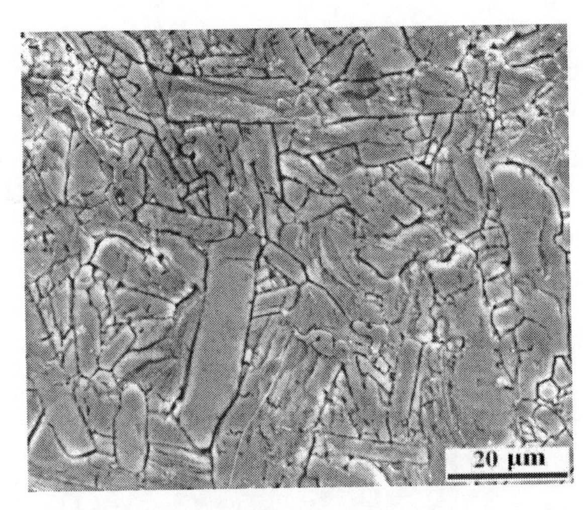

Figure 23. SEM micrograph showing the status of Ti_3AlC_2 matrix underneath the frictional oxide film [33].

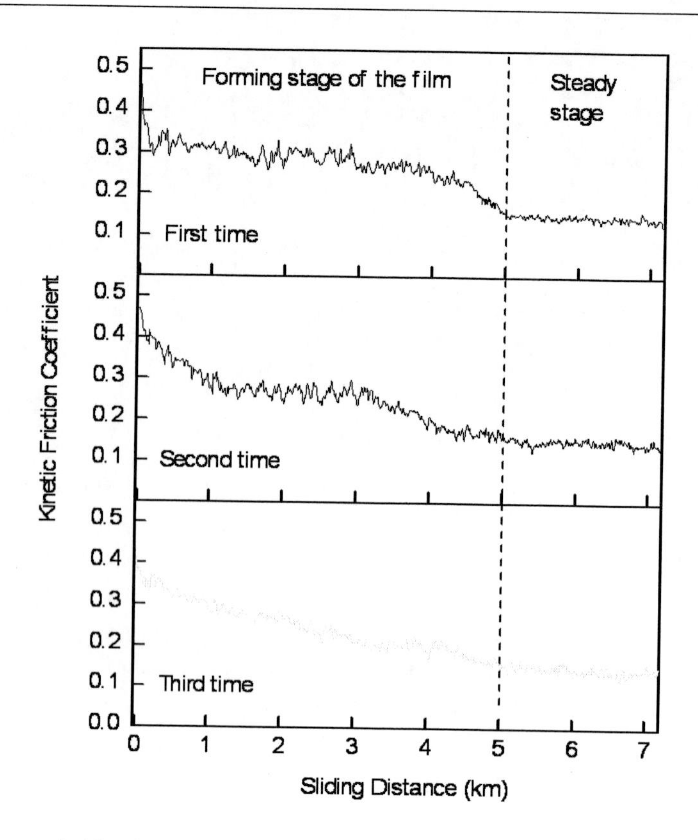

Figure 24. The change in kinetic friction coefficient of Ti_3AlC_2 sliding against the low carbon steel counterpart, after removed the frictional oxide film from the Ti_3AlC_2 worn surface [33].

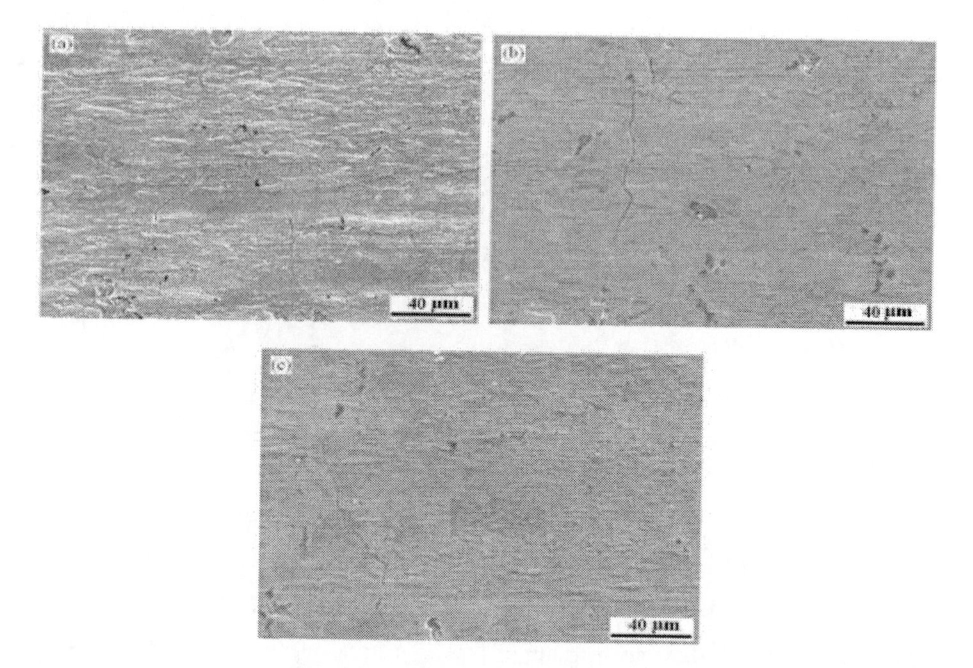

Figure 25. Ti_3AlC_2 worn surface status formed repeatedly after removed the frictional oxide film; (a) firstly removed, (b) secondly removed and (c) thirdly removed [33].

Figure 25(a-c) shows the Ti_3AlC_2 friction surface appears after the first time, the second time, and the third time test, respectively. Essentially, the frictional oxide films reformed in every test processes are identical to that shown in Figure 19(a). It is quite clear that the coefficient of friction has a larger value in the initial stage of the friction process whenever the oxide film is removed. Then, it reduces gradually with the oxide film forming, and finally stabilizes to a steady value. This evidence demonstrates that the quite low friction coefficients, as shown in Figure 17, can be measured only in the presence of a frictional oxide film, which was generated predominantly from the Ti_3AlC_2 friction surface.

4.2. Discussion

Essentially, the titanium and aluminum oxides in the frictional oxide film should be identical to the oxides generated in air at high temperatures [34-35]. In this chapter, we focus on the questions: What is the real morphology of the oxide film and how does it play a lubricating role during the friction behavior process?

To determine the phase state of the oxide film on Ti_3AlC_2 friction surface, XRD analysis has been used (Figure 26). In comparison with Ti_3AlC_2 virginal surface, the X-ray diffraction pattern of the oxide film showed a small change in intensities of the Ti_3AlC_2 phase peaks, while the diffraction background had a small increase, there is no appearing new peak belonging to the either titanium oxide, aluminum oxide, or ferric oxide. This suggested that the oxide film is amorphous. Otherwise, the diffraction peaks of these oxide phases should appear in XRD measurement.

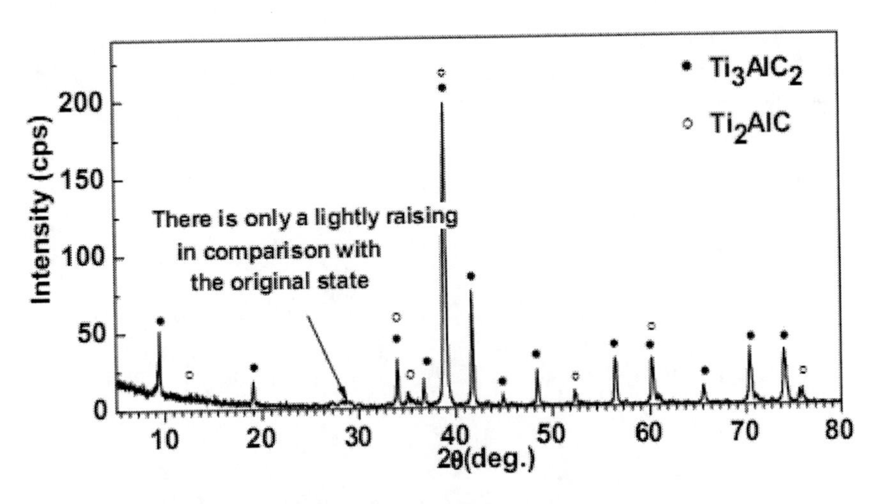

Figure 26. Typical XRD pattern of the Ti_3AlC_2 worn surface [33].

To make a clear understanding of the real status and behavior of the oxide film during sliding friction, the wear debris was carefully collected, and then analyzed by the EDS. The result showed that the chemical compositions of the wear debris were essentially identical to that of the oxide film, indicating that the wear debris is mainly the product of the oxide film, not the spall generated directly from the Ti_3AlC_2 matrix or the low carbon steel disk. This means that the oxide film served as an isolator between the Ti_3AlC_2 surface and the steel disk surface. However, the oxide film surface actually was considerably rougher than the Ti_3AlC_2

matrix surface. A rougher surface would have a larger coefficient of friction in terms of the tribological principle, but the result in our experiment was reversed. A reasonable inference is that the oxide film is in a liquid stage during the sliding friction, and the image observed after the friction test is only a freezing stage of the oxide film. In fact, if the oxide film is not in the liquid stage, it could not form the unusual surface pattern, like plastered viscous slurry.

In order to validate the fusibility of the oxide film, the fusion point of the wear debris was identified via the TG and DSC analysis. Figure 27 is the analyzed result. There is an endothermic peak in the range of 238-314 °C, indicating that the oxide film would be fused when the temperature of the friction surface reaches or exceeds 238 °C. To further validate this point, the collected wear debris was cold-pressed into a flake and then sintered in a muffle furnace. The sintering temperature was selected as 300 °C and held for 10 min, according to the TG and DSC analysis result. Figure 28 (a and b) are images of the wear debris before and after sintering. Figure 28 (b) shows that most of the wear debris is fused and conglutinated together, demonstrating that the oxide film would be in a fused state so long as the friction surface temperature reaches 300 °C. Such a temperature can be easily reached in the course of the friction behavior, at least, in some areas where the asperities generate drastic impact; otherwise, the oxide could not be generated. Clearly, such a liquid oxide film must have a significant lubricating effect on the friction surface, and consequently contribute to the low friction surface, and consequently contribute to the low friction coefficient, and small wear rates for either Ti_3AlC_2 or its friction counterpart.

Since the forming of the oxide film depends upon an adequate temperature on the friction surface, a higher sliding speed, which generates more frictional heat, would be more advantageous for the formation of the oxide film, and it thus leads the friction coefficient to reduce accordingly. Indeed, this has been shown in Figure 17, where the friction coefficient reduces with the increase of the sliding speed. However, due to the extrusion and scrape actions between the friction surfaces, the thickness of the oxide film is restricted, and the redundant oxides will be removed as wear debris. Thus, the wear rate of Ti_3AlC_2 is correspondingly increased with the sliding speed, because the redundant oxides are removed continuously as the wear debris, which has been shown in Figure 17.

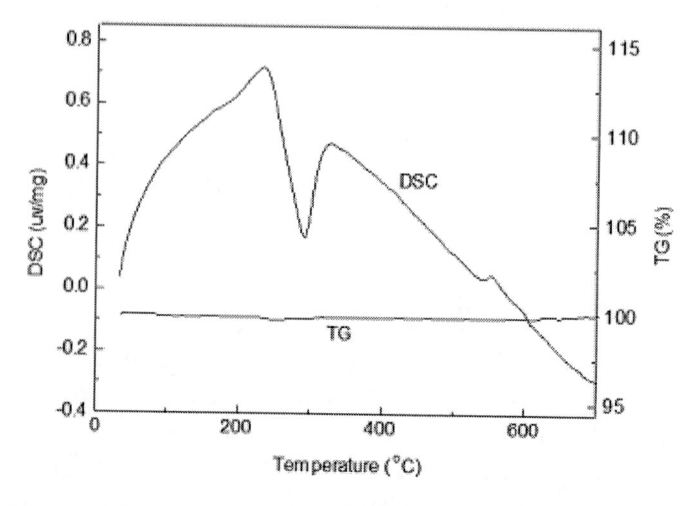

Figure 27. Analysis curves of the thermo-gravimetric (TG) and differential scanning calorimetry (DSC) for the wear debris [33].

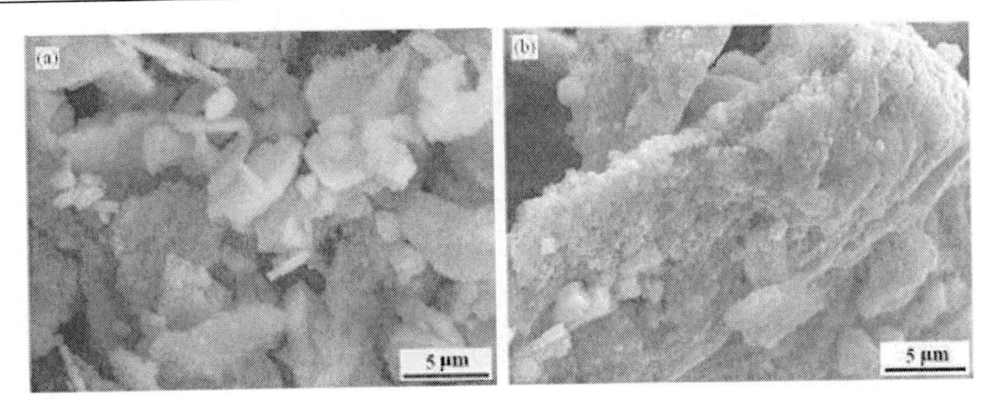

Figure 28. Change in the configuration of the wear debris before and after heating at 300 °C; (a) before and (b) after [33].

The transition behavior and the reducing of the fluctuation in the kinetic friction coefficient with the forming of the oxide film, as shown in Figure 24, is additional evidence. The results reveal that the oxide film can be formed and acts as a significant lubricating material.

Additional studies are needed to investigate the mechanism and influencing factors on the formation of the oxide film. In metal-to-metal friction, the oxidation behavior has always been ascribed to the high flash temperature because of the drastic impact of asperities [15]. However, for most ceramics, they have good oxidation resistance, and it is quite difficult for them to be oxidized during the friction process. In our experiment, the oxidation behavior occurrence was considered in correlation with the actual presence of the oxide film. There are two possibilities. First, the instantaneous temperature at some local areas of the friction surface would be quite high due to the drastic impact of asperities between the friction surfaces, and such high temperature is sufficient to induce the oxidation of Ti_3AlC_2. The second possibility is that the oxidation of Ti_3AlC_2 may not be independent during sliding friction. At high sliding speed, the wrapped oxygen gas and the mechanical alloying between the transferred materials and the Ti_3AlC_2 surface may accelerate Ti_3AlC_2 to be oxidized and further form a frictional oxide film. Thus, it may not need to reach the critical value at which the oxidation of the Ti_3AlC_2 occurs in a static state. In fact, the second possibility should be given more attention in tribological research. Additional research is in progress about this point.

4.3. Conclusions

The friction and wear behavior and the mechanism of Ti_3AlC_2 sliding dryly against low carbon steel have been investigated. The following conclusions can be drawn:

(1) The highly pure, dense polycrystalline bulk Ti_3AlC_2 has quite a low coefficient of friction and wear rate in comparison with most advanced ceramics.
(2) The low coefficient of friction and the wear rate are attributed to the presence of a frictional oxide film generated from the Ti_3AlC_2 friction surface.

(3) The oxide film consists of an amorphous mixture of Ti, Al and Fe oxides, which are in a fused state on the friction surface during the sliding friction.
(4) Increasing the sliding speed results in the reducing of the friction coefficient and the increasing of the Ti_3AlC_2 wear rate, for the increased frictional heat induces more oxides to be generated and consumed.

REFERENCES

[1] M.W. Barsoum, The $M_{N+1}AX_N$ phases: A new class of solids. *Prog. Solid State Chem.* 28 (2000) 201-281.

[2] T. El-Raghy and M.W. Barsoum, Processing and mechanical properties of Ti_3SiC_2: I, reaction path and microstructure evolution. *J. Am. Ceram. Soc.* (1999) 82 [10]: 2849-2854.

[3] J.F. Li, W. Pan, F. Sato, and R. Watanabe, Mechanical properties of polycrystalline Ti_3SiC_2 at ambient and elevated temperatures. *Acta. Mater.* (2001) 49: 937-945.

[4] T. El-Raghy, A. Zavaliangos, M.W. Barsoum, S. Kalidinidi, Damage mechanisms around hardness indentations in Ti_3SiC_2. *J. Am. Ceram. Soc.* 80 (1997) 513-516.

[5] S.B. Li, L.F. Cheng, L.T. Zhang. Identification of damage tolerance of Ti_3SiC_2 by hardness indentations and single edge notched beam test. *Mater. Sci and Tech.* (2002) 18 [2]: 231-233.

[6] M.W. Barsoum, L. Farber and T. El-Raghy. Dislocations, kink band, and room-temperature plasticity of Ti_3SiC_2. *Met. Mater. Trans. A* (1999) 30: 1727-1738.

[7] T. El-Raghy, A. Zavalliangos, M.W. Barsoum *et al.* Damage mechanisms around hardness indentations in Ti_3SiC_2. *J. Am. Ceram. Soc.* (1997) 80 [2]: 513-516.

[8] I.M. Low, S.K. Lee, and B.R. Lawn. Contact damage accumulation in Ti_3SiC_2. *J. Am. Ceram. Soc.* (1998) 81 [1]: 225-228.

[9] M.W. Barsoum, T. El-Raghy, C.J. Rawn *et al.* Thermal properties of Ti_3SiC_2. *J. Phys. Chem. Solids* (1999) 60: 429-439.

[10] M.W. Barsoum, H.I. Yoo, I.K. Polushina *et al.* Electrical conductivity, thermopower and Hall effect of Ti_3AlC_2, Ti_4AlN_3 and Ti_3SiC_2. *Phys. Rev. B* 62 (2000):10194-10198.

[11] S. Myhra, J.W.B. Summers, E.H. Kisi. Ti_3SiC_2 - a layered ceramic exhibiting ultra-low friction. *Mater. Lett.* (1999) 39: 6-11.

[12] T. El-Raghy, P. Blau, M.W. Barsoum. Effect of grain size on friction and wear behavior of Ti_3SiC_2. *Wear* (2000) 238: 125-130.

[13] Z.M. Sun, Y.C. Zhou. Tribological behavior of Ti_3SiC_2-based material. *J. Mater. Sci. Tech.* (2002) 18: 142-145.

[14] S.C. Lim, M. F. Ashby and J. H. Brunton, The effects of sliding conditions on the dry friction of metals. *Acta Metall.* 37 (1989) 767-772.

[15] S.C. Lim and M. F. Ashby, Wear-mechanism maps. *Acta Metallurgica* 35 (1987) 1-24.

[16] S. Wilson and A.T. Alpas, Thermal effects on mild wear transitions in dry sliding of an aluminum alloy. *Wear* 225-229 (1999) 440-449.

[17] D.A. Rigney Transfer, mixing and associated chemical and mechanical processes during the sliding of ductile materials. *Wear*, 2000; 245: 1-9.

[18] D.A. Rigney, Comments on the sliding wear of metals. *Tribology International*, 1997; 30 [5]: 361-367.

[19] H.X. Zhai, Z.Y. Huang, Y. Zhou, Z.L. Zhang, Y.F. Wang, M.X. Ai, Oxidation layer in sliding friction surface of high-purity Ti_3SiC_2. *J. Mater. Sci.* 2004; 39: 6635-6637

[20] Z.F. Zhang, Z.M. Sun, H, Hashimoto, T, Abe, A new synthesis reaction of Ti_3SiC_2 from Ti/$TiSi_2$/TiC powder mixtures through pulse discharge sintering (PDS) technique. *Mat. Res. Innovat*, 2002; 5: 185-189.

[21] A. Zhou, C. Wang, Y. Huang, A possible mechanism on synthesis of Ti_3AlC_2. *Mater. Sci. Eng. A*, 352, 333-39 (2003).

[22] N.V. Tzenov, M.W. Barsoum, Synthesis and characterization of Ti_3AlC_2. *J. Am. Ceram. Soc.*, 83 [4] 825–32 (2000)

[23] H.X. Zhai, Z.Y. Huang, Y. Zhou, Z.L. Zhang and S.B. Li. Ti_3AlC_2 — a soft ceramic exhibiting low friction coefficient. *Mater. Sci. Forum*, (2005) 475-479: 1251-1254.

[24] H.X. Zhai, Z.Y. Huang. Instabilities of sliding friction governed by asperity interference mechanisms. *Wear* 257 (2004) 414-422.

[25] Z.Y. Huang, H.X. Zhai, M.L. Guan, W. Zhou, M.X. Ai, Y. Zhou and S.B. Li. Oxide-film-dependent tribological behaviors of Ti_3SiC_2. *Wear* 262 (2007) 1079- 1085.

[26] H.X. Zhai, Z.Y. Huang, M.X. Ai, Y. Zhou *et al.* Tribological behaviors of bulk Ti_3SiC_2 and influences of TiC impurities, *Mater Sci and Eng A*. 435-436 (2006) :360-370.

[27] H.X. Zhai, Z.Y. Huang, Y. Zhou, Z.L. Zhang, Y.F. Wang, Frictional layer and its anti-friction effect in high-purity Ti_3SiC_2 and TiC-contained Ti_3SiC_2. *Key Eng. Mater.* 280-283 (2005) 1347-1352.

[28] Z.Y. Huang, H.X. Zhai, Y. Zhou, Y. F. Wang, Z.L. Zhang, Sliding Friction behavior of bulk Ti_3SiC_2 under different normal pressures. *Key Eng. Mater.* 280-283 (2005) 1353-1356.

[29] S.B. Li, L.F. Cheng, L.T. Zhang. Oxidation behavior of Ti_3SiC_2 at high temperature in Air. *Mater. Sci and Eng. A*. (2002) 341 [1-2]: 112-120.

[30] X.H. Wang, Y.C. Zhou. Oxidation behavior of Ti_3A1C_2 at 1000-1400℃ in air. *Corros Sci*. (2003) 45: 891-907.

[31] J. R. Barber, Thermoelastic instabilities in the sliding of conforming solids. *Proc. Roy. Soc. A*, 312(1969) 381-394.

[32] H.X. Zhai, Z.Y. Huang, M.X. Ai, Y. Zhou, Z.L. Zhang and S.B. Li. Tribophysics properties of bulk polycrystalline Ti_3AlC_2. *J. Am. Ceram. Soc.* (2005) 88 [11]: 3270-3274.

[33] Z.Y. Huang, H.X. Zhai, W. Zhou, M.X. Ai. Tribological behaviors and mechanisms of Ti_3AlC_2. *Tribol. Lett.* 27 (2007): 129-135.

[34] M.W. Barsoum. Oxidation of $Ti_{n+1}AlX_n$ (n = 1-3 and X = C, N) Part II. Experimental results. *J. Elec. Soc.* (2001) 148[8]: 551-562.

[35] X.H. Wang, Y.C. Zhou. Oxidation behavior of Ti_3AlC_2 powders in flowing air. *J. Mater. Chem.* (2002) 12 [9]: 2781-2785.

In: MAX Phases: Microstructure, Properties and Applications ISBN 978-1-61324-182-0
Editors: It-Meng (Jim) Low and Yanchun Zhou

Chapter 14

HIGH TEMPERATURE OXIDATION CHARACTERISTICS OF CR_2ALC

Dong Bok Lee
School of Advanced Materials Science and Engineering, SungkyunkwanUniversity, Suwon 440-746, South Korea

ABSTRACT

The successful deployment of Cr_2AlC requiresthe understanding of the high-temperature oxidation characteristics of Cr_2AlC such as theoxidation kinetics,the mechanism, and the oxide scales formed. Hence, the high-temperature isothermal and cyclic oxidation behavior of Cr_2AlC is described. Cr_2AlCoxidizes according to the following equation during the isothermal oxidation between 900 and 1300°C in air, *i.e.*, $Cr_2AlC + O_2 \rightarrow$ (Al_2O_3oxide layer) +(Cr_7C_3sub-layer) + (CO or CO_2gas). The oxide scale consists primarily of the Al_2O_3 barrier layerthat forms by the inward diffusionof oxygen. The consumption of Al to form the Al_2O_3 leads to the enrichment of Cr immediately below the Al_2O_3 layer, resulting in the formation of the Cr_7C_3sublayer. At the same time, carbon escapes from Cr_2AlC as CO or CO_2gas into the air.During the cyclic oxidation between 900 and 1100°C in air, Cr_2AlCsimilarly oxidizes according to the equation; $Cr_2AlC + O_2 \rightarrow$ (Al_2O_3oxide layer) + (Cr_7C_3sublayer) + (CO or CO_2gas). However, during the cyclic oxidation at 1200 and 1300 °C, Cr_2AlCoxidizes according to the equation, $Cr_2AlC + O_2 \rightarrow$ ($Al_2O_3/Cr_2O_3/Al_2O_3$ triple oxide layers) + (Cr_7C_3sub-layer) + (CO or CO_2gas), because the thermo-cyclingsfacalitate the formation of the intermediate Cr_2O_3-rich layer between the outer Al_2O_3-rich layer andthe inner Al_2O_3-rich layer.

1. ISOTHERMALOXIDATIONBEHAVIOR OF CR_2ALC

Cr_2AlC isa relatively new ternary carbide. Like metals, itis relatively soft, damage tolerant,thermal shock resistant, has good electrical and thermal conductivities, and good machinability. Like ceramics, it displays excellent heat and chemical resistance, elastic stiffness, maintains strength at high temperatures, and has a high melting point.These unique combinations of metallic and ceramic properties make it a potential candidate as high-temperature structural components. Hence, it is important to understand the oxidation behavior of Cr_2AlC [1-6].

For the oxidation study described here,Cr (< 45 μm, 99.9 % purity) and graphite (~ 10 μmϕ, 99.95% purity) powders were mixed at a molar ratio of Cr:C = 2:1 in a SPEX™ shaker mill for 10 min in an Ar atmosphere, pressed into a cylindrical green pellet by uniaxial pressing at 20 MPa. The pellet was then heated to 1550°C for 1 hr under a vacuum of 1.3 Pa for conversion into CrC_X(x=0.5). The partially sintered bulk CrC_X was ground, and screened to sizes < 45 μm in diameter. The $CrC_X$and Al (<45 μmϕ, 99.7% purity) powders were mixed at a molar ratio of 2:1 in a SPEX™ shaker mill for 10 min in an argon atmosphere, and hot pressed at 1300°C to 19 mmϕ x 10 mm under a pressure of 25 MPa for 1 hr in a flowing argon gas. During hot pressing, a reaction occurred between the CrC_X and Al powders to produce bulk Cr_2AlC. The average grain size of the synthesized Cr_2AlC was about 17 μm. The measured density was 5.2 g/cm^3, with a relative density of more than 99% of the theoretical value.

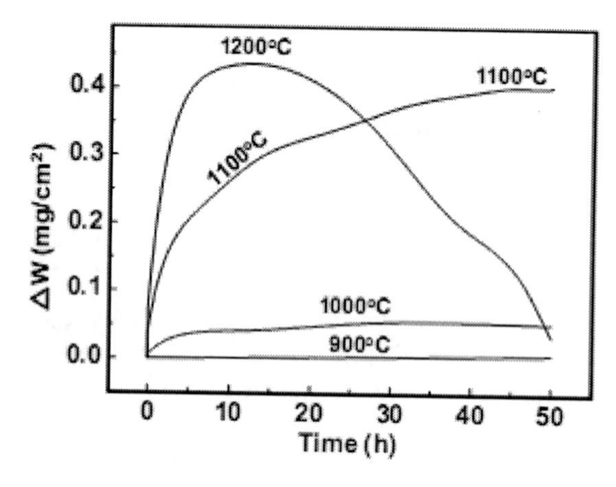

Figure 1.Oxidation curves of Cr_2AlC in air at 900, 1000, 1100, and 1200 °C.

Figure1shows the weight gain versus the oxidation time curves of Cr_2AlC at temperatures between 900 and 1200°C in air. The total weight gains wererelatively small owing to the formation of the protective alumina scale on the surface, reflecting the good oxidation resistance of Cr_2AlC. Chromium in Cr_2AlC does not easily oxidize owing to its thermodynamic nobility.It is well known that chromia has a high vapor pressure at high temperatures. The oxidation rates increased with increasing oxidation temperature. Factors that affect the oxidation curves are; (1) carbon escape from Cr_2AlC in the form of CO or CO_2,(2) preferential oxidation of Al in Cr_2AlC, (3) partial breakage or spalling and subsequent healing of the alumina scale formed, especially at higher temperatures, and (4) experimental error arisen by the quite slow oxidation rate, especially at lower temperatures. The 1200°C curve displays weight losses after about 10 h, due mainly to the evaporation of carbon. The higher the temperature is, the more severe the vaporization of carbon is. Similarly, the 1100°C curve negatively deviated from the familiar parabolic rate law. The overall oxidation rates of Cr_2AlC are roughly comparable with Al_2O_3-forming kinetics. Several types of Al_2O_3 have been reported for alloys forming alumina layers. At high temperatures the metastable,transientoxides γ , δ,and θ-Al_2O_3 transform to the α-Al_2O_3 according to the following sequence [7]:

γ-Al_2O_3(cubic spinel)→δ-Al_2O_3(tetragonal)→θ-Al_2O_3(monoclinic)→α-Al_2O_3 (rhombohedral)

Forβ-NiAl around 800 °C,the metastable (γ, δ)-Al_2O_3 are rapidly converted into θ-Al_2O_3 having a cation-vacancy network, and thereby the scale grows by outward transport of Al through both cation lattice and surface diffusion. In this way voids can be produced at the oxide-metal interface, resulting in a deterioration of the adherence of the oxide layer on the substrate. Less dense and faster growing θ-Al_2O_3 scales having a blade-like or whisker-like morphology transforms into highly stoichiometric and slower growing α-Al_2O_3 at higher temperatures, or in the later stage of oxidation. For example, theθ→ α-transition occursfor β-NiAl at 1000°C in less than 10 h, being accompanied by a change in transport properties of the scale. The growth of the α-layer is mainly governed by anion-diffusion, so that the previously formed voids are filled with oxide. This improves the adhesion of the oxide layer to the substrate. Therefore, the formation of the α-layer is preferred [8].

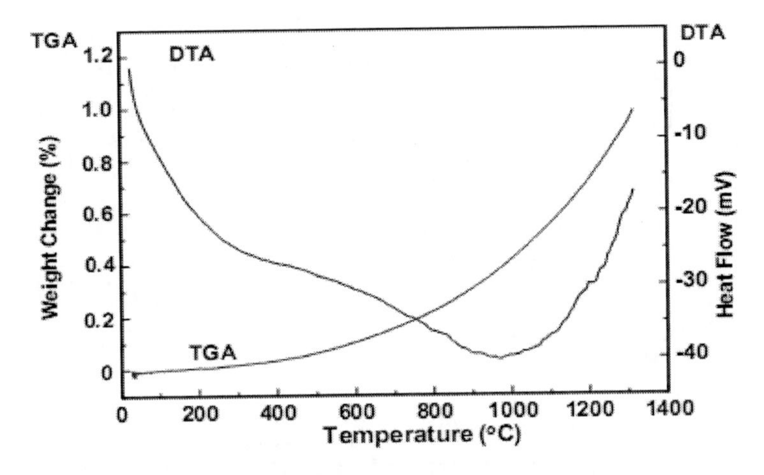

Figure 2.TG-DTA curves of Cr_2AlC, which were obtained during heating from room temperature to 1315 °C in air with a heating rate of 10 °C/min.

Figure 2 shows the TG-DTA analytical results for Cr_2AlC. The TGA curve indicates that the oxidation rate increased in proportion to the test temperature. The DTA curve indicates that an endothermic reaction was predominant below 970°C due to heating of the sample, whereas the exothermic reaction was predominant above 970°C due to oxidation of Cr_2AlC. This indicates that Cr_2AlCbegan to oxidize noticeably above 970°C. On the other hand, the TG-DSCcurves for the oxidation of Cr_2AlC powders indicated that the oxidation began to take place by the up-take of oxygen from 800°C [3].

Table 1 lists the average thickness of the oxide scale formed on Cr_2AlC. Generally, thin scales formed, because the continuous, stable α-Al_2O_3 layer decreased the oxidation rate and the transport process in the growing scale.The scale thickened progressively with increasing temperature and time. The samples listed in Table 1 were investigated by the XRD technique, as follows.

Table 1.Average thickness of the oxide scale formed on Cr_2AlCas a function of the oxidation temperature and time

	900°C/80 h	1000°C/30 h	1100°C/480 h	1300°C/15 h	1300°C/100 h
Thickness (μm)	0.8	1.0	3.4	8.8	27.1

Figure3 shows the XRD patterns of the oxide scales formed on Cr_2AlC. In Figure3(a), the major phase was the matrix with a minor Cr_7C_3 phase. No Al_2O_3 peakswere detected,owing to its thinness. In Figure3(b), the Cr_2AlC peaks werestrong, because the α-Al_2O_3scale and the Cr_7C_3sub-layer were still thin. As the oxidation temperature and time were increased, the intensities of the diffraction peaks for Cr_2AlC became decreased with a corresponding increase in the relative peak intensities for Cr_7C_3 and α-Al_2O_3.

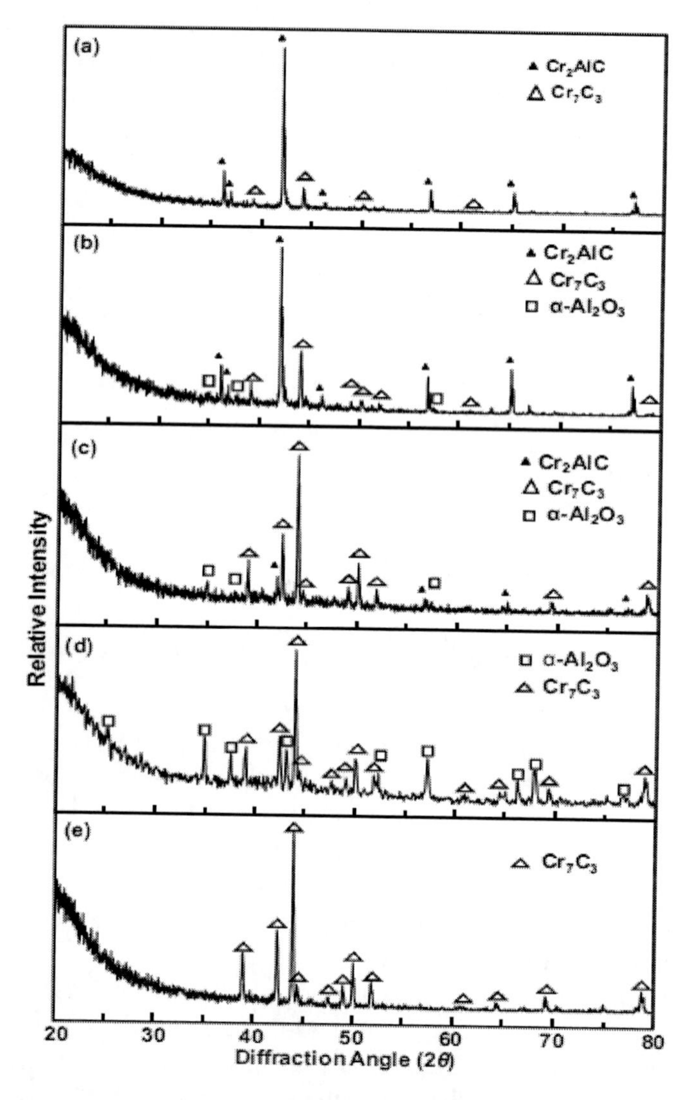

Figure 3.XRD patterns of the oxide scales formed on Cr_2AlC. The air-oxidation conditions were; (a) at 900 °C for 80 h, (b) at 1000 °C for 30 h, (c) at 1100 °C for 480 h, (d) at 1300 °C for 15 h,(e) at 1300 °C for 100 h.

In Figure 3(c), the major phase was the Cr_7C_3sub-layer that formed immediately below the α-Al_2O_3 surface layer. The diffraction pattern of α-Al_2O_3 was not so distinct owing to its partial spallation. Generally, the main weakness of the α-Al_2O_3 scales is their susceptibility to cracking and spalling. Scale failure might be caused by the buildup of growth stress caused by the anisotropic volume expansion during oxidation and thermal stress caused by differences in the thermal expansion coefficients between the matrix and the oxide upon coolingfrom the oxidation temperature. In Figure 3(d), the Cr_2AlC matrix phase disappeared completely, because a thick Cr_7C_3sub-layer formed and the slowlygrowing α-Al_2O_3 surface layer was partially spalled off.In Figure 3(e), only Cr_7C_3 was detected because the Al_2O_3 surface layer hadcompletely spalled off. The absence of chromium oxides in Figure 3 is due to the followings. (1) $Cr_2O_3$is below the detection limit of the XRD method, (2) Cr_2O_3 evaporates considerably as CrO_3 (g) above 1000°C,and, the most likely, (3) Cr_2O_3dissolves in Al_2O_3.

$Cr_2O_3$is thermodynamically less stablethanα-Al_2O_3.Also, Cr atoms in Cr_2AlC are bound to the carbon atoms asCr_2C layers by covalent-ionic bonding, which are interleaved with layers of Al. This strong Cr-C interaction decreases the activity of Cr in Cr_2AlC. Hence, aluminum oxidizesselectively to form the Al_2O_3 layer.The oxidation resistance of Cr_2AlC depends on the continuity and compactness of the Al_2O_3 layer.The preferential oxidation of Al causes Al-depletion underneaththe surface, promoting the formation of the Cr_7C_3sub-layer immediately below the Al_2O_3 layer.Cr_7C_3 is hard, and its stability arises from strong covalent-ionic bonding of Cr-C-Cr chains. The strong Cr-C bonding in the Cr_2AlC matrix and in the Cr_7C_3sub-layer makes Cr difficult to oxidize into a Cr_2O_3layer. Cr_2O_3, when formed to a small amount on the surface, tends to form a Cr_2O_3-Al_2O_3solid solution,because both oxides have the corundum structure.

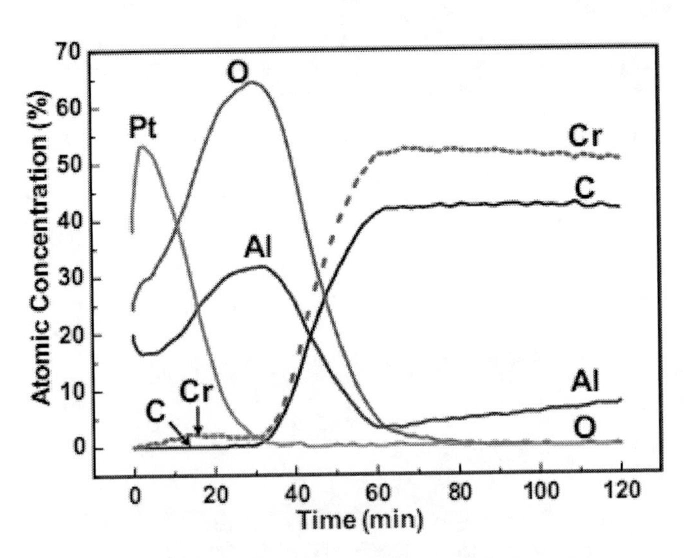

Figure 4.AES depth profiles of Cr_2AlC taken after oxidation at 1100 °C for 5 min in air. The penetration rate is 18 nm/min for the reference SiO_2.

Figure4 shows the AES depth profiles of oxidized Cr_2AlC, which were obtained in order to understand the oxidation mechanism during the initial oxidation stage. This inert marker experiment was performed by the sputter-deposition of a thin Pt film on top of the Cr_2AlCprior to its oxidation. It can be seen that some Al diffused out toward the surface, although most of the Al was oxidized by the inwardly-transported O^{-2} ions to Al_2O_3. This is consistent with the fact that alumina grows primarily through the inward progression of O^{-2} ions along the oxide grain boundaries, although there is some component of outward growth as well [9]. $Al_2O_3$is highly stable, and exhibits low diffusivities for both cations and anions.There was a small amount of dissolved Cr ions in the Al_2O_3.Carbon diffused out from the surface, while oxygen diffused inward.

Figure 5(a) shows the SEM top view of the oxide scale formed on Cr_2AlC after oxidation at 900°C for 50 h in air. Micrometer-size oxide blades were scattered over the thin surface oxide layer. Figure 5(b) shows another area where oxide blades were present in high density. The non-uniformity of these oxide blades originated mainly from the anisotropy of the matrix grains, and the variation in the purity of each matrix grain. The EDS spectrum shown in Figure 5(c) indicates that the oxide blades were Al_2O_3. These blades were frequently found, even though their amount varied for each oxidation condition. The Cr peak may have originated from the matrix. Chromium accelerates the transformation of metastable aluminasto stable α-Al_2O_3[9].The Pt peak originated from Pt deposited for SEM investigations.

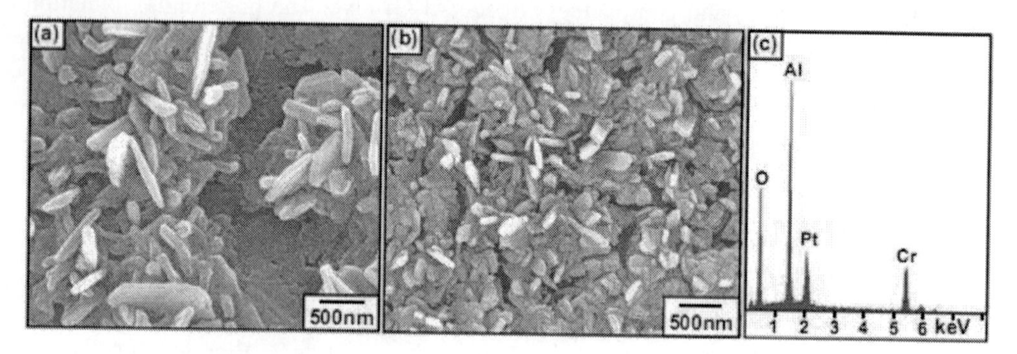

Figure 5.Oxide scales formed on Cr_2AlC after oxidation at 900 °C in air. (a) SEM top view (50 h), (b) SEM top view (80 h), (c) EDS spectrum of (b).

Figure 6 shows the SEM fracture surface of the oxide layer that formed after oxidation at 1000 °C for 50 h. The oxide layer with a thickness of 1 μm consisted of tiny, round Al_2O_3 crystallites, which later grew to coarser ones as oxidation progressed.

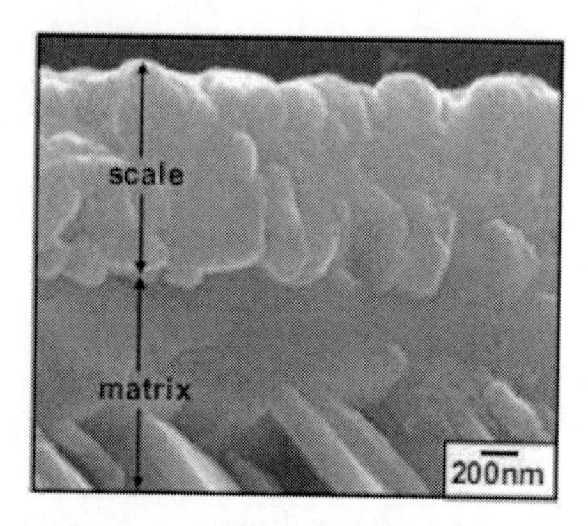

Figure 6.SEM fracture surface of the oxide scale formed on Cr_2AlC after oxidation at 1000 °C for 50 h in air.

Figure 7(a) shows some oxide blades that formed on the surface after oxidation at 1100°C for 50 h. The flat, background oxide layer was enlarged in Figure 7(b). Here, sub-microcrystalline Al_2O_3 grains can be seen, which are mainly due to their slow growth rate even at 1100°C. When oxidized at 1100°C for 480 h, the Al_2O_3 oxide layer that was 3.4 μmthick partiallyspalled out during the cooling stage, which exposed the Cr_7C_3sub-layer, asshown in Figure 7(c). A trace of the polyhedral Al_2O_3 grains was imprinted on the Cr_7C_3sub-layer. The EDS spectrum shown in Figure 7(d) indicates that the Cr_7C_3 layer shown in Figure 7(c) contained some dissolved oxygen.

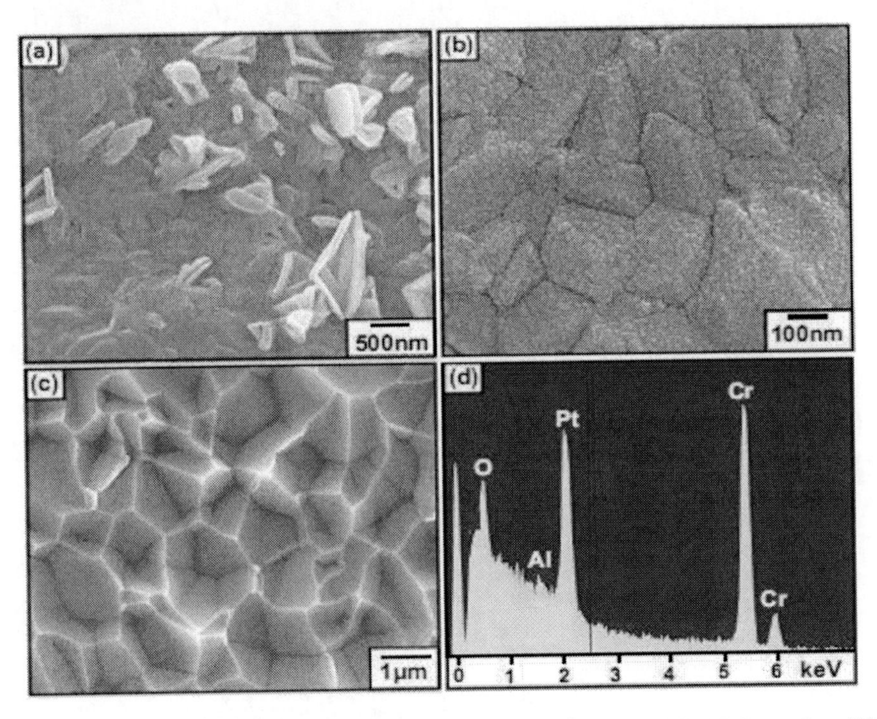

Figure 7.Oxide scales formed on Cr_2AlC after oxidation at 1100 °C in air. (a) SEM top view (50 h),(b)enlarged SEM top view of the α-Al_2O_3 grains shown in (a), (c) SEM top view of the Cr_7C_3 layer (480 h), (d) EDS spectrum of (d).

Figure 8 shows the EPMA results of the oxide scale formed on Cr_2AlC after oxidation at 1200°C for 50 h. The Al_2O_3scale was 5.7 μmthick.There was very little Cr and C in the Al_2O_3 layer.The Cr_7C_3 sub-layer with approximately 3~7 μm in width was depleted in Al, and enriched in Cr.The rate of Cr_7C_3 formation was not constant for every matrix grain due to the layered structure of Cr_2AlC. Cracks developed mainly in the lower part of the Al_2O_3 layer as well as in the Cr_7C_3 sub-layerbecause of the escape of carbon, the loss of aluminum in the Cr_7C_3 sub-layer, the build-up of growth stress during scaling, and the thermal stress occurring during cooling. Cracking and partial spallation of the Al_2O_3 scale can facilitate oxygen penetration into the matrix to a certain extent. The volume expands when the alumina scaleforms.It is noted that thelinear thermal expansion coefficients are(7.2~8.6) × 10^{-6} K^{-1} for Al_2O_3, 6.7 × 10^{-6} K^{-1} for Cr_2O_3, 10.6 × 10^{-6} K^{-1} for Cr_7C_3, and 13.3 × 10^{-6} K^{-1} for Cr_2AlC. On cooling, compressive stress istherefore generated in the alumina scale, while tensile stress arises in Cr_7C_3. Hence, it was easy to crack the mechanically hard Cr_7C_3 layer under tensile stress. The larger the temperature change, the bigger the thermal stress.

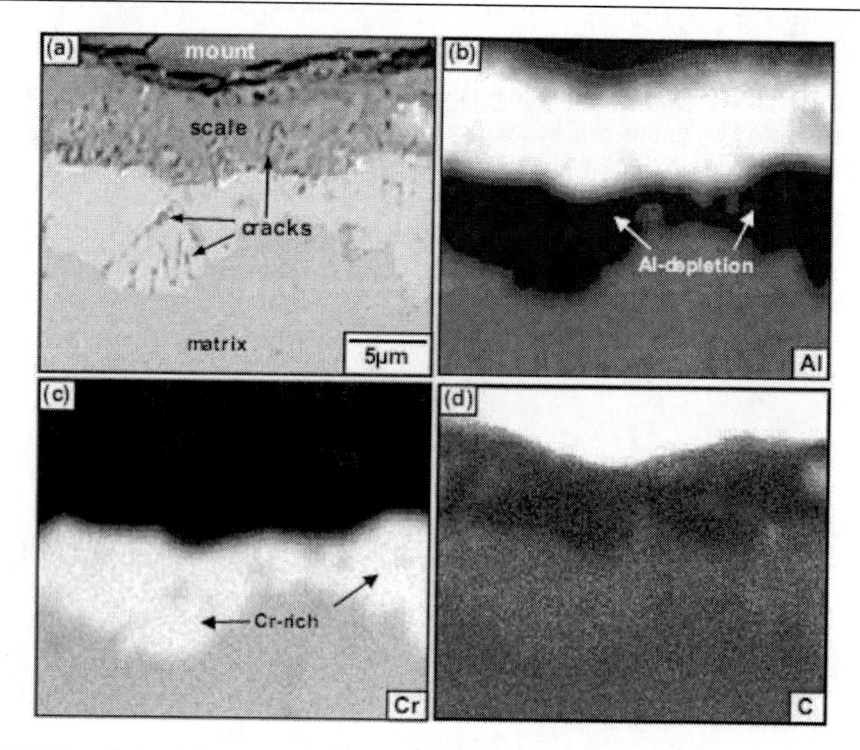

Figure 8.EPMA analysis of Cr_2AlC after oxidation at 1200 °C for 50 h in air. (a) cross-sectional back-scattered electron (BSE) image, (b) Al map, (c) Cr map, (d) carbon map.

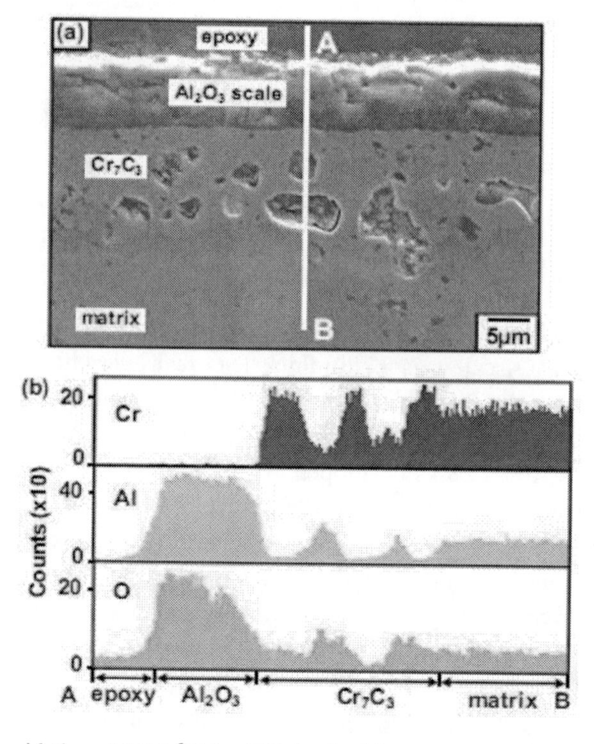

Figure 9.Cr_2AlC after oxidation at 1300 °C for 15 h in air. (a) SEM cross-sectional image,(b) EDS line profiles.

Figure 9 shows the SEM results of the oxide scale formed on Cr_2AlC after oxidation at 1300°C for 15 h.The Al_2O_3 and Cr_7C_3layers are about 9 μm and 15~20 μm thick, respectively.Inside the Cr_7C_3 layer, round or irregularly-shaped voids (dark) androundislands(dark)were seen. The roundislands were rich in Al and oxygen that diffused inward. Oxygen can easily diffuseinward because the alumina scale tended to flake-off. The consumption of Al on the surface depletes Al in the Cr_7C_3 layer. Into this layer, the supply of Al from Cr_2AlC via the outward diffusion of Al is limited, owing to the chemical stability of the Cr_2AlC compound. The limited supply of Al, and the loss of carbon result in the formation of voids in the Cr_7C_3sub-layer. Hence, the oxidation of Cr_2AlC resulted in the formation of an Al_2O_3 layerat the surface, and an underlying Cr_7C_3 layer containing voids and Al_2O_3islands.

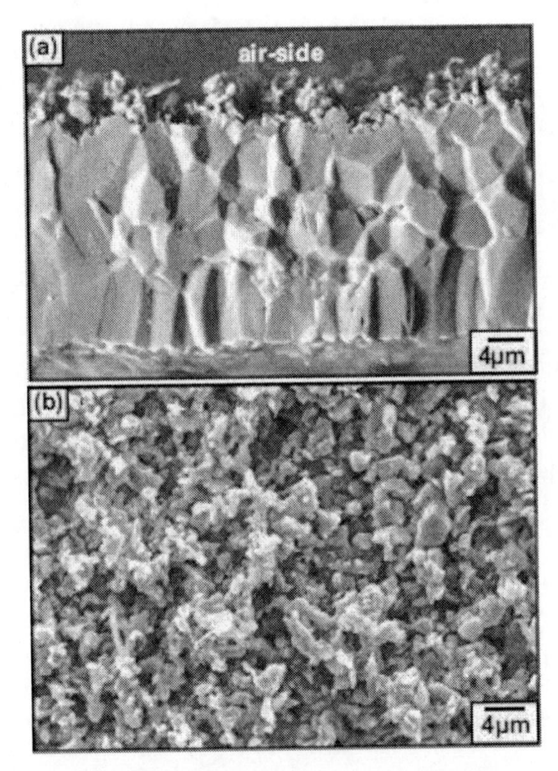

Figure 10.SEM image of the Al_2O_3 scale formed on Cr_2AlC after oxidation at 1300 °C for 240 h in air.(a) BSE image of the fracture surface, (b) top view.

Figure 10 shows the SEM results of the oxide scale formed on Cr_2AlC after oxidation at 1300°C for 240 h.Figure10(a) shows the fracture surface of the α-Al_2O_3 layer that had spontaneously spalled off from the oxidized specimen.SinceAl_2O_3 grows exceedingly slow and the supply of Al from Cr_2AlC is limited, the thickness of the α-Al_2O_3 layer is only ~ 27 μm. The outer grains were rather equiaxed, whereas the inner grains were more columnar. This resembled the typical morphology of the α-Al_2O_3 oxide scale that formed on the Al_2O_3-forming metal alloys.Figure 10(b) shows the surface morphologyof the outerAl_2O_3grains. Thesegrains that presumably formed during the initial oxidation stage were fine,despite long time oxidation at a high temperature.

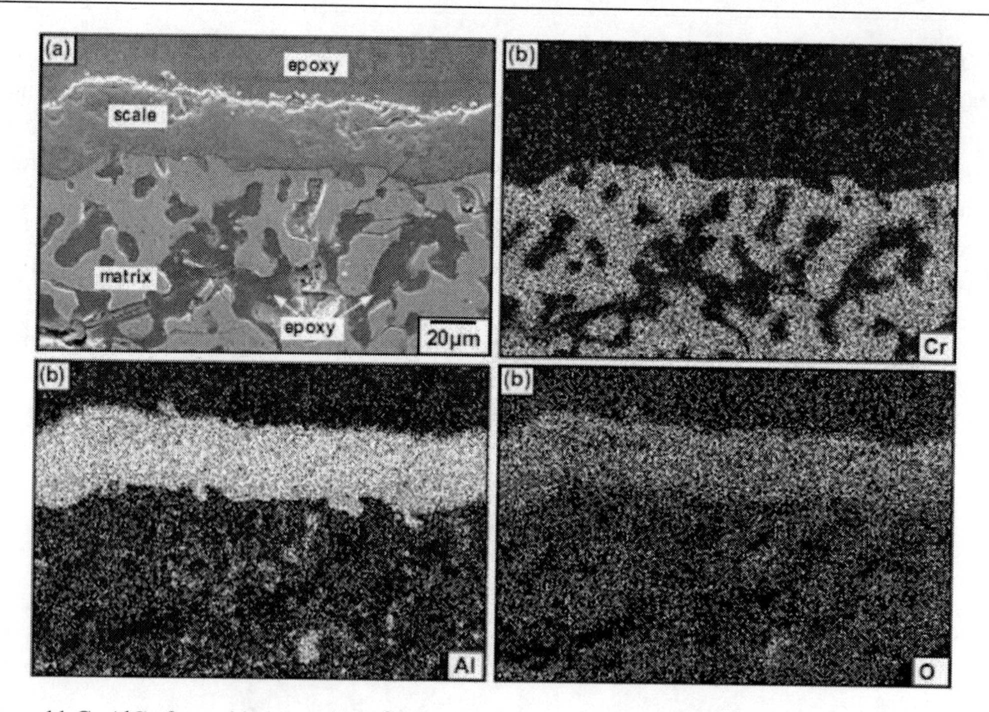

Figure 11.Cr_2AlC after oxidation at 1300 °C for 336 h in air. (a) SEM cross-sectional image,(b) EDS mappings of Cr, Al and oxygen.

Figures 11(a) and (b) show the cross-sectional image and corresponding EDS mappings of Cr_2AlC after oxidation at 1300°C for 336 h in air, respectively. Most of the scale spalled off. Locally, the scale was adherent, as shown in Figure 11(a). Despite the lengthy oxidation at 1300 °C, the Al_2O_3 layer was only ~ 30 μm thick. The underlying Cr_7C_3 layer hadexcessive voids with some Al_2O_3islands. Voids grew and destroyed the Cr_7C_3 layeras oxidation progressed. In Figure 11(a), voids were filled with mounting epoxy. The growth of the Al_2O_3 oxide layerwas determined by the rate of Al supply from the Cr_2AlC matrix,and the adherence of the Al_2O_3 layer. Since Cr_2AlC below the Cr_7C_3 layer does not easily decomposeowing to its chemical stability, the supply of Al through theCr_7C_3 layer toward the Al_2O_3scale during oxidation is limited. Hence, the Al_2O_3layersformedon the surface were relatively thin, and excessive voids formed in the Cr_7C_3 layer.

2. CyclicOxidationBehavior of Cr_2AlC

Figure 12 shows the cyclic oxidation curves at 1000, 1100, 1200 and 1300 °C in air. Each data point indicates one thermal cycle. The test cycles involved exposing the specimens for 2 h, cooling them quickly to room temperature, measuring the weight changes, and returning them to the furnace. The heating and cooling rateswere 350 and 150 °C/min, respectively. The specimens during cyclic oxidation were given a total exposure of 100 h (50 cycles). The displayed weight changes are the sum of the weight gain due to scaling and the weight losses due to not only partial scale spallation but also carbon escape from Cr_2AlC in the form of CO or CO_2gas. At 1000 °C, the weight changes were almost nil, indicating excellent oxidation resistance. At 1100°C, small weight gains were continuously measured, indicating that the rate of scaling was faster than the combined rate of spallation and carbon loss. Visual inspection indicated that little scale spallation occurred, which was mainly attributed to the formation of a quite thin scale. At 1200 °C, small initial weight gains occurred, followed by small weight losses, due to partial scale spallation and carbon loss at the later stage of oxidation. At 1300 °C, big weight losses occurred, due mainly to local spallation of the thickened oxide scale and an increased rate of carbon loss. Generally, the cyclic oxidation resistance of Cr_2AlC was acceptable at least up to 1100 °C, owing to the formation of thin, protective oxide scales.

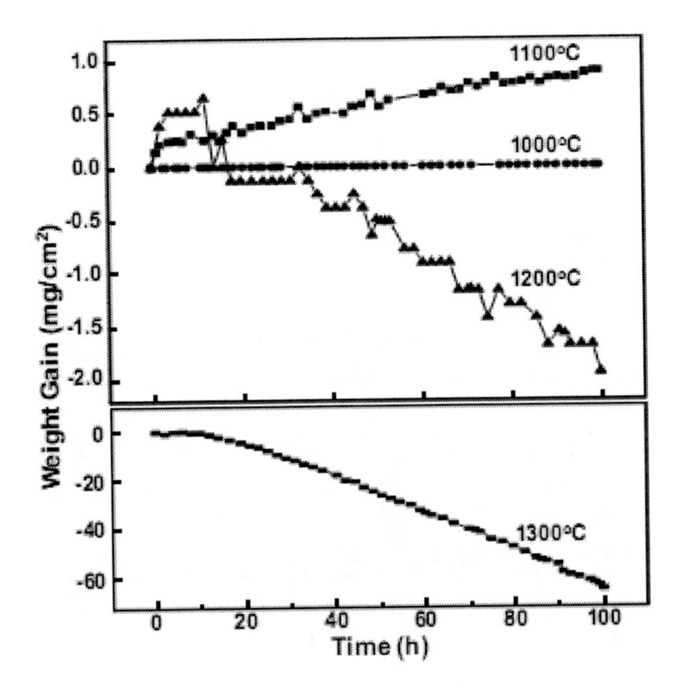

Figure 12.Weight change vs. oxidation time curves of Cr_2AlC during cyclic oxidation for 100 h between 1000 and 1300 °C in air.

Figure13 shows the XRD patterns of Cr_2AlC after cyclic oxidation for 100 h in air. In Figure 13(a),the Cr_2AlC matrix peaks were the strongest, owing to the small extent of oxidation. Cr_2AlCoxidized tothe α-Al_2O_3 oxide layer and the Cr_7C_3 sub-layer. In Figure 13(b),the phases were identified as α-Al_2O_3, Cr_7C_3, and Cr_2AlC, in decreasing order of intensity. The Cr_2AlC peaks became weak, owing to the growth of the Al_2O_3 layer and the Cr_7C_3 sub-layer.In Figure 13(c),Cr_2O_3, α-Al_2O_3, and Cr_7C_3 were detected. However, no matrix peaks were detected, because the X-rays were not able to penetrate beyond the Cr_7C_3sub-layer. From 1200 °C, Cr_2O_3 began to appear. The growth of Cr_2O_3 is primarily caused by Cr ions moving outward along the grain boundaries [9]. Although Cr_2O_3 grows faster than α-Al_2O_3 and has a high evaporation rate, both oxides are protective. Hence, the total weight changes at 1000, 1100, and 1200°C were small (Figure 12).As shown inFigure 13(d),oxides of Cr_2O_3, Al_2O_3 and Cr_7C_3 formed at 1300°Cbut their diffraction patterns were weak because of the rough oxide surface. No matrix peaks were seen because of the thick surface oxide layers.

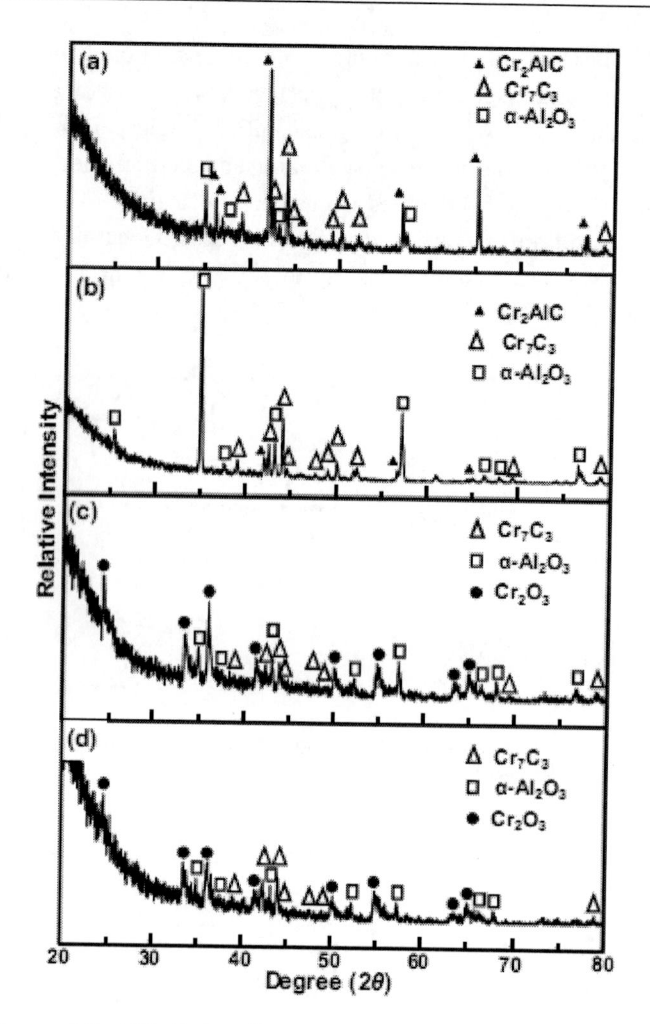

Figure 13. XRD patterns of Cr_2AlC after cyclic oxidation for 100 h in air. (a) at 1000 °C, (b) 1100 °C,(c) 1200 °C, (d) 1300 °C.

Figure 14 shows the SEM results of Cr_2AlC after cyclic oxidation at 1000 °C for 100 h. The α-Al_2O_3 layer was dense, adherent, and 1.3 μm-thick (Figure14(a)). The Al-depletion and Cr-enrichment in the Cr_7C_3sub-layer were not distinctin Figure14(b), due to the small extent of oxidationand well-known limited spatial resolution of the EDS employed.

Figure15shows the analytical results of Cr_2AlC after cyclic oxidation at 1100 °C for 100 h. The α-Al_2O_3 layer was adherent, and4.5μm-thick (Figure15(a)). Here, cracks developed around the scale-matrix interface, because of carbon loss from Cr_2AlC, growth stress due to the different volume expansion of the phases, and thermal stress due to the mismatch in their thermal-expansion coefficients. The EPMA mappings shown in Figs. 15(b)-(e) indicate the Cr-free Al_2O_3 layer and the (Al-depleted, Cr-enriched)-Cr_7C_3 sub-layer. Figure15(f) shows the fracture surface of the Al_2O_3 layer. Itsupper grains were rather equiaxed, whereas itslower grains were coarse and columnar.These grainsresembled the typical morphology of the α-Al_2O_3 oxide layer.Although theoxidation was carried out at 1100 °C, the Al_2O_3oxide layer was thin. Figure15(g) shows the morphology of the surface scale. The fine Al_2O_3oxides had a blade-like and sometimes equiaxed shape.

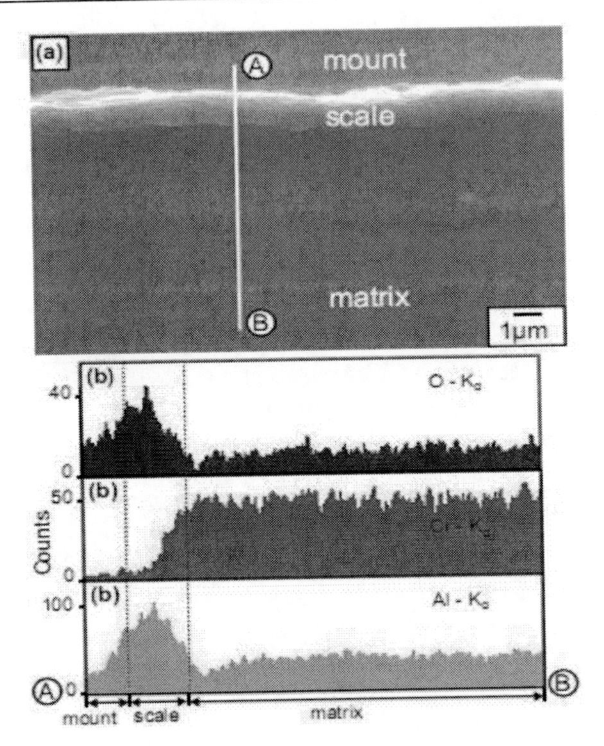

Figure 14. Cr_2AlC after cyclic oxidation at 1000 °C for 100 h in air. (a) SEM cross-sectional image,(b) EDS line profiles along A-B.

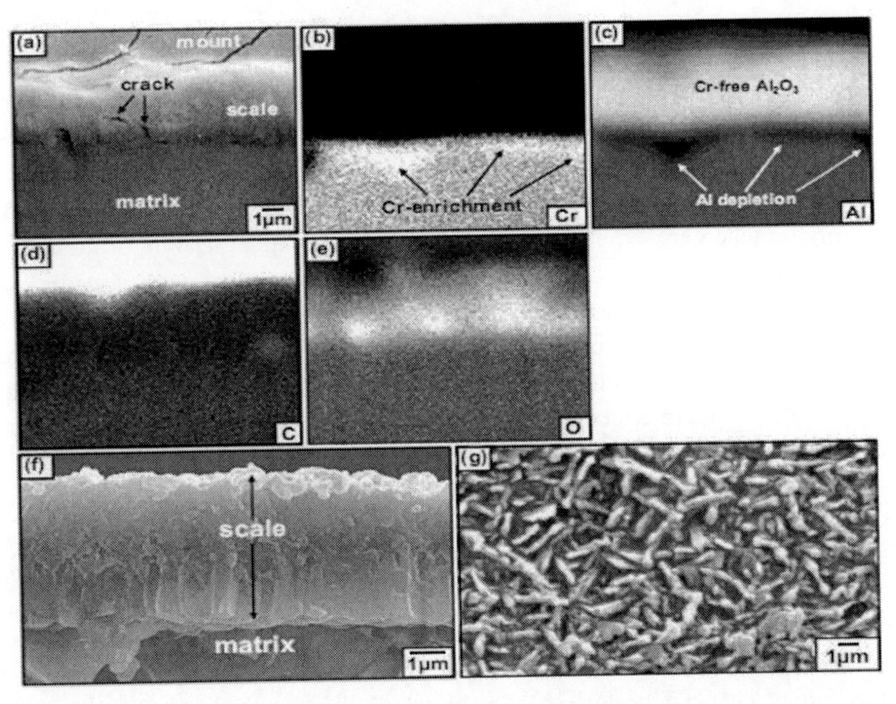

Figure 15.Cr_2AlC after cyclic oxidation at 1100 °C for 100 h in air. (a) EPMA cross-sectional image, (b) Cr map, (c) Al map, (d) C map, (e) oxygen map, (f) SEM fractograph, (g) SEM top view of the scale.

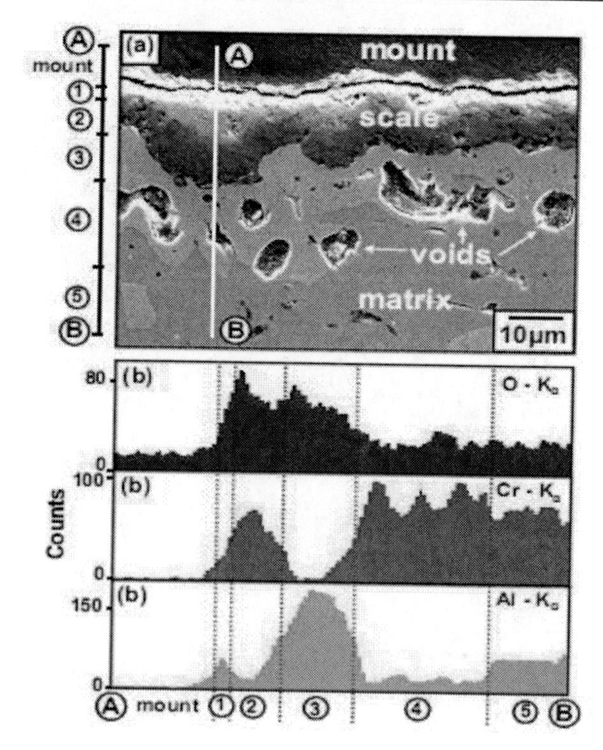

Figure 16. Cr_2AlC after cyclic oxidation at 1200 °C for 100 h in air. (a) SEM cross-sectional image, (b) EDS line profiles. (1= outer Al_2O_3-rich layer, 2= intermediate Cr_2O_3-rich layer, 3= inner Al_2O_3-rich layer, 4= Cr_7C_3 sub-layer, 5= matrix).

Figure16 shows the SEM results of Cr_2AlC after cyclic oxidation at 1200 °C for 100 h. Figure16(a) can be divided into an outer Al_2O_3-rich layer, an intermediate Cr_2O_3-rich layer, an inner Al_2O_3-rich layer, an Al-depleted Cr_7C_3sub-layer, and the matrix. Voids were scattered in the Cr_7C_3sub-layer, owing to the consumption of Al and Cr in the oxide scale, together with the evaporation of carbon from Cr_2AlC. Voids provide the channels for the diffusion of cations and oxygen, and act as stress concentrators, so as to adversely affect the scale integrity or adherence. The partial breakage and subsequent healing of the scale above 1200 °C led to the formation of the triple-layered oxide scale. In the case of cyclic oxidation above 1200 °C in air, the more active aluminum oxidized preferentially, and grew to form the outer Al_2O_3-rich layer. As the oxidation progressed, the oxide layer became thicker, and susceptible to breakage and spalling. Then, the underlying (Al-depleted, Cr-enriched) layer was oxidized to form the intermediate Cr_2O_3-rich layer. This caused the Cr-depletion and Al-enrichment underneath. This area reacted with oxygen to become the inner Al_2O_3-rich layer. Oxygen can penetrate along the short-circuit diffusion paths such as cracks. Between the inner Al_2O_3-rich layer and the matrix, the Cr_7C_3sub-layer formsowing to shortage of the Al-supply from the matrix. It is noted that the single-layered Al_2O_3oxide formedduring the isothermal oxidation of Cr_2AlC between 1000 and 1300 °C in air. Once a continuous, stable Al_2O_3 oxide layer developed on Cr_2AlC during isothermal oxidation, the Al_2O_3 oxide layer kept growing, without the formation of Cr_2O_3. During cyclic oxidationabove 1200 °C, the scale morphology changed because of the effect of thermal cycling, which caused scale failure owing to the buildup of growth stress during the scaling and the thermal stress which occurred upon heating and cooling.

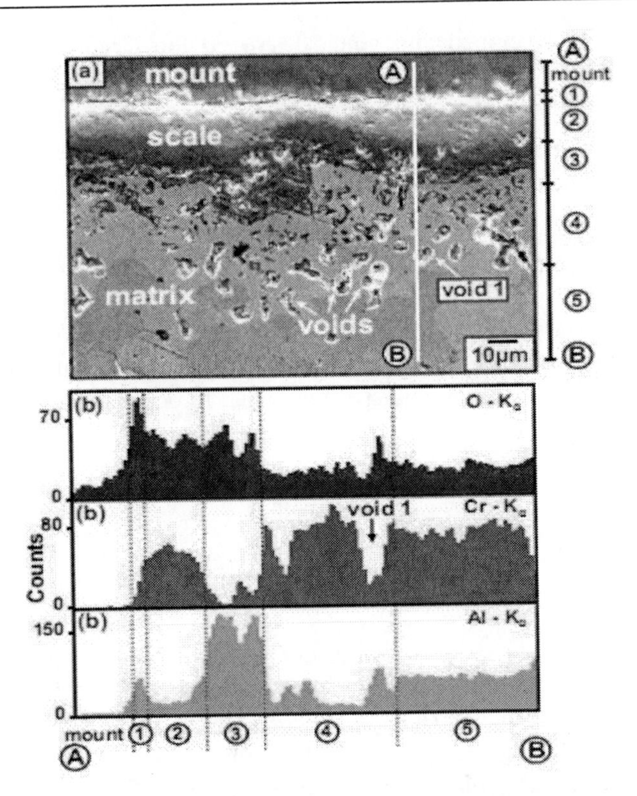

Figure 17.Cr_2AlC after cyclic oxidation at 1300 °C for 100 h in air. (a) SEM cross-sectional image,(b) EDS line profiles. (1= outer Al_2O_3-rich layer, 2= intermediate Cr_2O_3-rich layer, 3= inner Al_2O_3-rich layer, 4= Cr_7C_3 sub-layer, 5= matrix).

Figure17shows the SEM results of Cr_2AlC after cyclic oxidation at 1300 °C for 100 h. The outer Al_2O_3-rich layer (2.4 μm-thick), intermediate Cr_2O_3-rich layer (15.6 μm-thick), inner Al_2O_3-rich layer (15.1 μm-thick), and Al-depleted Cr_7C_3sub-layer (31.1 μm-thick) were seen. At 1300 °C, excessive voids formed in the Cr_7C_3sub-layer, and the cyclic oxidation resistance decreased significantly.Cracks developed in the scale due to the carbon loss, the build-up of growth stress caused by the anisotropic volume expansion of Cr_2O_3 and Al_2O_3, and the thermal stresses which occurred upon cooling. The larger the temperature change, the bigger the thermal stress. Hence, severe cracking and continuous spallation of the scale occurred at 1300 °C.

CONCLUSIONS

During the isothermal oxidation tests between 900 and 1300°C in air, Cr_2AlCoxidized according to the equation, $Cr_2AlC + O_2 \rightarrow Al_2O_3 + Cr_7C_3 + (CO \text{ or } CO_2)$. The oxide scale consisted primarily of Al_2O_3. The consumption of Al to make the Al_2O_3 layer led to the enrichment of Cr immediately below the Al_2O_3 layer, resulting in the formation of the underlying Cr_7C_3 layer. Carbon escaped from Cr_2AlC into the air. As the oxidation temperature and time were increased, the Al_2O_3 oxide layer and the underlying Cr_7C_3 layer containing voids thickened. The Al_2O_3 oxide layer was susceptible to cracking and partial spalling. Small cracks developed and voids destroyed the Cr_7C_3 sub-layer, when serious oxidation occurred. The oxidation resistance of Cr_2AlC was good up to 1200°C due to the formation of the Al_2O_3 barrier layer. However, Cr_2AlC could not survive lengthy oxidation at 1300°C.

The cyclic oxidation behavior between 1000 and 1300°C for up to 100 h in air was explained. At 1000 and 1100 °C, Cr_2AlC displayed excellent cyclic oxidation resistance by forming a less than 5 μm-thick, adherent Al_2O_3 oxide layer and a narrow Cr_7C_3 sub-layer. The α-Al_2O_3 that formed was Cr-free, and the Cr_7C_3 sub-layer was depleted in Al.At 1200 and 1300 °C, an outer Al_2O_3-rich layer, an intermediate Cr_2O_3-rich layer, an inner Al_2O_3-rich layer, and a Cr_7C_3 sub-layer formed above the matrix. Voids formed mainly in the Cr_7C_3 sub-layer.From 1200 °C, scale cracking and spalling began to occur locally to a small extent. At 1300 °C, the cyclic oxidation resistance deteriorated owing to the formation of voids and the spallation of the scales.

REFERENCES

[1] Lin, Z. J.; Zhou, Y. C.; Li, M. S.; Wang, J. Y.*Z. Metallkd.*2005, 96, 291-296.

[2] Lin, Z. J.; Li, M. S.; Wang, J. Y.; Y. C. Zhou, *Acta Mater.*2007, 55, 6182-6191.

[3] Tian, W.; Wang, P.; Kan, Y.; Zhang G.*J. Mater. Sci.*2008, 43, 2785-2791.

[4] Lee, D. B.; Nguyen, T. D.; Han, J. H.; Park, S. W. *Corros. Sci.* 2007, 49, 3926-3934.

[5] Lee, D. B.; Park, S. W. *Oxid. Met.* 2007, 68, 211-222.

[6] Lee, D. B.; Nguyen, T. D. *J. AlloysCompd.* 2008,464, 434-439.

[7] Grabke, H.J.;Brumm, M.W.;Wagemann, B. In *Oxidation of Intermetallics*;Ed.Grabke, H.J.;Schütze, M.; WILEY-VCH:NY, 1997; pp79-84.

[8] Klumpes,R., Marée,C.H.M., Schramm E.and de Wit, J.H.W. In *Oxidation of Intermetallics*;Ed.Grabke, H.J.;Schütze, M.; WILEY-VCH:NY, 1997; pp99-108.

[9] Birks, N.; Meier, G.H.; Pettit, F.S. *Introduction to the High-Temperature of Metals;*2nd ed., Cambridge University Press, England, 2006.

INDEX

A

B

C

D

O

P

R

S

T

U

V

X

Z